Carbonylation

Direct Synthesis of
Carbonyl Compounds

Carbonylation

Direct Synthesis of Carbonyl Compounds

H. M. Colquhoun
ICI Chemicals and Polymers Ltd.
Runcorn, England

D. J. Thompson
ICI Specialty Chemicals
Wilmington, Delaware

and

M. V. Twigg
European Vinyls Corporation International
Brussels, Belgium

Plenum Press • New York and London

Library of Congress Cataloging-in-Publication Data

Colquhoun, H.M.
    Carbonylation : direct synthesis of carbonyl compounds / H.M.
Colquhoun, D.J. Thompson, and M.V. Twigg.
        p.   cm.
    Includes bibliographical references and index.
    ISBN 0-306-43747-3
    1. Carbonyl compounds.   I. Thompson, D.J.   II. Twigg, M.V.
III. Title.
QD305.A6C623   1991
547'.036--dc20   91-10817
                                                            CIP

This book is to provide general information concerning its subject
matter; it is not intended to provide detailed procedural instructions.
The reader should consult appropriate sources for details of
experimental methods and statutory safety regulations. Any person using
information on equipment, methods, or materials given in this book
should do so only after seeking the advice of a suitably qualified
person. While all reasonable care has been used in compiling this book,
the authors and publisher accept no responsibility for the accuracy of
the contents, nor for any misuse of the book, including its use as a
procedural manual. Many of the topics covered in this book are the
subject of patents, and therefore nothing contained herein should be
construed as granting a license or permission to use any invention or
process that may be protected by a patent.

ISBN 0-306-43747-3

© 1991 Plenum Press, New York
A Division of Plenum Publishing Corporation
233 Spring Street, New York, N.Y. 10013

Printed in the United States of America

Preface

Since the publication of our earlier book on transition metal mediated organic synthesis,* there has been a widespread increase of interest in this topic, and transition metal based methodology has become firmly established in many areas of organic chemistry. The direct, catalytic formation of organic carbonyl compounds using carbon monoxide as the source of the carbonyl group has seen exceptional progress, and this carbonylation chemistry is being used increasingly in research and on a larger scale for fine chemicals production.

In view of these developments, there is a need for a modern, practically oriented book dealing with transition metal based carbonylation chemistry. The present monograph should help fulfill this need, since it is intended specifically to foster the adoption of catalytic carbonylation as a general tool in synthetic organic chemistry. It deals exclusively with reactions involving the interconversion of carbon monoxide and organic carbonyl compounds, and although the majority of the reactions discussed involve *catalytic* formation of carbonyl compounds, potentially valuable syntheses requiring *stoichiometric* quantities of transition metal are also included. In addition, a chapter is devoted to the remarkably useful reverse transformation (decarbonylation), in which an organic carbonyl group is *eliminated* in the form of carbon monoxide.

The main part of the book is organized around the nature of the *organic product being discussed*, and includes syntheses of aldehydes, ketones, carboxylic acids, anhydrides, esters, amides, imides, carbamates, carbonates, ureas, and isocyanates, as well as a variety of their cyclic analogues. The overall format follows that of our earlier book in providing

---

* H. M. Colquhoun, J. Holton, D. J. Thompson, and M. V. Twigg, *New Pathways for Organic Synthesis: Practical Applications of Transition Metals*, Plenum, New York (1984).

mechanistic background, details of specific synthetic applications, and an outline of the experimental approaches involved. Both established reactions and emerging chemistry are covered, and extensive reference is made to the original literature, where more detailed information can be found. Chapters concerned with the properties of carbon monoxide itself, necessary safety precautions, and preparations of the more important catalysts and reagents referred to in the text are also included. The book thus provides a basic source of reference, and a guide to the literature, for all important aspects of laboratory scale carbonylation chemistry.

Contents

Chapter 1

Introduction

In the late nineteenth and early twentieth centuries the discoverers of carbon–carbon bond-forming reactions which depended on the presence of carbonyl groups (Michael, Claisen, Knoevenagel, *et al.*) laid the foundations of modern synthetic organic chemistry, and, despite an explosion in the number of new reagents and reactions in recent years, the carbonyl group still remains one of the most versatile functionalities available to the organic chemist.[1] The importance of this simple group derives not only from its own inherent reactivity, being susceptible to both nucleophilic attack at carbon and electrophilic attack at oxygen, but also from the polarizing effect it has on neighboring atoms and functional groups, particularly its ability to stabilize an adjacent carbanion by charge delocalization into the $C = O$ double bond.

However, the immense utility of carbonyl compounds as starting materials is scarcely matched, in conventional organic chemistry, by a corresponding wealth of clean, high-yielding *syntheses* of compounds of this type. Aldehydes, for instance, are traditionally accessible by carefully controlled oxidation or reduction procedures of relatively poor efficiency, and direct routes to carbonyl compounds such as esters, acyl halides, or anhydrides, from non-carbonyl-containing precursors, do not figure prominently in the classical literature.

The purpose of the present monograph is to show how transition metal based reagents and catalysts can be used to bring *carbon monoxide* into reaction with many organic compounds, often under very mild conditions. Indeed, there is now a growing realization that carbon monoxide can be regarded for many synthetic purposes as an "incipient carbonyl group," which can be introduced directly into a number of different sites in an organic molecule, and that such reactions can often be achieved in good yield and with high selectivity. As in our earlier book, [2] the emphasis is on novel synthetic transformations of genuine value to the organic chemist, providing single products or easily separable mixtures of products in reasonable yield.

The text is generally organized according to the nature of the *organic product* being synthesized, rather than by reference to catalyst or reaction type. Syntheses of virtually every class of organic carbonyl compound— including aldehydes, ketones, carboxylic acids, acid halides, anhydrides, esters, amides, imides, ureas, carbonates, carbamates, isocyanates, and a wide variety of cyclic compounds—are described, together with examples of their application to specific synthetic problems. Chapters are also devoted to the electronic structure and properties of free and coordinated carbon monoxide, the reaction mechanisms encountered in transition metal mediated carbonylation chemistry, the preparation of catalysts, necessary safety precautions, equipment design, and general experimental procedures.

Reactions of carbon monoxide yielding *non*-carbonyl compounds, such as the Fischer–Tropsch hydrocarbon synthesis, [3] the formation of tertiary alcohols from carbon monoxide and alkylboranes, [4] and the manufacture of methanol from synthesis gas (a mixture of carbon monoxide and hydrogen), [3] are not included in this book. Neither is there much detailed discussion of large-scale processes based on carbonylation chemistry such as the manufacture of acetic acid from methanol, or 1-butanal from propene, since many excellent reviews of the industrial aspects of carbonylation chemistry are already available. [3, 5] The emphasis here is rather on laboratory scale synthesis. However, it should be noted that industrial process research has played a major role in the development of catalytic carbonylation, and that careful studies of such processes have provided the basis for much of our present mechanistic understanding.

The first well-defined carbonylation reaction was discovered just over fifty years ago by Otto Roelen, [6] working in the laboratories of Rührchemie AG. During an investigation into the mechanism of the high-pressure, cobalt-catalyzed Fischer–Tropsch synthesis of hydrocarbons from carbon monoxide and hydrogen (a mechanism that has still to be completely clarified) Roelen observed that addition of ethene to the usual feed-gas mixture of carbon monoxide and hydrogen led to formation of propanal in high yield [Eq. (1.1)].

$$H_2C{=}CH_2 \; + \; CO \; + \; H_2 \; \xrightarrow{\text{HCo(CO)}_4} \; CH_3CH_2CHO \qquad (1.1)$$

This reaction (subsequently named "hydroformylation" as it resulted in the addition of hydrogen to one end of the $C=C$ double bond and a formyl group to the other), proved to be completely independent of the heterogeneous Fischer–Tropsch synthesis, and Roelen suggested that the new process might be a homogeneous reaction, catalyzed by the then recently discovered cobalt tetracarbonyl hydride $HCo(CO)_4$. This view has since been abundantly confirmed,[7] and carbonylation chemistry remains to this day an area of catalysis uniquely dominated by homogeneous processes. At present, hydroformylation (mainly of propene to a mixture of butanal and 2-methylpropanal) annually accounts for some 8 million tons of products worldwide. Butanal is either hydrogenated directly to butan-1-ol for solvent applications, or else is subjected to aldol condensation followed by hydrogenation of the resulting C_8 aldehyde. The latter process yields 2-ethylhexanol, which is used principally in the manufacture of plasticizers for PVC.

Roelen's discovery was quickly followed by the work of Reppe and co-workers at IG Farben, who, in an extensive research program between 1939 and 1945, showed that many types of organic carbonyl compound (notably carboxylic acids and esters) could be obtained from unsaturated hydrocarbons via stoichiometric or catalytic reactions involving metal carbonyl complexes.[8] This work led directly to an efficient nickel carbonyl catalyzed process for acrylic acid production from acetylene [Eq. (1.2)],[9] though this coal-based chemistry has now been largely superseded by a process based on the oxidation of petrochemically derived propene.[10]

$$HC\equiv CH \ + \ CO \ + \ H_2O \ \xrightarrow{\ Ni(CO)_4\ } \ H_2C=CHCOOH \qquad (1.2)$$

Some 25 years after the initial research into carbonylation was carried out, a comprehensive review[11] indicated that, although many new reaction types had been described, the essentials of carbonylation chemistry remained very much as Reppe and Roelen had discovered them in the 1930s and 1940s. Thus, with few exceptions, reactions still required the use of high temperatures (100–300°C) and pressures (100–1000 bar), expensive autoclave equipment, large quantities of dangerously toxic, volatile, and unstable catalysts [$Ni(CO)_4$, $Fe(CO)_5$, or $HCo(CO)_4$], and instead of a single major product, usually gave a complex mixture of compounds requiring separation. Consequently, although forming the basis of several large-scale industrial processes, carbonylation was scarcely to be contemplated as a basis for fine-chemicals production, and as a laboratory-scale research tool for synthetic organic chemistry it was probably not considered at all.

The second 25 years of carbonylation chemistry have, however, seen dramatic changes in this picture, particularly with the work of

Wilkinson,[12] Heck,[13a] and Tsuji,[13b] which led to the discovery of stable but extremely active catalysts based on organophosphine complexes of rhodium and palladium. In addition, there have been significant advances involving the application of new techniques such as phase transfer to existing catalytic systems based on cobalt or iron.[14] As a result, many carbonylation reactions can now be carried out at low temperatures (below 100°C), at pressures close to atmospheric, using only very small quantities (typically 0.1–1 mole% on substrate) of involatile, air-stable catalyst precursors such as $Pd(PPh_3)_2Cl_2$ or $RhCl(CO)(PPh_3)_2$ which are converted to the active catalytic species *in situ*. Moreover, the scope and understanding of carbonylation chemistry have grown to such an extent that it can now be regarded, like catalytic hydrogenation, as one of the most generally useful techniques of synthetic organic chemistry, with a reasonably well-developed set of guidelines for choice of catalysts, reaction conditions, and work-up procedures. In many cases, the functional group tolerances of catalysts and reactions have been examined, and selectivities between different functional groups established, so that reactions using new substrates can be carried out with a degree of confidence impossible only a few years ago.

The need for a new monograph on carbonylation chemistry, drawing together established science and recent advances in the field, is emphasized by the fact that new results are now published, not only in journals devoted

Scheme 1.1

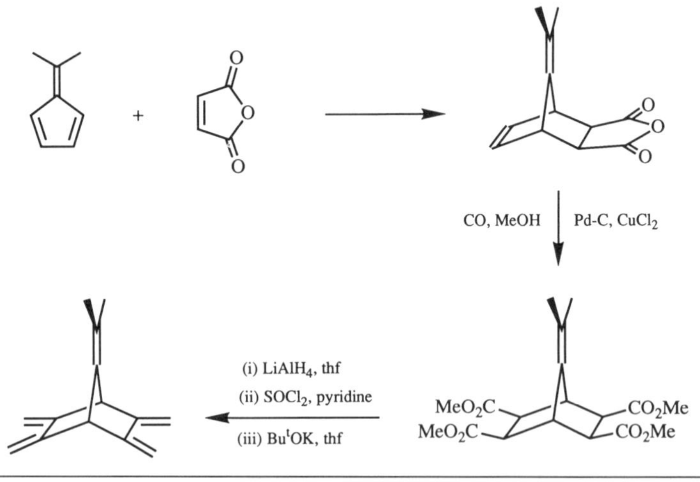

to organometallic and catalytic chemistry, but also in the synthetic organic literature. Here carbonylation may be used for only one step in a multi-stage synthesis, where it simply represents the best approach available. A good example of this is shown in Scheme 1.1, where a novel bis-diene for multiple Diels–Alder reactions is generated in high yield,[15] using palladium-catalyzed oxidative carbonylation of an alkene[16] as a key step in the synthesis. Selectivity for carbonylation at the *endocyclic* (disubstituted) double bond is essentially 100%.

Similarly, a new approach[17] to the 2-alkyl cyclopentenone skeleton which occurs in many natural products depends on the cobalt-catalyzed carbonylation of an epoxide. As originally reported,[18] this type of reaction was carried out at 130°C under 240 bar pressure of carbon monoxide, but with only slight modifications to the catalyst system, the more recent results (Scheme 1.2) were achieved at atmospheric pressure and at temperatures only slightly above ambient.

Scheme 1.2

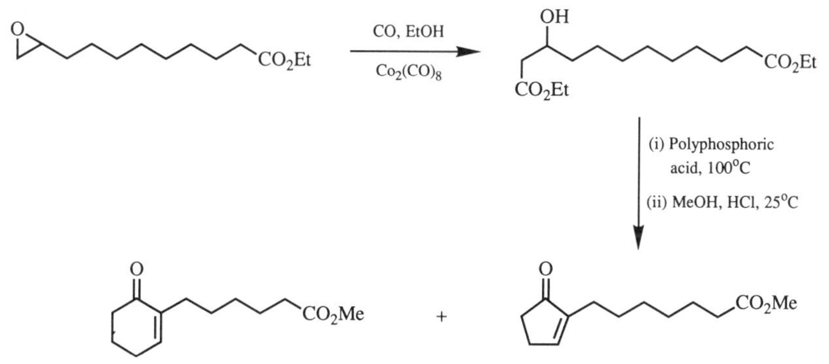

Even after 50 years of continuous development, it is clear that catalytic carbonylation chemistry has not yet reached its full potential. Indeed, the pace of discovery is if anything quickening, with the introduction of new *double* carbonylation reactions[19] leading to α-keto carboxylic acids, -esters, and -amides, new cobalt-based catalysts for carbonylation of aryl halides[20] (perhaps providing an alternative to highly effective but expensive palladium-based systems), and new types of substrate such as aryl triflate esters,[21] which effectively allow phenols to be used as reactants in place of iodoarenes.

The versatility and increasing sophistication of carbonylation as a synthetic tool will undoubtedly lead to its much greater application in the production of complex organic molecules, both in the research laboratory and on a larger scale for fine chemicals manufacture. The present monograph sets out to describe most of the established and emerging chemistry that could prove valuable in such a context.

Chapter 2

Reaction Mechanisms in Carbonylation Chemistry

2.1. Introduction

The "mechanism" of a homogeneous catalytic reaction can be viewed at two levels. First, one may identify a series of relatively stable, though not necessarily isolable, metal complexes whose sequential interconversion leads to a cyclic and therefore catalytic process. This is the level at which reaction mechanisms are most often treated in the present book. At the same time, each "elementary reaction" such as oxidative addition or reductive elimination must obviously have its own intimate mechanism, but the details for many catalytic processes are still controversial, and discussion at this level is therefore restricted to Section 2.5, where the elementary reactions themselves are described.

2.2. The Carbon Monoxide Molecule

The conventional valence bond description of carbon monoxide involves two canonical forms: a "carbenelike" structure, in which divalent carbon is linked to oxygen by a double bond, and a "dinitrogenlike" form, in which both atoms carry a lone pair and are linked by a triple bond. The latter canonical form, which is by far the more important, leads to the assignment of formal charges $O(+)$ and $C(-)$, as shown in Fig. 2.1.

Figure 2.1. Valence bond description of carbon monoxide.

This very simple picture is consistent with many of the observed physical properties of carbon monoxide (Table 2.1), including the very high $C-O$ bond energy (1076 kJ mol^{-1}, the highest of any diatomic molecule), the extremely short bond length (1.128 Å), and the direction, though not the magnitude of the dipole moment (1.2×10^{-20} esu). That the observed dipole is so slight presumably reflects a considerable polarization of electron density within the triple bond, away from carbon and towards the more electronegative oxygen, thereby neutralizing much of the formal charge separation.

The *chemistry* of carbon monoxide, particularly its coordination chemistry, is, however, more often rationalized on a qualitative *molecular orbital* (MO) basis than in valence bond terms. An energy-level MO diagram for carbon monoxide, with occupied-orbital energies determined by photoelectron spectroscopy,[22] is shown in Fig. 2.2. Assignment of the filled orbitals 4σ and $5\sigma^*$ to lone pairs on oxygen and carbon, respectively,

Table 2.1. Physical Properties of Carbon Monoxide

Melting point	$-205\,°C$ (1 bar)
ΔH_{fusion}	0.84 kJ mol^{-1} (at $-205\,°C$)
Boiling point	$-191.5\,°C$ (1 bar)
$\Delta H_{\text{vaporization}}$	6.03 kJ mol^{-1} (at $-191.5\,°C$)
Vapor pressure equation (liquid)	$\log_{10} p = a - (b \log_{10} T) - c/T$ ($a = 13.718$, $b = 2.393$, $c = 432.8$)
Density (gas)	1.25 g liter^{-1} (at $0\,°C/1$ bar)
$\Delta G^0_{\text{formation}}$	-137.34 kJ mol^{-1}
$\Delta H^0_{\text{formation}}$	-110.59 kJ mol^{-1}
$\Delta S^0_{\text{formation}}$	198.0 J deg^{-1} mol^{-1}
Bond energy	1076 kJ mol^{-1}
Bond length	1.128 Å
Dipole moment	1.2×10^{-2} D
Force constant	19.02 mdyn Å^{-1}
Ionization potential	1.35 MJ mol^{-1}
Electron affinity	less than -0.17 MJ mol^{-1}
Proton affinity	-0.594 MJ mol^{-1}

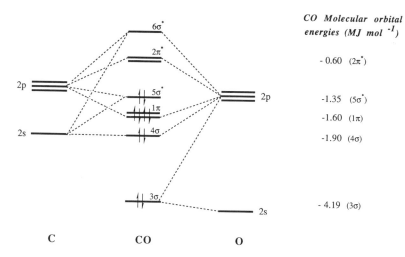

Figure 2.2. Molecular orbital description of carbon monoxide.

and orbitals 3σ and 1π (the latter being doubly degenerate) to the triple bond, allows an approximate correlation to be made between valence bond and MO descriptions.

2.3. *Reactivity of Carbon Monoxide*

Despite the presence of a formally divalent carbon atom, carbon monoxide is not in fact a particularly reactive molecule. This is largely a kinetic rather than a thermodynamic phenomenon, reflecting the low polarity, high ionization potential, and low electron affinity of the molecule (see Table 2.1). The energy required to convert the $C \equiv O$ triple bond to a double bond is not exceptionally high, and can often be generated by the formation of two additional bonds to carbon. Reaction of carbon monoxide with chlorine, for instance, involves breaking a $Cl-Cl$ bond ($+244\ kJ\ mol^{-1}$), converting $C \equiv O$ to $C = O$ ($+272\ kJ\ mol^{-1}$), and forming two $C-Cl$ bonds ($-660\ kJ\ mol^{-1}$). In bond enthalpy terms the formation of $COCl_2$ is therefore favored by about $144\ kJ\ mol^{-1}$, and even though an unfavorable entropy term reduces the free energy change somewhat, the experimentally determined value is still a substantial $-73\ kJ\ mol^{-1}$. The reaction is nevertheless extremely slow in the absence of photolysis or catalysis (in practice a charcoal catalyst is used), and indeed the relative kinetic inertness of carbon monoxide means that much of its chemistry depends on the use of either extreme conditions, energetic reagents, or, as described extensively in the present monograph, some form of catalysis.

Perhaps the simplest examples of such catalysis are found in the reactions of carbon monoxide with protic reagents such as alcohols or secondary amines, affording esters or amides of formic acid. These reactions are catalyzed by alkoxide $[RO]^-$ or amide $[R_2N]^-$ anions, respectively, and, as shown in Scheme 2.1, the key step is nucleophilic attack on carbon monoxide by the catalyst. The strongly basic alkoxyacyl or aminoacyl anion that is formed then undergoes rapid proton transfer from alcohol or amine, thus regenerating the catalytic species.

Scheme 2.1

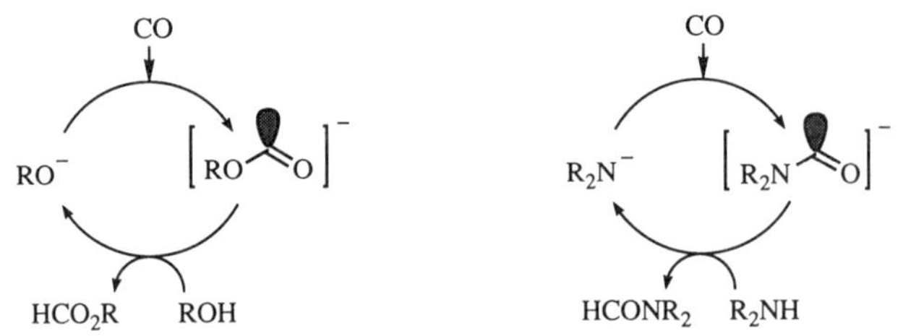

The aminoacyl anions in the catalytic cycle of Scheme 2.1 can also be generated as stoichiometric reagents in the form of lithium derivatives, which are stable at low temperatures $(-75°C)$ and react with a variety of electrophiles,[23] including alkyl halides, aldehydes, and ketones (Scheme 2.2).

Scheme 2.2

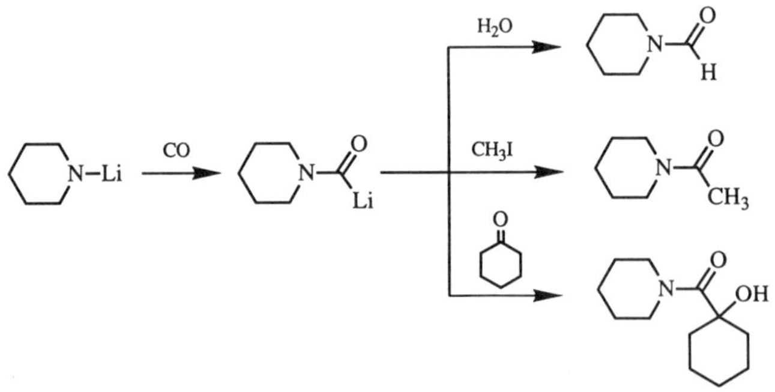

Organolithium reagents, like lithium amides, react with carbon monoxide at low temperatures (around $-100°C$) via formal nucleophilic attack, to give acyl lithium species which can be trapped cleanly by reaction with alkyl or silyl halides. Other trapping agents may also be used, so that aldehydes and ketones afford α-hydroxyketones, and esters give good yields of α-diketones (Scheme 2.3). In the absence of such trapping agents, however, complex mixtures of products tend to be obtained on hydrolytic work-up. Thus, although reaction of phenyl lithium with carbon monoxide yields benzophenone as the major product, appreciable quantities of benzoin, benzil, benzpinacol, benzoylmethanol, and benzhydrol, are also formed.[24]

Scheme 2.3

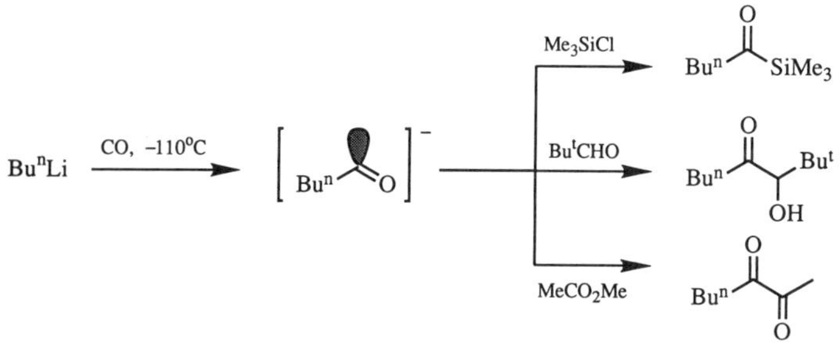

A rather different type of reaction involves one-electron transfer to carbon monoxide from alkali metals in liquid ammonia or ether solvents. The first-formed products are salts of dihydroxyacetylene $M^+_2[OC\equiv CO]^{2-}$, presumably generated via dimerization of the radical anion $[CO]^-$, and such salts undergo further transformations on heating in air,[25] to give a fascinating series of delocalized oxocarbon dianions $[C_nO_n]^{2-}$ ($n = 2-6$), as shown in Scheme 2.4.

Since the highest filled molecular orbital in carbon monoxide is the weakly antibonding, carbon-centered, $5\sigma^*$ orbital, it would be reasonable to expect electrophilic reagents to attack the CO molecule at carbon, and indeed, powerful electrophiles such as carbenium ions[26] and diborane[27] do react at this position, to give acylium ions $[R_3CCO]^+$ and "borane carbonyl," $H_3B \cdot CO$, respectively. However, carbon monoxide is an extremely weak σ-donor and does not form stable complexes with more conventional Lewis acids such as BF_3 or aluminum chloride, though there

Scheme 2.4

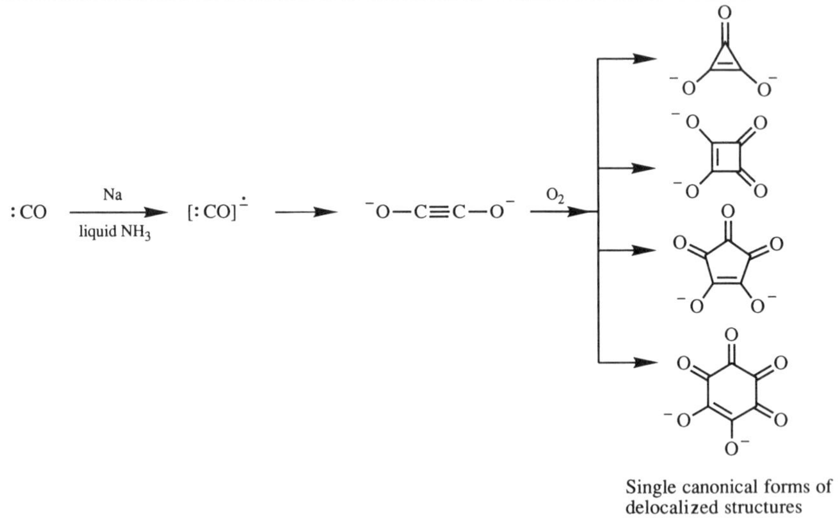
:CO
Na
liquid NH3
[:CO]
O—C≡C—O
O2
Single canonical forms of delocalized structures

is good evidence for a transient protonated species ("the solvated formyl cation") in acid-catalyzed reactions of carbon monoxide with aromatics (the Gattermann–Koch synthesis; Scheme 2.5).[28]

Scheme 2.5

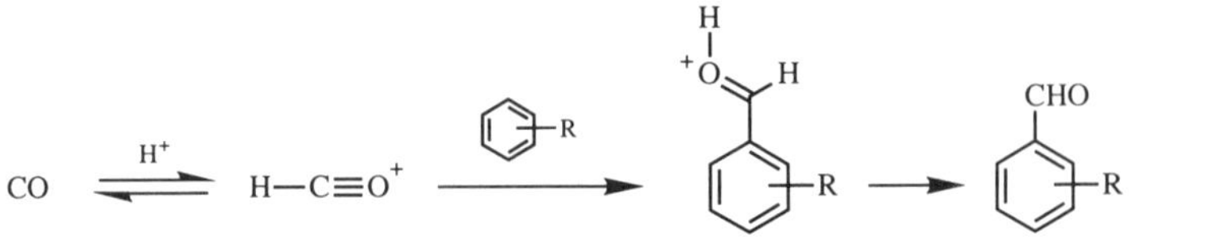
CO
H+
H—C≡O+
R
CHO

In contrast to main-group Lewis acids, *transition metals* and their ions form a vast array of complexes containing coordinated carbon monoxide. For example, the acute toxicity of carbon monoxide stems from its ability to complex, and thus block, the oxygen-binding iron centers of hemoglobin. It has long been recognized that this characteristic behavior of transition metals results from the presence of filled or partly filled *d* orbitals, which are of the correct symmetry and energy to interact with the

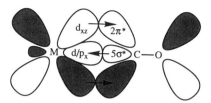

Figure 2.3. Synergic bonding model for coordination of carbon monoxide to a transition metal.

vacant, antibonding $2\pi^*$ orbitals of carbon monoxide, as shown schematically in Fig. 2.3.

The resulting *synergic* type of bonding, in which carbon monoxide donates electron density from the weakly antibonding $5\sigma^*$ orbital and accepts electrons into its strongly antibonding $2\pi^*$ orbitals, results in a reduction in the order of the $C-O$ bond, and transition metal carbonyl complexes invariably show a shift in the $C-O$ stretching frequency from the value in free carbon monoxide ($2155\,\text{cm}^{-1}$) towards lower wave number, a shift that increases in magnitude with increasing negative charge on the metal. The isoelectronic complexes $Ni(CO)_4$, $[Co(CO)_4]^-$, $[Fe(CO)_4]^{2-}$, and $[Mn(CO)_4]^{3-}$ thus show infrared $C-O$ stretching frequencies (B_{1u}) at 2046, 1890, 1730, and $1670\,\text{cm}^{-1}$, respectively, consistent with increasing occupancy of the antibonding $2\pi^*$ orbitals of the carbonyl ligand as the charge on the metal increases.[29] This ability of carbon monoxide to behave as a strong "π acid" is responsible for the stabilization of unusually low oxidation states in transition metal carbonyl complexes (0 to -3), since the build-up of electron density at the metal center, produced by conventional ligand-to-metal σ donation, is more than offset by metal-to-ligand electron transfer via the π system.

2.4. *The Chemistry of Coordinated Carbon Monoxide*

An interesting point that emerges from the above molecular orbital analysis is that coordination of carbon monoxide to a transition metal involves removal of electron density from the HOMO ($5\sigma^*$) and addition of electron density to its LUMO ($2\pi^*$). An analogous pattern of electron redistribution is of course found when the carbon monoxide molecule is raised to its first excited state, which might lead one to expect some enhancement of reactivity on coordination to a transition metal. This expectation is certainly fulfilled, but the nature and degree of enhanced reactivity vary considerably with the type of metal center involved. A metal in a high formal oxidation state and/or with strong π acid coligands acts as a strong σ acceptor but only a weak π donor. It is thus much more likely to promote nucleophilic attack on coordinated carbon monoxide than is a metal in a low oxidation state having strongly electron-donating coligands. In fact, although carbon monoxide itself is already somewhat susceptible to nucleophilic attack (see for example Schemes 2.1–2.3), coordination to a metal can greatly increase this susceptibility. A major reason for this is that the negative charge resulting from addition of a nucleophile to *coordinated* carbon monoxide is no longer centered solely at the carbonyl carbon but is delocalized onto the metal and thence to any other π-acid ligands that may be present. Reactions are now known in which the mildest of nucleophiles (water, alcohols, etc.) add directly to a carbonyl ligand (see, for example, Scheme 2.6) under strictly neutral conditions.[30]

Scheme 2.6

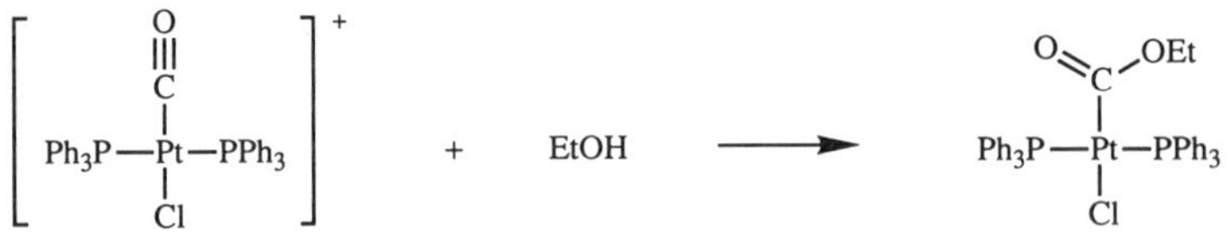

As expected, the most electrophilic carbonyl ligands are found in cationic, high oxidation state complexes, although even neutral metal carbonyls can react readily with strong nucleophiles[31] such as dialkyl-amides, alkoxides, and carbanions, as shown in Scheme 2.7.

Scheme 2.7

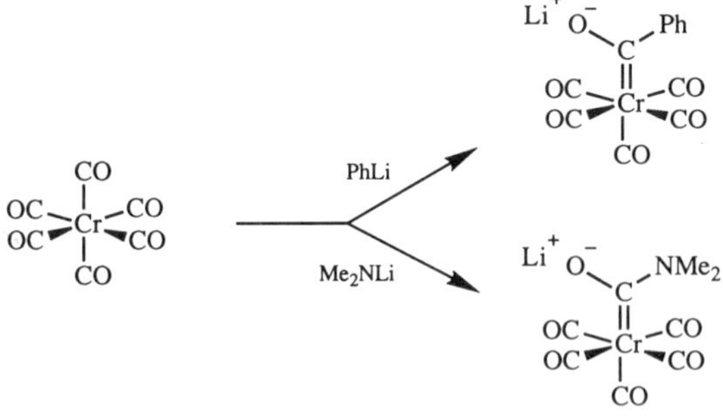

Perhaps the most characteristic reaction of coordinated carbon monoxide, however, is the so-called *insertion* process, in which a carbonyl ligand undergoes concerted *intramolecular* attack by another ligand, typically an alkyl, aryl, or other 1-electron ligand (Scheme 2.8). Although more properly referred to as a "metal-to-ligand migration" or "ligand combination reaction," the term "insertion" provides a convenient shorthand for the process and is widely used.[32]

Scheme 2.8

The importance of the insertion reaction in CO-based synthetic organic chemistry can scarcely be overemphasized since nearly all catalytic carbonylation reactions, and many stoichiometric syntheses, rely on insertion of carbon monoxide. Even in main-group carbonylation chemistry, such as that based on organoboranes,[33] migration of an alkyl group from the metal or metalloid to a carbonyl ligand provides the key C−C bond-forming step, as illustrated in Scheme 2.9. In practice, however, organoborane methodology is more often used to obtain tertiary alcohols than to synthesize organic carbonyl compounds, and hence much of borane-based carbon monoxide chemistry lies outside the scope of this book.

Scheme 2.9

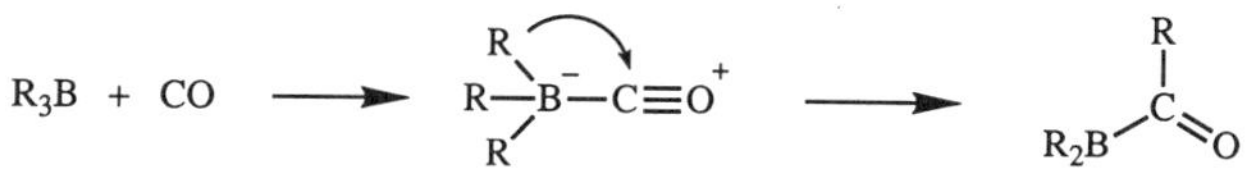

In homogeneous, transition metal catalyzed carbonylation, the most common mechanistic pattern involves metal–carbon bond formation followed by insertion of carbon monoxide, giving an acyl-metal intermediate $[RCOML_n]$ which eventually goes on to liberate the organic carbonyl product and regenerate the catalyst complex in its original form. The very many types of catalytic carbonylation reaction that have been discovered (affording aldehydes, ketones, acids, esters, amides, anhydrides, etc.) arise from variations in the method of forming the initial M–R bond and of decomposing the postinsertion metallo-acyl complex. The field of transition metal carbonylation catalysis may in fact be almost entirely rationalized, as described in Section 2.5, on the basis of a relatively small number of "elementary reactions," which combine in a variety of ways to generate a series of catalytic cycles.

2.5. Elementary Reactions in Organo-Transition Metal Chemistry

2.5.1. Oxidative Addition

One of the most pronounced characteristics of a transition metal is its ready access to a number of different oxidation states. It can thus reversibly acquire electrons from, or supply electrons to, its environment under relatively mild conditions. If this "environment" is a σ-antibonding orbital of an approaching molecule, then progressive transfer of electron density to this orbital from a metal orbital of matching symmetry (together with some degree of electron transfer *to* the metal from the corresponding σ-bonding orbital) can result in formation of a three-center metal–ligand bond, and eventually in complete dissociation of the original ligand σ bond (Scheme 2.10). At the end of the reaction the nonbonding "metal" electrons have been used to break a σ bond in the incoming molecule, with formation of two new bonds to the resulting fragments. Since the two previously nonbonding electrons are now involved in bonding, the metal has increased its formal oxidation state by two units, and the overall reaction is referred to as "oxidative addition."

Scheme 2.10

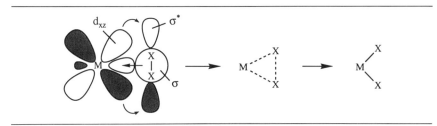

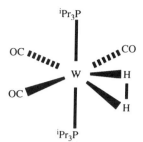

Figure 2.4. Structure of an η^2-dihydrogen complex.

This picture of concerted bond breaking and bond making is undoubtedly idealized,[34] and indeed many oxidative additions (especially of bonds to halogens) are believed to proceed via more complex ionic[35] or caged-radical[36] mechanisms. Strong evidence for at least one major concerted process has recently been discovered, however, in the form of a stable dihydrogen complex,[37] where the addition of dihydrogen to a metal has been arrested at the intermediate three-center bond stage (Fig. 2.4). Many complexes of this type are now known, including several that were formerly assigned conventional polyhydride structures.

A number of different bond types are capable of undergoing oxidative addition to low-valent transition metal complexes. Perhaps the most widely recognized are $H-H$, $C-H$, $Si-H$, $S-H$, $X-H$ (X = halogen), $N-H$, $O-H$, $C-C$, $C-X$, $X-X$, and $C-O$, and in carbonylation chemistry cleavage of the dihydrogen, hydrogen–halogen, and carbon–halogen bonds are of particular importance. For example, many catalytic carbonylation reactions are initiated by oxidative addition of a $C-X$ bond to a palladium(0) or rhodium(I) complex, the rate of addition almost invariably decreasing along the sequence $C-I > C-Br \gg C-Cl \ggg C-F$, as might be predicted from the $C-X$ bond energies, which *increase* in the same order. Oxidative addition of dihydrogen to Rh(I)[38] and Co(I)[39] is the penultimate step in two commercially important processes for aldehyde production, and addition of $Ar-X$ to Pd(0) and H_2 to Pd(II) are both implicated in the catalytic carbonylation of aryl and heteroaryl halides to aldehydes.[40]

Since both the coordination number and electron configuration of the metal increase on oxidative addition, the reaction is clearly forbidden for electronically and/or coordinatively saturated complexes. However, many apparently "saturated" complexes such as the 4-coordinate, 18-electron species Pd(PPh$_3$)$_4$, undergo reversible dissociation in solution[41] to give unsaturated 16e- or 14e-complexes, which are often highly susceptible to oxidative addition. The two-coordinate, 14e complex Pd(PPh$_3$)$_2$, for example, although not an isolable material, is a key intermediate in an important series of catalytic reactions leading to aromatic carbonyl compounds. It is formed *in situ* by dissociation either of $Pd(PPh_3)_4$ or of $Pd(PPh_3)_2(CO)_n$, and undergoes facile oxidative addition with bromo- and iodo-arenes (Scheme 2.11).[41]

Scheme 2.11

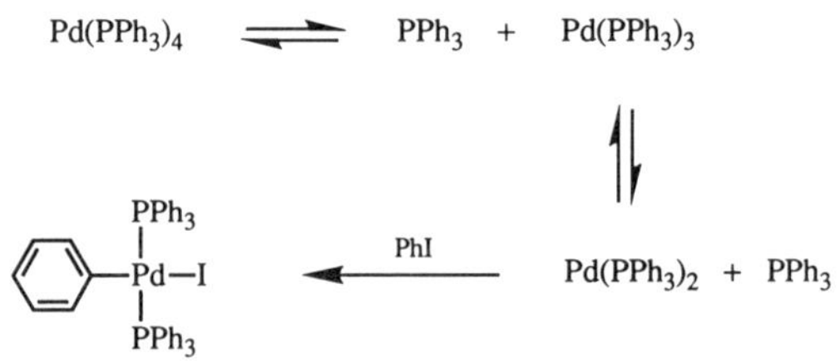

Finally in this section, it should be noted that oxidative addition is by no means confined to transition metal chemistry. Grignard reagents, for example, are formed in this way from magnesium and organic halides, and halogens add readily to triphenyl stibine to give pentavalent $SbPh_3X_2$. Indeed, few of the individual "elementary reactions" of homogeneous catalysis are entirely unknown among the main group elements, but the key to catalysis lies in the unique ability of *individual* transition metals to undergo a sufficient variety of elementary reactions to generate a cyclic, catalytic process.

2.5.2. Nucleophilic Attack by the Metal

Although oxidative addition to a metal raises its formal oxidation state by two units, the process also, paradoxically perhaps, increases the total number of electrons associated with the metal by the same number. Oxidative addition cannot therefore occur if the metal center is already electronically saturated. Such complexes can, however, if they are not also *coordinatively* saturated, act as metal-centered *nucleophiles* towards alkyl halides and other species containing electrophilic centers (Scheme 2.12). The formal oxidation state of the metal again increases by two units, but the coordination number increases by only one and the 18-electron configuration remains unchanged.

Scheme 2.12

$[Co(CO)_4]^-$ + CH$_3$—I $\longrightarrow$ OC—Co(CO)(CO)(CO) with CH$_3$ + I$^-$

2.5.3. Reductive Elimination

This unimolecular decomposition pathway is simply the reverse of oxidative addition, and involves the loss of two "one-electron ligands" from a metal center, the two ligands combining to give a single elimination product. A concerted elimination clearly requires the combining ligands to be in the *cis* configuration (although by no means all reductive eliminations are concerted), and in the product the coordination number and formal oxidation state of the metal are both reduced by two units.

In carbonylation chemistry, reductive elimination is frequently the "product-releasing" step of a catalytic cycle, so that, for example, decomposition of an acyl hydrido complex of Co(III) releases the product aldehyde during alkene hydroformylation,[38] as shown in Scheme 2.13a. The recently discovered double-carbonylation of aryl halides to keto esters similarly depends on reductive elimination of the OC−CO bond from a Pd(II) complex (Scheme 2.13b).[42] Some less obviously related reactions

Scheme 2.13

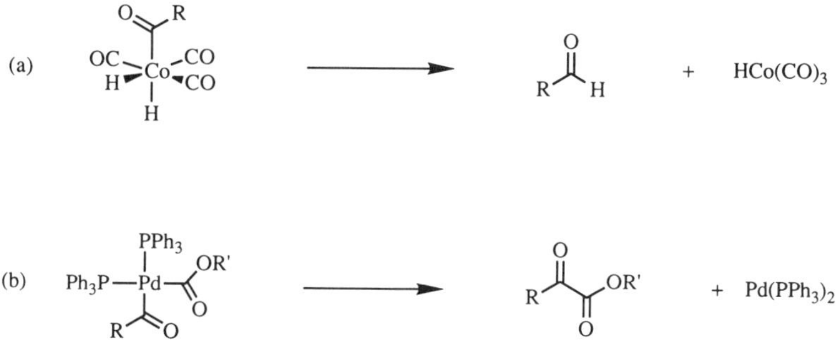

(a) [acyl hydrido Co complex] $\longrightarrow$ R—C(=O)—H + HCo(CO)$_3$

(b) [Pd complex with PPh$_3$] $\longrightarrow$ R—C(=O)—C(=O)—OR' + Pd(PPh$_3$)$_2$

can also be usefully considered as reductive eliminations. For example, certain cyclic ketones can be obtained by the reaction of strained alkenes with carbon monoxide, a key intermediate being the metallacyclopentane formed by coupling of two alkene ligands.[43] The relationship of alkene coupling to reductive elimination is clearly shown in Scheme 2.14, where the metal–alkene units are represented in valence bond terms as metalla-cyclopropane rings.

Scheme 2.14

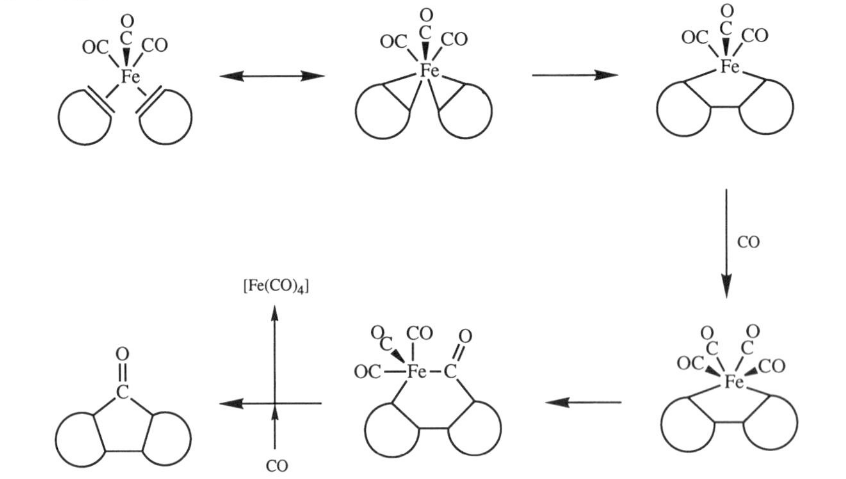

2.5.4. Insertion

The importance of carbon monoxide "insertion" into a metal–carbon bond was emphasized in Section 2.3, but in carbonylation chemistry several other reactions in which a one-electron ligand migrates from the metal to an unsaturated ligand are now well established. Thus, in hydroformylation or hydroesterification (see Section 2.6), migration of hydride to a coordinated alkene generates an alkyl ligand that subsequently inserts (i.e., migrates to) coordinated carbon monoxide (Scheme 2.15). Similarly, alkyne ligands can insert into both $M-H$ and $M-C$ bonds, and the resulting vinyl ligands will themselves migrate to coordinated alkenes, alkynes, or carbon monoxide. Even acyl groups can migrate from the metal to coordinated alkenes forming keto–alkyl ligands. Many different types of organic molecule can thus be assembled, often with a high degree of chemical specificity, via sequences of different insertion reactions.

Not unreasonably, the concerted mechanism of ligand migration requires a *cis* configuration of the combining ligands, and in keeping with

Scheme 2.15

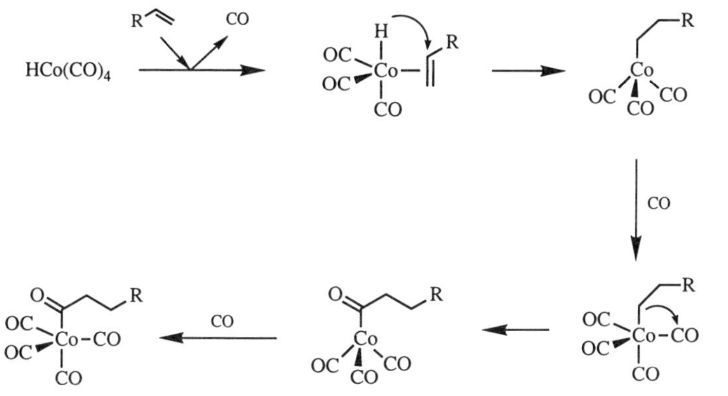

a concerted process, insertion is normally highly stereospecific. Insertion of carbon monoxide in particular has been shown to proceed with complete retention of configuration at the migrating carbon atom, which is consistent with the "front-side" attack implied by concerted migration (Scheme 2.16).[44] A further corollary of concerted migration is that insertion of alkenes or alkynes into $M-H$ or $M-C$ bonds should produce *syn* addition to the double or triple bond, and indeed this has been amply confirmed,[45] as shown for example in Scheme 2.17. Unless subsequent

Scheme 2.16

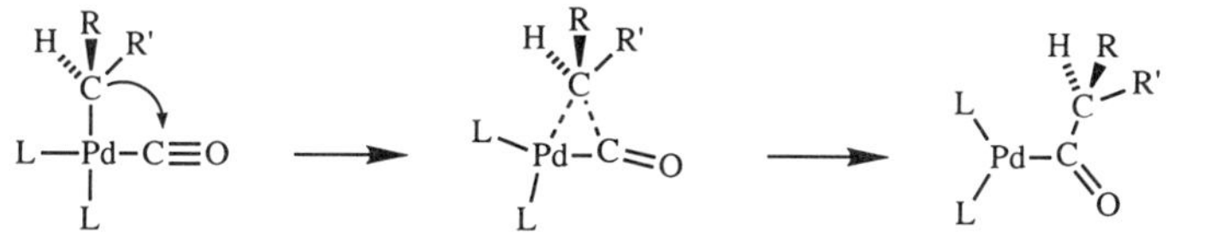

Scheme 2.17

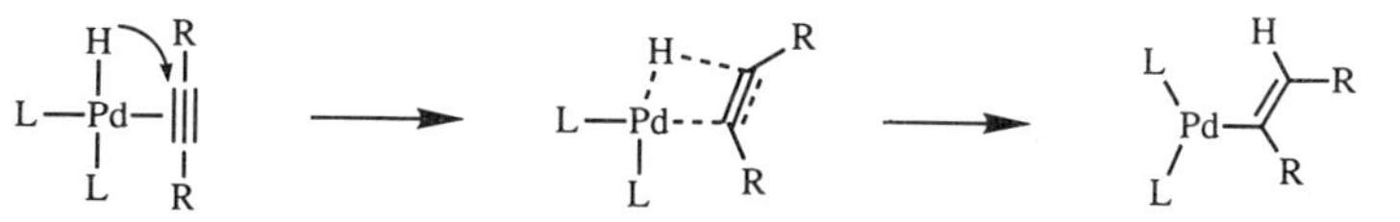

isomerization processes intervene, insertion reactions can generate new organic ligands with a high degree of both geometric and stereochemical specificity.

There are, however, a number of insertion processes for which assignment of a concerted mechanism is inappropriate. The reactions of silyl transition metal complexes with aldehydes, ketones, or small-ring cyclic ethers, for example, are believed[46] to proceed via heterolytic cleavage of the silicon–metal bond, followed by nucleophilic attack of the metal anion on the resulting silyloxonium cation (Scheme 2.18).

Scheme 2.18

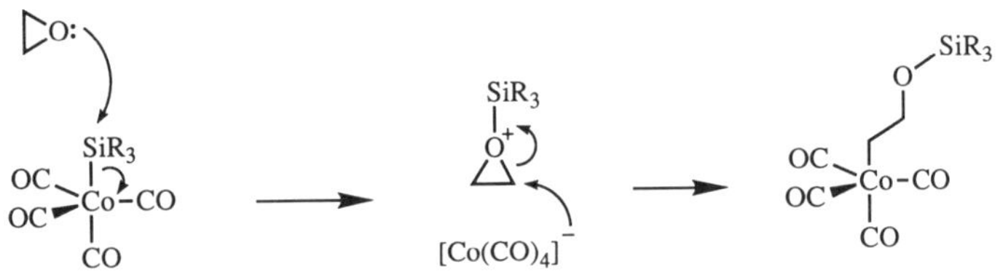

2.5.5. α- and β-Eliminations

These elementary reactions are simply "reverse-insertions," i.e., ligand-to-metal migrations, the description "α-" or "β-" referring to the number of carbon atoms from the metal at which ligand fragmentation occurs. Thus, reversal of the carbon monoxide insertion reaction involves migration of an alkyl or aryl ligand from the α-carbon to the metal as shown in Scheme 2.19a, and is therefore an α-elimination. Reversal of an alkene-hydride

Scheme 2.19

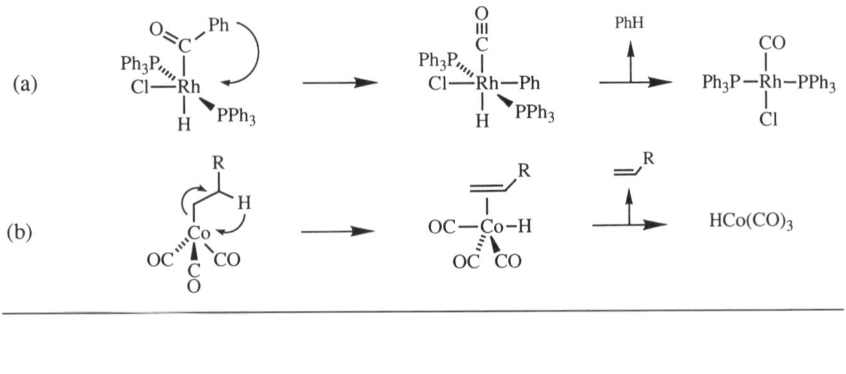

insertion, however, cleaves the alkyl ligand at the β-carbon (Scheme 2.19b) and is thus a β-elimination. Both α- and β-eliminations increase the coordination number of the metal by one, so that coordinatively saturated, kinetically stable complexes are not susceptible to this type of process. The presence of strongly bound ligands such as chelating phosphines and carbon monoxide in coordinatively saturated complexes can indeed completely inhibit β-elimination of hydride from alkyl ligands. An alternative and equally successful strategy for obtaining stable transition metal alkyls has been to avoid β-hydrogens altogether by using benzyl or methyl ligands, and although α-hydride elimination (generating a hydrido carbene complex) has recently been established as a viable process,[47] particularly where the earlier transition metals are involved, it does not seem to play a significant role in carbonylation chemistry.

2.5.6. *Nucleophilic Addition to a Ligand*

The enhanced susceptibility of coordinated carbon monoxide (relative to the free molecule) towards nucleophilic attack has already been discussed in Section 2.3, but it should also be noted that the facility to delocalize charge over both metal and ligands results in a similarly enhanced reactivity to nucleophiles for many unsaturated ligands including alkenes, alkynes, and arenes.

2.5.7. *Reductive Displacement*

Like reductive elimination and β-hydride elimination, reductive displacement is often the product-forming step of a catalytic cycle. It involves reductive cleavage of a metal–ligand bond and is an especially characteristic reaction of metal–acyl complexes under basic conditions. Depending on the system concerned, this type of reaction can result in formation of *inter alia* carboxylic acids, esters, amides, anhydrides, and acyl fluorides, as shown in Scheme 2.20.

Scheme 2.20

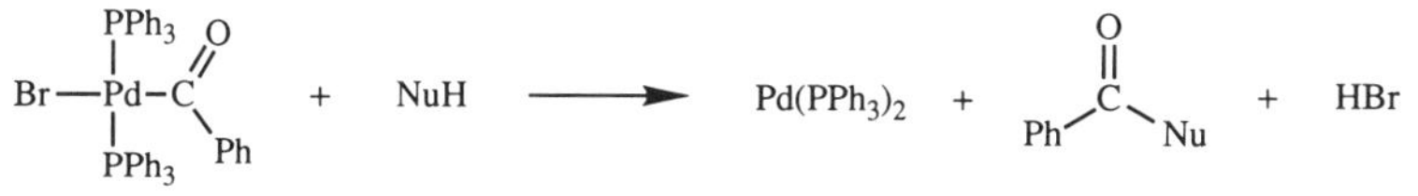

(Nu = RO, HO, R_2N, F, RCOO, etc.)

The detailed mechanism of reductive displacement varies from system to system, so that whereas reductive elimination of acyl halide, followed by hydrolysis or alcoholysis, occurs in some rhodium-catalyzed carbonylations (Scheme 2.21a),[48] recent mechanistic studies of certain palladium-based syntheses[42] suggest that alcoholysis may occur at the metal *before* reductive elimination takes place (Scheme 2.21b). Two further possibilities must also be considered, however, particularly for reductive cleavage of acyl cobalt complexes. The propensity of $[Co(CO)_4]^-$ to behave as a leaving group means that, under strongly basic conditions, direct nucleophilic cleavage of the cobalt–acyl bond becomes feasible (Scheme 2.21c). Under neutral or acidic conditions, however, this type of cleavage reaction is probably replaced by a two-step process involving *addition* of the protonated nucleophile to the acyl ligand followed by β-hydrogen elimination from the $M-C-O-H$ unit (Scheme 2.21d).

Scheme 2.21

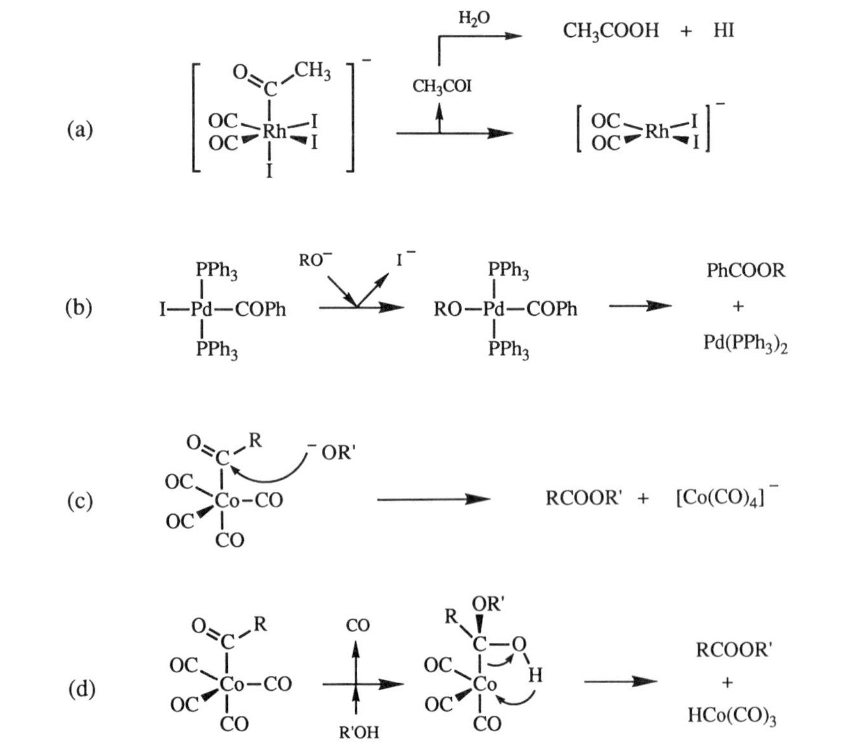

2.5.8. Ligand Dissociation and Replacement

A key step in any catalytic cycle is the simple metathetical replacement of one ligand by another. Examples are the replacement of a halide ligand by a carbanion or alkoxide anion, and the coordination of carbon monoxide to a metal, which almost invariably requires displacement of another ligand, if only a solvent molecule. The facile exchange of neutral ligands such as phosphines, alkenes, and carbon monoxide at a kinetically labile metal center is in fact a prerequisite for effective homogeneous catalysis, and its occurrence in any catalytically active system can almost be taken for granted.

2.6. Mechanistic Patterns in Carbonylation Catalysis

By far the greater proportion of the carbonylation chemistry, both stoichiometric and catalytic, which is described in later chapters, can be rationalized on the basis of the eight "elementary reactions" cited in Section 2.5. Although some mechanistic discussion is given in sections dealing with individual syntheses, it is useful to exemplify here the ways in which elementary reactions can combine to generate different types of catalytic reaction.

2.6.1. Direct Carbonylation

This simplest of all carbonylations $(RX + CO \rightarrow RCOX)$ is perhaps best exemplified by the rhodium-catalyzed conversion of iodomethane to

Scheme 2.22

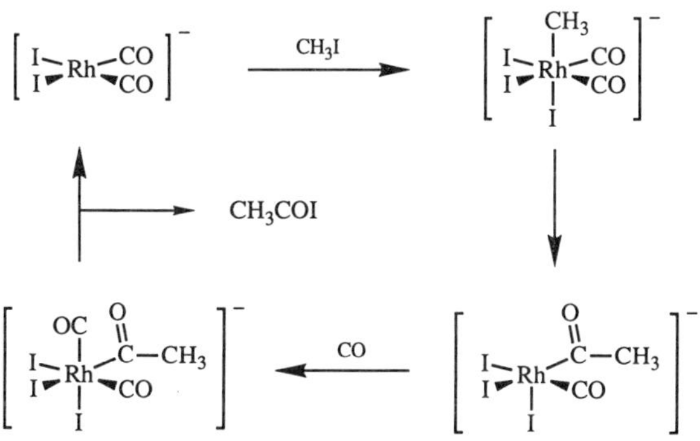

acetyl iodide,[48] a reaction that is crucial to the success of the low-pressure "Monsanto" process for acetic acid production from methanol.[49] Iodomethane (which in the acetic acid process is generated in a pre-equilibrium between methanol and HI) undergoes oxidative addition to the Rh(I) complex $[Rh(CO)_2I_2]^-$, and as shown in Scheme 2.22, methyl migration to a carbonyl ligand is followed by coordination of free carbon monoxide and reductive elimination of acetyl iodide with reformation of the initial Rh(I) complex. Hydrolysis of acetyl iodide by the water formed in the preequilibrium then yields acetic acid and regenerates HI. In the absence of water, however, iodomethane can be catalytically converted to acetyl iodide in high yield.[48]

2.6.2. *Substitutive Carbonylation*

Direct carbonylation of an organic halide is in fact a rather rare synthetic process, and systems in which halide ion is replaced by a nucleophile during the reaction $(RX + CO + [Nu]^- \rightarrow RCONu + [X]^-)$ are much more frequently encountered. This is a particularly versatile type of reaction in that it provides a wide range of "carbonyl anion equivalents," $([NuCO]^-)$ allowing the synthesis of many carboxylic acid derivatives directly from organic halides. As an example, the elementary steps involved in palladium-catalyzed carbonylation of a bromoarene are shown in Scheme 2.23, where oxidative addition is followed by insertion of carbon monoxide and reductive cleavage by whichever nucleophile ($[Nu]^- = [HO]^-$, $[RO]^-$, $[R_2N]^-$, $[RCOO]^-$, or $[F]^-$) is present.

Scheme 2.23

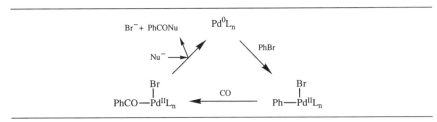

It has been found that by working at higher pressures of carbon monoxide, it is possible to produce doubly carbonylated products (α-keto esters and amides) from this type of reaction.[42, 50] Extensive mechanistic studies[42] have shown that such products arise, not by double CO insertion into the Pd−Ar bond, but by formation of a cationic acyl/carbonyl complex which is attacked at the carbonyl ligand by an external nucleophile. Reductive elimination then yields the product keto-acid derivative (ester or amide) and regenerates Pd(0) (Scheme 2.24).

Scheme 2.24

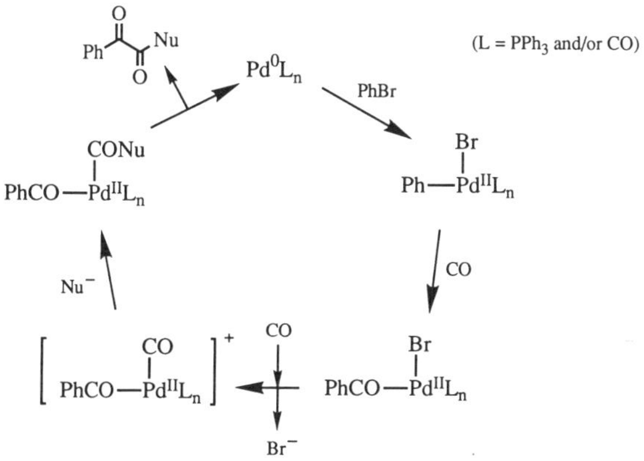

Substitutive carbonylation of aliphatic halides is also possible, but here a different type of catalyst is necessary because of the tendency of alkyl palladium complexes to "short-circuit" the catalytic cycle by undergoing β-hydride elimination before carbon monoxide insertion can occur. Platinum–phosphine complexes such as $Pt(PPh_3)_2Cl_2$ are, however, now known to catalyze such aliphatic carbonylations, probably via the same type of cycle as in Scheme 2.23, although more vigorous conditions are required.[51] The success of this type of catalyst depends on the fact that alkyl platinum complexes are much less prone to β-elimination, possibly because, for relativistic reasons,[52] the M−C σ-bond energy is significantly greater in third-row transition metal complexes than in analogous second-row derivatives.

A rather different approach to carbonylation of aliphatic halides involves the anionic complex $[Co(CO)_4]^-$, which, being a kinetically stable 18-electron species, attacks the halide by nucleophilic displacement rather than by undergoing oxidative addition. The probable catalytic cycle for cobalt-catalyzed carbonylation of benzyl chloride[53] is shown in Scheme 2.25.

Scheme 2.25

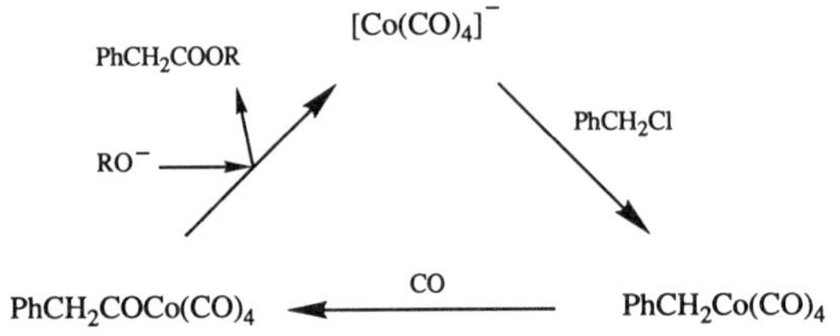

2.6.3. *Additive Carbonylation*

As noted in Chapter 1, this class of reaction includes hydroformylation ($RCH=CH_2 + CO + HY \rightarrow RCH_2CH_2COY$; $Y = H$), the first catalytic carbonylation reaction to be discovered. Replacing hydrogen in this synthesis by water or an alcohol yields carboxylic acids ("hydrocarboxylation") or esters ("hydroesterification"), respectively, and although different catalysts may be required, the general mechanistic pattern is very similar in all three cases. The reaction is initiated by alkene insertion into a metal hydride, and the resulting alkyl ligand (which may be linear or branched depending on the direction of insertion) migrates to coordinated carbon monoxide. In cobalt-catalyzed hydroformylation, outlined in Scheme 2.26, oxidative addition of hydrogen to the metal is followed by reductive elimination of aldehyde, and the original metal hydride complex is thus regenerated.[54]

Scheme 2.26

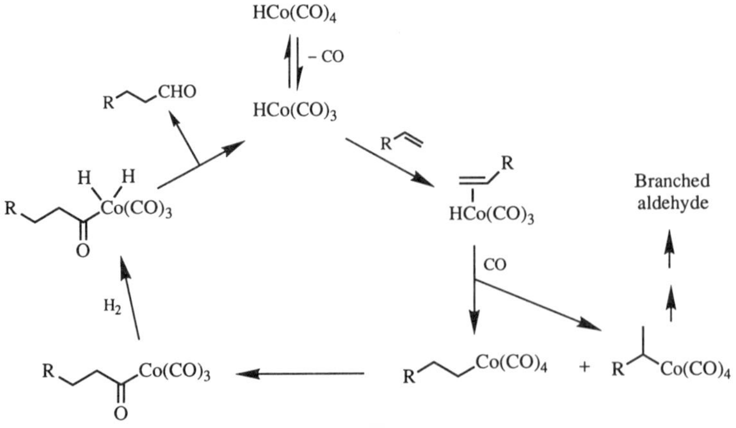

In hydroesterification, however, the exact nature of the product-releasing step is less clear. Under strongly basic conditions nucleophilic cleavage followed by reprotonation of the metal could occur, but, as discussed in Section 2.4.7, under neutral conditions addition of alcohol to the acyl carbonyl followed by β-hydride elimination from the O—H group (Scheme 2.27) seems likely to be the preferred pathway.

Scheme 2.27

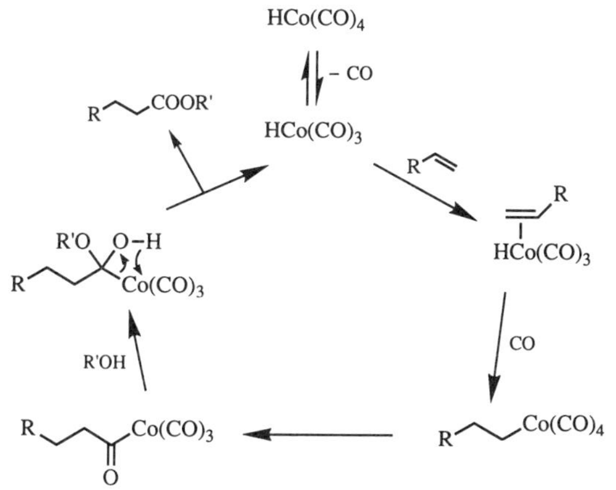

Although outside the main theme of this book, which is the use of *transition metal* catalysts and reagents in carbonylation chemistry, it should be noted here that an entirely different form of alkene carbonylation is possible under very strongly acidic conditions. Protonation of the alkene affords a carbenium, ion which, as mentioned in Section 2.3, is one of the few types of conventional Lewis acid that will react directly with carbon monoxide. The resulting acylonium ion can be trapped by addition of water, and this reaction forms the basis of the "Koch process" for manufacture of carboxylic acids (Scheme 2.28).[26] Transition metal ions such as

Scheme 2.28

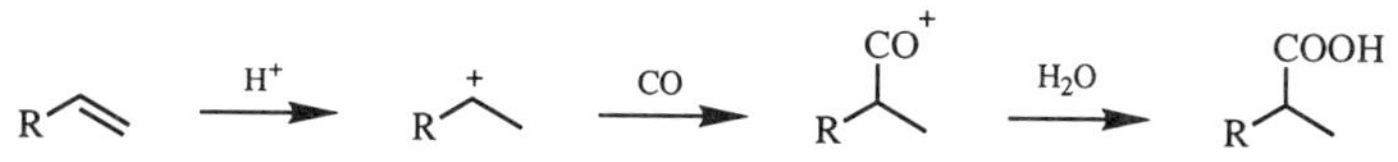

copper(I) and silver(I) do promote the reaction,[55] but their role is uncertain and they may well serve only to increase the effective concentration of carbon monoxide in solution by forming loosely bound, soluble carbonyl complexes of the type $[M(CO)_n]^+$.

2.6.4. Multicomponent Carbonylations

One of the best illustrations of the ability of homogeneous catalysts to assemble complex organic structures by sequential insertion reactions is the hydroxybutenolide synthesis shown in Scheme 2.29. Although some details of the catalytic cycle are speculative, this scheme, based on known elementary reactions, at least provides a rationale for the remarkable result that no fewer than five separate molecules (CH_3I, $HC \equiv CR$, H_2O, and $2 \times CO$) can be combined in selective fashion (yields of up to 70% can be attained) under essentially ambient conditions.[56]

Scheme 2.29

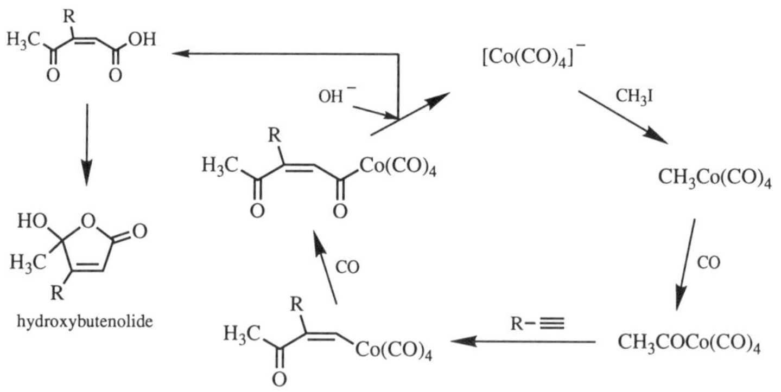

2.6.5. Oxidative Carbonylation

A number of potentially useful carbonylation reactions, particularly those based on organopalladium chemistry, result in the transition metal undergoing a reduction in oxidation state (for example, Pd(II) → Pd(0), as in Scheme 2.30). In order to achieve a catalytic reaction, it is necessary to reoxidize the metal back to its original state, and for palladium this can often be achieved by addition of a copper(II) salt to the reaction, which then becomes stoichiometric in copper but catalytic in palladium (see Section 7.3 for further examples).

Scheme 2.30

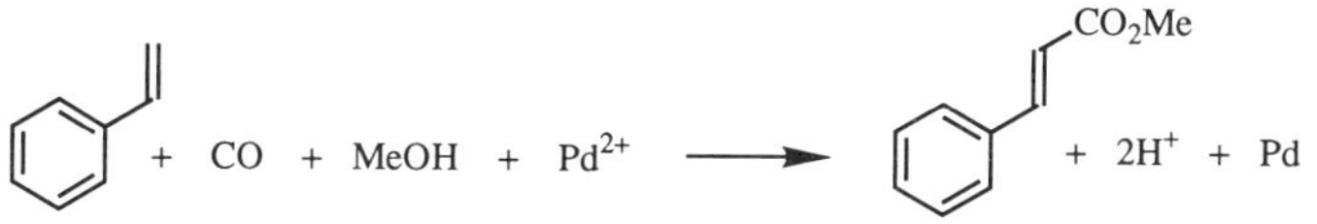

2.6.6. Decarbonylation

Since most of the elementary reactions involved in carbonylation chemistry are readily reversible, it is not unreasonable that metal complexes that catalyze carbonylation of organic compounds should also catalyze their *decarbonylation* $(RCOX \rightarrow RX + CO)$.[57] The key step in this surprisingly useful transformation is fragmentation (α-elimination) of an acyl ligand to give an alkyl or aryl ligand and coordinated carbon monoxide. As an example, the complete cycle for catalytic decarbonylation of an aldehyde by $RhCl(CO)(PPh_3)_2$ is given in Scheme 2.31, and many other synthetically valuable decarbonylation reactions are discussed in Chapter 11.

Scheme 2.31

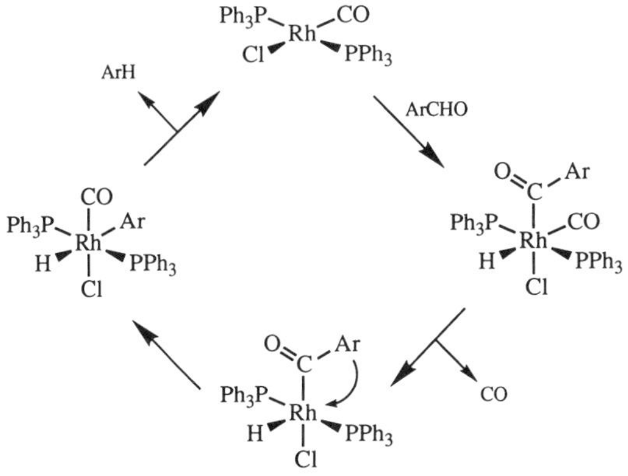

Chapter 3
Practical Aspects

3.1. Introduction

This chapter is concerned with the practical aspects of carbonylation chemistry. In view of the very active catalyst systems now available, autoclave equipment is only occasionally required and there is rarely any need for the extremes of pressure and temperature synonymous with older industrial carbonylation processes. Small-scale atmospheric pressure procedures can in fact often be carried out in conventional laboratory glassware. In procedural terms, carbonylation reactions can be most closely compared with hydrogenations, both types of reaction involving gas handling and transition metal catalysis. Although carbon monoxide requires particularly careful handling in view of its toxicity and the potentially explosive nature of its mixtures with air, when used correctly it is a versatile reagent that can be employed safely in most laboratories.

This chapter contains information about carbon monoxide, details of atmospheric-pressure carbonylation techniques, and a short section on commercially available high-pressure autoclave equipment. There is also a section on the volatile metal carbonyls and a general discussion of the properties of transition metal based carbonylation catalysts, complementing the preparative information given in Chapter 12. Here emphasis is given to properties relevant to storing and handling these materials. To complete the chapter there is a brief review of the methods used to isolate and purify the products of carbonylation reactions, although in practice conventional purification techniques (chromatography, recrystallization, distillation, etc.) normally require only minor modification, to ensure the removal of catalyst residues.

3.2. Carbon Monoxide

The nature of the bonding in carbon monoxide, its physical properties, and its general chemical reactivity are discussed in the previous chapter. Here we focus on properties associated with its use in the laboratory.

3.2.1. General Considerations

Carbon monoxide is a colorless, odorless gas, liquefying at $-191.5°C$ (compare with dinitrogen, which liquefies at $-196°C$). It is readily available, at pressures of up to about 100 bar, in a variety of cylinder sizes so that, when necessary, high-pressure experiments can be run at 80 bar without the need for a compressor. The main cause of the acute toxicity of carbon monoxide is its high affinity for hemoglobin. By strongly coordinating to the iron center it prevents oxygen transport, thus effectively causing asphyxia when sufficient quantities are taken into the lungs. It is stated[58] that 400–500 ppm of carbon monoxide in air can be breathed for an hour without appreciable effect, while levels above 1000 ppm are dangerous, and at more than 4000 ppm the gas causes death within an hour. The previously cited reference gives a Threshold Limit Value (TLV) of 50 ppm. The very real danger associated with high concentrations of carbon monoxide is that, as with dinitrogen, there are no obvious warning signs of oncoming asphyxiation. Prolonged exposure to lower concentrations causes shortness of breath and headache followed, when exposure is severe, by confusion, dizziness, and impaired hearing and vision. Carbon monoxide must therefore always be used in an efficient fume cupboard, keeping emissions into the laboratory atmosphere to an absolute minimum.

Detectors based on chemical reaction, such as "Draeger tubes," provide a quick and reliable means of estimating carbon monoxide concentration. They are straightforward to use and have the advantage of requiring no electrical maintenance. Concentrations of carbon monoxide in a fume cupboard can be checked with them whenever it is thought necessary, but since each tube can be used for only a single determination, frequent use becomes costly. Therefore, while chemical detectors are convenient for single concentration measurements, if carbon monoxide is in routine use it is recommended that an electronic detection system be installed to provide continuous monitoring of the laboratory atmosphere.

There are a number of commercial devices available for detecting carbon monoxide. The older "explosive meter" types used industrially for detecting explosive gas mixtures in coal mines are inexpensive but are rather less convenient to use than more modern gas-leak detectors, which use efficient catalytic or electrochemical sensors. Both of the latter types of detector provide reliable performance, although they may require periodic recalibration as they age. Even small portable units are now sensitive to as little as 1 ppm of carbon monoxide, and these can be used for personal monitoring, for example when clipped to a belt. If equipped with a probe tube they also can be used as point source leak detectors for valves and connectors.

More detailed information about commercially available hazardous-gas detectors and selective carbon monoxide detectors can be found in product literature provided by the larger gas suppliers, chemical companies, and specialist electronic equipment manufactures. Some names and addresses are given in Appendix 3.

The limits of flammability for carbon monoxide at atmospheric pressure and room temperature are, on a volume basis, 12.5% (lower) and 74.2% (upper) so that under normal circumstances, when working with small amounts of the gas in an efficient fume cupboard, there should be minimal risk of fire or explosion. Moreover, under these conditions there is unlikely to be a build-up of carbon monoxide in the working environment.

3.2.2. Handling Carbon Monoxide

As discussed in the previous section, all operations involving carbon monoxide must be carried out in an efficient fume cupboard. For atmospheric pressure reactions, conventional rubber tubing or the clear plasticized PVC tubing (for example "Tygon R3603") now widely used in laboratories may be employed, with appropriate securing clips, for delivering carbon monoxide to the reaction system. Where higher pressures are to be used, stainless steel tubing and couplings are required.

Carbon monoxide of 99.5% minimum purity is available in small lecture bottles (typically containing about 30 liters of gas at standard temperature and pressure). These provide a particularly convenient source of the gas for occasional laboratory scale atmospheric pressure carbonylations, since not only are they small enough to be mounted inside the fume cupboard when in use (and also stored there at other times), but they also represent a considerably smaller hazard than do conventional cylinders, which contain vastly more gas. Although use of a lecture bottle may appear to limit the scale of reaction, the quantity of carbon monoxide available (about 1.3 moles) is sufficient to carry out, for example, twenty carbonylations each using 20 g of a substrate with a molecular weight of 300 (assuming complete utilization of carbon monoxide). Even at a more realistic 50% utilization level, a respectable amount of chemistry can still obviously be achieved with a single lecture bottle.

Information on available cylinder types, valves, and pressure regulators is best obtained from manufacturers or suppliers, together with the most recent safety recommendations—names and addresses of some suppliers are given in Appendix 2. In general, carbon monoxide under ambient conditions demands no special materials considerations, and brass fittings are satisfactory. With lecture bottles either brass or stainless steel needle control valves can be used, together with PTFE sealing washers. Since needle valves do not regulate pressure when connected to a closed system, a bubbler, lute, or other relief device must always be used, to avoid a build-up of pressure in the equipment. Small, chromium-plated brass, single-stage pressure regulators are now available for lecture bottles, and their use when carrying out carbonylations is strongly recommended. These are available, for example, from the Aldrich Chemical Company, and are provided with a needle valve on the outlet. For atmospheric pressure reactions, the most convenient unit is one fitted with a delivery pressure gauge reading 0–30 psig. An important advantage of these regulators is that the cylinder pressure reading on the inlet gauge gives a good indication of the amount of gas remaining in the cylinder.

3.2.3. Impurities and Purification

The impurities present in carbon monoxide depend on its origin and history. Industrially it is separated from synthesis gas (a mixture of hydrogen and carbon oxides), which is itself manufactured by the steam reforming of hydrocarbons. Natural gas is the most widely used feedstock.[59] Synthesis gas itself can be useful for carbonylation reactions involving both carbon monoxide and hydrogen, and commercially available cylinders provide gases in a range of CO/H_2 ratios.

Depending on the purification process used, commercial carbon monoxide may contain small quantities of hydrogen, carbon dioxide, methane, dinitrogen, and one or more of the noble gases. The major impurities in laboratory grades are usually dinitrogen ($\sim 0.5\%$) and methane ($\sim 0.2\%$). Carbon monoxide that is stored in iron cylinders over prolonged periods may contain traces of iron pentacarbonyl. This can be removed by passing the gas over heated silica or copper turnings, although such purification is not normally required when using high-grade carbon monoxide supplied in lecture bottles, and only rarely with the lower grades of gas. In fact the most troublesome impurity is generally oxygen, which, although present at very low levels, can cause progressive catalyst deactivation.

Oxygen can be removed from carbon monoxide by passing the gas through a short bed of reduced copper catalyst or manganous oxide, as in the purification of inert gases such as dinitrogen or argon.[60] Elevated temperatures are not necessary. Such purification systems are, however, somewhat inconvenient to install for individual preparations so that purchasing high-purity carbon monoxide is usually more attractive than purification of a lower-grade gas. Purification systems, if used, should be flushed free of carbon monoxide before reactivation of the absorbent at high temperature, since under these conditions any nickel or iron in the system can promote disproportionation of carbon monoxide to carbon and carbon dioxide.

3.2.4. Solubility and Gas Volume Measurement

The solubility of carbon monoxide in common organic solvents is, like that of most gases, very low when compared to solid or liquid reagents. Gas solubilities can be expressed in a number of different ways, and the literature in this area may be confusing. The Bunsen absorption coefficient (α), for example, is the volume of gas, reduced (assuming ideal gas behavior) to $0°C$ and 1 atm partial pressure, which dissolves in a unit volume of the solvent at the temperature of the measurement. The Ostwald absorption coefficient (L), on the other hand, is simply the volume of gas dissolved per unit volume of solvent at the temperature and pressure of the measurement. In the physical chemistry literature, the mole fraction of dissolved gas (X_2) is often used, though precise density measurements are required to calculate this parameter from solubility coefficients. Only rarely are gas solubilities reported in conventional molar concentration units (M). These are, however, included in Table 3.1. The relationships between the various gas solubility parameters are given in Eq. (3.1)–(3.5),

$$L = V_2(t)/V_1(t) \tag{3.1}$$

$$\alpha = 273.15 V_2(t)/V_1(t)\,Tp \tag{3.2}$$

$$L = \alpha Tp/273.15 \tag{3.3}$$

$$X_2 = [(RT/LpV_M) + 1]^{-1} \tag{3.4}$$

$$M = X_2/[MV(1 - X_2)] \tag{3.5}$$

where $V_2(t)$ is the volume of gas dissolved at temperature $t°C$, $V_1(t)$ is the volume of solvent at temperature $t°C$, p is the partial pressure of gas (atm), X_2 is the mole fraction of dissolved gas, T is the absolute temperature ($t + 273.15$), R is the gas constant (0.08205 liter atm K^{-1} mol^{-1}), V_M is the molar volume of pure solvent at $t°C$ (liters), and the units of $V_2(t)$ and $V_1(t)$ are the same.

As shown in Table 3.1, the equilibrium concentrations of carbon monoxide in organic solvents at room temperature and 1 atm partial pressure are in the region of 10^{-2}–10^{-3} molar, and the value for water is at the lower end of this range. The solubility of carbon monoxide tends to

Table 3.1. Solubility of Carbon Monoxide in Selected Solvents at 25 °C

Solvent	Molar volume	Solubility				Ref.
		$X_2{}^a$	L^b	α^c	C^d	
n-Heptane	146.46	17.24×10^{-4}	0.287	0.263	11.71×10^{-3}	*f*
Cyclohexane	108.75	9.91×10^{-4}	0.223	0.204	9.12×10^{-3}	*g*
Methylcyclohexane	128.35	12.41×10^{-4}	0.237	0.217	9.68×10^{-3}	*h*
Benzene	89.41	6.68×10^{-4}	0.183	0.168	7.47×10^{-3}	*f*
Toluene	106.86	8.11×10^{-4}	0.186	0.170	7.59×10^{-3}	*h*
Decalin^e	154.10	9.17×10^{-4}	0.143	0.131	5.96×10^{-3}	*k*
Perfluoroheptane	227.33	38.75×10^{-4}	0.419	0.384	17.11×10^{-3}	*f*
Perfluorobenzene	115.79	21.20×10^{-4}	0.449	0.411	1.35×10^{-3}	*l*
Carbon tetrachloride	97.09	8.76×10^{-4}	0.221	0.203	9.03×10^{-3}	*i*
Chloroform	80.94	6.43×10^{-4}	0.194	0.178	7.94×10^{-3}	*j*
Chlorobenzene	102.27	6.47×10^{-4}	0.155	0.142	6.33×10^{-3}	*i*
1,4-Dichlorobutane^e	111.35	5.97×10^{-4}	0.129	0.127	5.36×10^{-3}	*j*
1,2-Dichloroethane	69.79	3.88×10^{-4}	0.136	0.134	5.56×10^{-3}	*j*
Acetone	74.01	7.72×10^{-4}	0.255	0.234	10.44×10^{-3}	*i*
Methanol	40.73	3.76×10^{-4}	0.226	0.207	9.24×10^{-3}	*m*
Ethanol	58.68	4.84×10^{-4}	0.202	0.185	8.26×10^{-3}	*m*
n-Propanol^e	74.79	5.50×10^{-4}	0.177	0.165	7.36×10^{-3}	*n*
i-Propanol^e	76.55	6.04×10^{-4}	0.190	0.132	7.89×10^{-3}	*n*
Isobutanol	92.88	6.52×10^{-4}	0.172	0.158	7.03×10^{-3}	*o*
Nitrobenzene	102.72	3.72×10^{-4}	0.089	0.081	3.63×10^{-3}	*p*
Dimethylformamide^e	77.04	1.40×10^{-4}	0.044	0.041	1.82×10^{-3}	*q*
Water	18.07	0.17×10^{-4}	0.023	0.021	0.95×10^{-3}	*r*

^{a} Mole fraction of dissolved carbon monoxide at 1 atm partial pressure of carbon monoxide.
^{b} Ostwald solubility coefficient; see text for definition.
^{c} Bunsen solubility coefficient; see text for definition.
^{d} Concentration in moles per liter at 1 atm partial pressure of carbon monoxide.
^{e} Refers to measurements at 20 °C.
^{f} C. Gjalgbaek, *Acta Chem. Scand.*, **6**, 623 (1952).
^{g} E. Wilhelm and R. Battino, cited in *Chem. Rev.*, **73**, 1 (1973).
^{h} L. Field, R. Battino, and E. Wilhelm, cited in *Chem. Rev.*, **73**, 1, (1973).
^{i} J. Horiuti, *Sci. Pap. Inst. Phys. Chem. Res. Tokyo*, **17**, 125 (1931).
^{j} W. J. Knebel and R. J. Angelici, *Inorg. Chem.*, **13**, 632 (1974).
^{k} M. Basato, J. P. Fawcett, and A. J. Poë, *J. Chem. Soc. Dalton Trans.*, 1350 (1974).
^{l} F. D. Evans and R. Battino, *J. Chem. Thermodyn.*, **3**, 753 (1971).
^{m} J. C. Gjaldbaek, *Kgl. Dan. Vidensk. Selsk. Mat. Fys. Medd.*, **13**, 24 (1948).
^{n} J. C. Gjaldbaek and E. K. Anderson, *Acta Chem. Scand.*, **2**, 683 (1948).
^{o} R. Battino, F. D. Evans, W. F. Danforth, and E. Wilhelm, *J. Chem. Thermodym.*, **3**, 743 (1971).
^{p} J. C. Gjaldbaek and E. K. Anderson, *Acta Chem. Scand.*, **8**, 1389 (1954).
^{q} Based on approximate data given in R. S. Kittila, *Dimethylformamide Chemical Uses*, E. I. Du Pont De Nemours & Co, 1967.
^{r} E. Wilhelm, R. Battino, and R. J. Wilcock, *Chem. Rev.*, **77**, 219 (1977).

be less in polar solvents. The concentration of dissolved gas decreases rapidly as the temperature approaches the boiling point of the solvent, although the solubility coefficient does not in fact change very greatly—the partial pressure of carbon monoxide decreases as that of the solvent vapor increases to that at the boiling point. Increasing the reaction temperature thus requires increased pressure to maintain the concentration of dissolved gas (following Henry's law), but since this requirement is greater the more volatile the solvent, there is clearly an advantage in using a high-boiling-point solvent, even when reactions are carried out under pressure.

When the solubilities given in Table 3.1 are compared with the concentration of substrate being carbonylated (typically about 1 molar), it is clear that carbon monoxide may be in stoichiometric deficit by a factor of several hundred. This huge differential highlights the need for efficient stirring, to promote gas transfer into the condensed phase by providing a high interfacial surface area. In many ways, carbonylations resemble catalytic hydrogenations—reaction rates can be high enough for the overall process to become limited by mass transfer of gas into the liquid phase.

For synthetic work, it is not always necessary to know how much carbon monoxide has been consumed, merely that no more is being taken up and that the reaction is complete. However, when making kinetic measurements, or optimizing reaction parameters such as temperature or rate of stirring, or defining conditions for decarbonylation reactions (see Chapter 11), it becomes necessary to measure gas volumes. This can be conveniently carried out with the type of gas burette (containing water or a light oil) normally used in conjunction with atmospheric pressure laboratory hydrogenators.[61] A simple gas burette is shown in Fig. 3.1; it is connected to the reaction system via flexible tubing, and is initially emptied of gas by raising the reservoir with the isolation tap open. Subsequently, when the apparatus has been evacuated and refilled with carbon monoxide (see below), gas is drawn into the burette by slowly lowering the reservoir. This process is repeated two or three times to remove all of the remaining air. The initial volume reading is taken with the two liquid levels at the same height (when the internal pressure equals atmospheric pressure), and this is repeated at intervals whenever the volume consumed is required. The apparent volume of gas used is the difference between the two burette readings, but in order to calculate the number of moles of carbon monoxide consumed, the temperature, barometric pressure, and vapor pressure of liquid in the burette all need to be taken into account. Measurements of carbon monoxide liberated during decarbonylations involve essentially the reverse of this procedure, the whole apparatus first being filled with dinitrogen (unless kinetic studies, for example, require the presence of a carbon monoxide atmosphere).

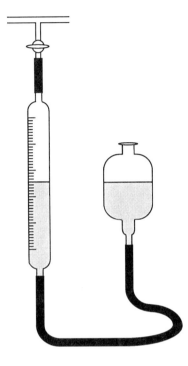

Figure 3.1. Gas burette for measurement of carbon monoxide volume.

3.3. Transition Metal Carbonyls

The properties of the common binary metal carbonyls are given in Table 3.2. An important consideration when using metal carbonyls is the very high toxicity of their vapors, since even the solid compounds may have appreciable vapor pressures and many can be sublimed. Solid materials, however, rarely represent anything like the same degree of hazard as the volatile liquid carbonyls, for which stringent operating procedures are required, taking full account of their toxicity and flammability.[62] The most important volatile carbonyls are $Ni(CO)_4$ and $Fe(CO)_5$. The variation of their vapor pressures with temperature is shown in Fig. 3.2, and some of their other physical properties are given in Table 3.3. The volatile carbonyl hydrides $HCo(CO)_4$ and $HMn(CO)_5$ are less often encountered. Nickel tetracarbonyl (TLV 0.05 ppm) and iron pentacarbonyl (TLV 0.1 ppm) are much the most toxic,[58] their TLVs being some three orders of magnitude lower than that of carbon monoxide. Failure to appreciate the hazards associated with these volatile compounds has, in the past, resulted in some serious accidents, and it is strongly recommended

Table 3.2. *Properties of Transition Metal Binary Carbonyls Arranged According to Their Position in the Periodic Table*[a]

$V(CO)_6$	$Cr(CO)_6$	$Mn_2(CO)_{10}$	$Fe(CO)_5$	$Co_2(CO)_8$	$Ni(CO)_4$
Blue solid	White crystals	Yellow crystals	Yellow liquid	Orange crystals	Colorless liquid
MW 219.0	MW 220.1	MW 390.0	MW 195.9	MW 342.0	MW 170.8
MP 60 °C	MP 150 °C	MP 154 °C	MP −20 °C	MP 51 °C	MP −19.3 °C
AS very	AS stable	AS moderately	BP 104 °C	AS moderately	BP 42.2 °C
			AS mildly		AS mildly
			$Fe_2(CO)_9$	$Co_4(CO)_{12}$	
			Brown plates	Black solid	
			MW 363.8	MW 571.9	
			MP 100 °C (d)	MP 60 °C	
			AS moderate	AS moderate	
	$Mo(CO)_6$	$Tc_2(CO)_{10}$	$Ru(CO)_5$	$Rh_2(CO)_8$	
	White crystals	White crystals	Colorless liquid	Unstable	
	MW 264.0	MW 476.1	MW 241.1		
	MP 146 °C	MP 159 °C	MP −22 °C		
	AS stable	AS stable	AS moderate		
			$Ru_3(CO)_{12}$	$Rh_4(CO)_{12}$	
			Orange crystals	Dark red crystals	
			MW 639.3	MW 747.8	
			MP 154 °C	MP 150 °C	
			AS stable	AS moderate	
				$Rh_6(CO)_{16}$	
				Black solid	
				MW 1065.6	
				MP 220 °C	
				AS moderate	

[a] AS, Air sensitive.

Table 3.2.   (Continued)

W(CO)$_6$	Re$_2$(CO)$_{10}$	Os(CO)$_5$	Ir$_2$(CO)$_8$
White	White	Colorless	Unstable
crystals	crystals	liquid	
MW 351.9	MW 625.5	MW 330.3	
MP 169 °C	MP 170 °C (d)	AS moderately	
AS stable	AS stable		

		Os$_2$(CO)$_9$	Ir$_4$(CO)$_{12}$
		Unstable	Yellow
			solid
			MW 1104.9
			MP 230 °C
			AS moderately

		Os$_3$(CO)$_{12}$	
		Yellow	
		crystals	
		MW 906.7	
		MP 224 °C (d)	
		AS stable	

Table 3.3.   Selected Properties of Nickel Tetracarbonyl and Iron Pentacarbonyl

Property	Ni(CO)$_4$	Fe(CO)$_5$
Color	Colorless	Yellow
Molecular weight	170.7	195.9
Boiling point (°C)	42.2[a]	104[f]
Melting point (°C)	−19.3[b]	−20.0[g]
Density (g cm^{-3})	1.31 (20 °C)[c]	1.46 (21 °C)[h]
Vapor pressure (mm Hg)	322 (20 °C)[d]	22 (20 °C)[g]
$\Delta H_{vaporization}$ (kcal mol^{-1})	6.50[e]	9.60[g]

[a] B. Suginuma and K. Satozaki, *Bull. Inst. Phys. Res. Tokyo*, **21**, 432 (1942).

[b] J. E. Spice, L. A. K. Staveley, and G. A. Harrow, *J. Chem. Soc.*, 100 (1955).

[c] L. Mond and R. Nasini, *Z. Phys. Chem.*, **8**, 150 (1891); for summary of related data see F. W. Laird and M. A. Smith, *J. Am. Chem. Soc.*, **57**, 266 (1935).

[d] J. S. Anderson, *J. Chem. Soc.*, 1653 (1930).

[e] Average value, for discussion see Ref. *b*.

[f] M. Trautz and W. Badstubner, *Z. Elektrochem.*, **35**, 799 (1929).

[g] A. J. Leadbetter and J. E. Spice, *Can. J. Chem.*, **37**, 1923 (1959).

[h] *Handbook of Chemistry and Physics*, 66th Edition, R. C. West (ed.). CRC Press, Boca Raton, Florida (1985).

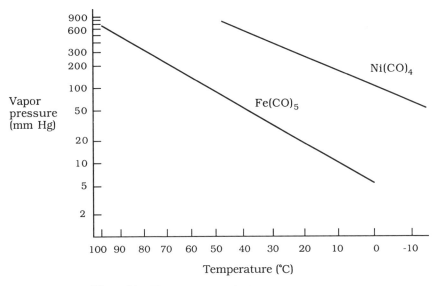

Figure 3.2. Vapor pressures of $Ni(CO)_4$ and $Fe(CO)_5$.

that volatile liquid metal carbonyls be used only where no alternative catalyst or reagent exists.

The effects of inhalation of nickel tetracarbonyl or iron pentacarbonyl may not be apparent for several hours.[62] Early symptoms include headache, giddiness, vomiting or nausea, breathlessness, and weakness of limbs. Subsequently (12–18 h) there may be severe breathlessness, pain during respiration, dry cough and cyanosis (bluish coloration of face and ears). In the event of an accident the person concerned should be kept warm, put to bed as soon as possible, and then kept lying down. Oxygen should be given and a doctor sent for at once. In acute cases, supervision of the patient over an extended period is necessary to detect the possible onset (often sudden) of pulmonary oedema. Moreover, it has been established that there is a correlation between continued exposure to low concentrations of nickel tetracarbonyl and cancer of the respiratory tract.[63]

When working with volatile metal carbonyls, medicinal oxygen, a stretcher, and blankets should be available, and before work begins the person in charge must consult the current safety literature from the suppliers for the latest safety/first aid recommendations. Whenever there is the slightest risk of escape of metal carbonyl vapor a self-contained compressed air respirator must be worn, and *on no account* should a charcoal-type mask be used.

Volatile metal carbonyls will burn in air, and explosive mixtures can be formed. This can occur spontaneously when pouring [particularly $Fe(CO)_5$] from one container to another, as a pyrophoric sludge is often found at the bottom of containers of liquid carbonyls exposed to air. It is therefore preferable to make transfers using thin tubing with application of differential pressures of nitrogen or carbon monoxide. These carbonyls are insoluble in water, but miscible with most organic solvents. They react vigorously with nitric acid or halogens, forming salts and carbon monoxide.[64] Bromine water has been widely used as a reagent for the controlled oxidative decomposition of residual metal carbonyls in used apparatus because it is one of the few convenient reagents that will destroy such residues smoothly under ambient conditions. With care, this procedure can also be used for disposing of larger quantities of liquid carbonyls.

The conditions under which volatile carbonyls of nickel and cobalt can be formed are surprisingly mild, so that appropriate precautions must be taken when contacting their compounds with carbon monoxide. The most significant problem in this respect is to avoid formation of nickel tetracarbonyl since, in the absence of donor ligands such as triphenylphosphine, carbonylation of nickel salts at room temperature and atmospheric pressure under only mildly reducing conditions[65] can lead to surprisingly rapid formation of $Ni(CO)_4$. Because of its volatility, $Ni(CO)_4$ is easily distilled from the reaction mixture, and this in fact forms the basis of one preparative method.[66] At elevated temperatures, both finely divided nickel metal[67-69] and aqueous solutions of nickel salts containing chelating amines[70] afford the carbonyl without additional reducing agent. In a number of carbonylation procedures $Ni(CO)_4$ is a stoichiometric by-product. Iron pentacarbonyl is, however, much less readily generated.

Although cobalt tetracarbonyl hydride, $HCo(CO)_4$, is easily formed and is highly toxic, it is also relatively unstable and quickly decomposes under normal conditions of temperature and pressure. Thus a laboratory preparation[71, 72] of $Co_2(CO)_8$ involves merely passing carbon monoxide through an alkaline suspension of $Co(CN)_2$ to generate, on acidification, the unstable hydride $HCo(CO)_4$. This decomposes in solution on standing at room temperature to give the dimeric binary carbonyl. Liquid $HCo(CO)_4$ has a boiling point[73] of 47°C, close to that of nickel tetracarbonyl, but its revolting odor means that in practice it is more easily detected. The corresponding manganese hydride $HMn(CO)_5$ is more stable. It is a liquid at room temperature with a relatively low vapor pressure (5 mm Hg at 25°C).[74] It is not readily formed by accident, and in any case manganese compounds find only occasional use in carbonylation chemistry.

Most metal carbonyl vapors have a musty smell [$HCo(CO)_4$ in particular has, as noted, a powerful and characteristically unpleasant odor, which warns of its presence], but any level that can be detected by the average person must be regarded as a dangerously high concentration.

3.4. Carbonylation Catalysts

Preparations of the most common carbonylation catalysts are given in Chapter 12, together with some of their individual properties. Here, more general aspects of ligands and organometallic and coordination compounds are considered. Most of the noncarbonyl ligands involved in homogeneous catalysis are good σ-electron donors, and many of them, like carbon monoxide, are quite effective π-acceptors. They invariably possess some toxic properties (sometimes, like carbon monoxide, they bind strongly to hemoglobin), but because they are often solids or liquids of relatively low vapor pressure they do not present the same level of hazard as carbon monoxide. Nevertheless, skincontact should always be avoided, and it is advisable to use such ligands in a fume cupboard. Before preparing catalysts, and especially when using unfamiliar ligands, current safety data should be carefully checked.

Phosphine ligands, PR_3, are generally oxygen sensitive, forming the corresponding phosphine oxide (R_3PO) more or less readily. The lower trialkyl derivatives are volatile and ignite spontaneously in air, and while liquid tributylphosphine is less reactive, it does slowly adsorb oxygen. Crystalline triphenylphosphine is much more stable and is in fact by far the most commonly used phosphine in carbonylation catalysis. It melts at $80\,^{\circ}C$, and any contaminating phosphine oxide is readily removed by recrystallization from ethanol. Recrystallized triphenylphosphine should always be used for catalyst preparations, and, if stored, the recrystallized material should be kept in dark bottles under dinitrogen to restrict oxidation to a minimum. Its oxidation appears to be promoted by sunlight, and is catalyzed by a number of transition metal species.

Phosphites, $P(OR)_3$, are weaker electron donors than the corresponding phosphines but they are still often good ligands, most probably because of their greater π-acceptor abilities. The commonly used phosphites such as $P(OMe)_3$ and $P(OPh)_3$ are liquid at room temperature and are usually more stable and easier to handle than liquid phosphines, but they find much less use in transition metal catalysis.

Most of the transition metal "catalysts" discussed in this book are catalyst *precursors*, since the actual complex added to a reaction is frequently reduced, dissociated, or otherwise transformed before entering the catalytic cycle. Such precursors are usually neutral, stable species, with only moderate solubility in the organic solvents used for carbonylations. Their stability allows ready isolation and purification, and their limited solubility is not generally a problem in use since only very small amounts (typically 0.1–1 mol %, relative to the reactants) are needed. Low oxidation state transition metal species such as $Pd(PPh_3)_4$ and Wilkinson's catalyst, $RhCl(PPh_3)_3$, are, however, somewhat light and oxygen sensitive. They should be stored in the dark under nitrogen, preferably in a refrigerator. Some metal carbonyls, especially $Co_2(CO)_8$, are best stored refrigerated in sealed ampoules under carbon monoxide to further inhibit degradation.

A few strongly reduced transition metal compounds are even pyrophyric and can cause ignition (possibly explosively) of flammable solvent/air mixtures. This must be borne in mind when handling residues from reactions involving heterogeneous catalysts, and homogeneous systems that may contain species very different from those originally added to the reaction. The rule is never to allow such residues to become dry; they should be transferred while moist to the appropriate collection container, prior to disposal or subsequent metal recovery.

The toxicological properties of most transition catalysts have not been fully evaluated, and all such materials should be handled carefully. Some people may become sensitized to a very small number of soluble chloroplatinum compounds, leading to a condition known as platinosis.[58, 75] This is discussed further in Chapter 12 (Sec. 12.3.2). Other platinum group metals do not cause such sensitization, nor do most of the more common transition metals (though, surprisingly, many people can become sensitized to zinc compounds). The present authors are not aware of complications arising from the use of palladium- or rhodium-based carbonylation catalysts described in this book, but it is always prudent to exercise caution.

3.5. Carbonylation at Atmospheric Pressure

Much of the apparatus used for atmospheric pressure carbonylations can be based on conventional laboratory glassware, and the procedures are largely derived from methodology developed over a number of years for carrying out catalytic hydrogenations and organometallic reactions under inert atmospheres. As noted previously, the solubility of carbon monoxide in most common solvents is very low compared with the concentration of the substrate being carbonylated, so it is necessary to constantly replace it in solution as it is consumed. Both rapid stirring and high pressures facilitate transfer of gas into solution.

3.5.1. Dynamic Systems

It is possible to carry out atmospheric pressure carbonylation reactions merely by bubbling a stream of carbon monoxide through the stirred reaction solution. A magnetic stirrer is usually adequate for small-scale reactions, and a reflux condenser would normally be fitted to the flask. Excess carbon monoxide is vented high into the fume cupboard via a small bubbler.

Initially, air is displaced by carbon monoxide, and the catalyst is then added before the flask is raised to reaction temperature. The use of a bubbler is important in preventing back-diffusion of air into the system, and in allowing the excess gas flow to be monitored and minimized. When the reaction is completed, the mixture is allowed to cool while maintaining the flow of carbon monoxide, and when cold the apparatus is flushed with nitrogen and then with a stream of air. As noted earlier, it is inadvisable to have carbon monoxide, air, and catalyst present together, especially at high temperatures, as flammable or explosive mixtures could be ignited by dry catalyst on the walls of the flask.

The procedure outlined maintains an atmosphere of carbon monoxide above the liquid in a straightforward fashion, but is really suitable only for carrying out occasional exploratory experiments. When larger-scale synthetic work is necessary, or where carbonylation reactions are carried out on a more routine basis, it is worth installing a slightly more sophisticated assembly such as that described below. The key elements of this system, which is shown in Fig. 3.3, are as follows:

1. A lecture bottle of carbon monoxide fitted with a regulator, pressure gauges, and needle valve.
2. A relatively large oil bubbler fitted with an internal ground glass float valve and vented high in the fume cupboard. Commercially available bubblers of this type tend, in the authors' experience, to be too small to cope with the gas flows occasionally required (for example, when purging the system), but the bubbler shown here is easily constructed by any professional glass-blower.
3. A two-way glass stopcock, allowing the system to be alternately evacuated and refilled with carbon monoxide.
4. A vacuum pump. Although a rotary oil pump is shown in Fig. 3.3, in practice even an efficient water pump operating at, say, 0.02 bar (and suitably protected against suck-back) can be acceptable.*
5. A stirred reaction vessel. This should be relatively thick walled to withstand repeated evacuation, and multinecked to allow for auxiliary equipment such as additional funnels, condensers, and a

---

* This is because repeated evacuation and refilling with carbon monoxide very effectively purges the system of air. Three evacuate/refill cycles would reduce the partial pressure of air to $(0.02)^3 = 8 \times 10^{-6}$ bar, equivalent to about 0.0001 mmol of O_2 per liter. Given a typical catalyst loading of 0.1–1 mmol and a total system volume of around 1 liter, any catalyst deactivation due to oxidation would clearly be insignificant.

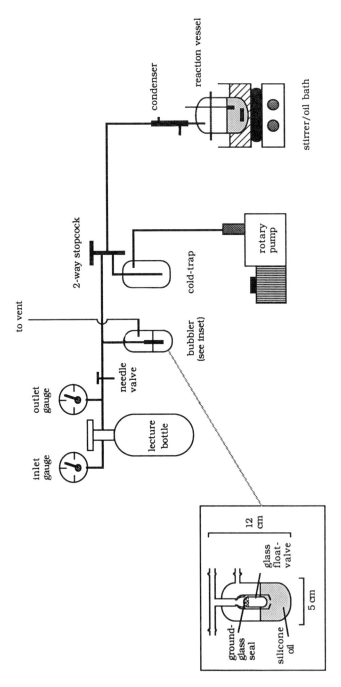

Figure 3.3. Simple apparatus for carbonylation at atmospheric pressure (inset shows details of nonreturn bubbler). The apparatus should be installed in a fume cupboard.

mechanical (as opposed to magnetic) stirrer. A flange-flask meets all of these criteria, and has the additional advantage over a normal flask that removal of insoluble reaction products and catalyst residues is much more straightforward.

In operation, a typical carbonylation procedure such as the conversion of an aryl halide to a benzoate ester, catalyzed by $Pd(PPh_3)_2Cl_2$, might be carried out as follows:

1. With the system assembled, the reaction vessel connected, and the two-way stopcock closed, the needle valve is opened to allow a slow stream of carbon monoxide to pass through the bubbler.
2. The system is evacuated and refilled with carbon monoxide. The float valve in the bubbler closes as soon as the evacuated system is opened to the gas supply, but reopens when atmospheric pressure is reestablished in the system.
3. The reactants and dry, degassed solvent are added to the flask, which is then cooled to $-78°C$. The catalyst is added.
4. The system is evacuated and refilled three times with carbon monoxide, and is then left open to the bubbler, with a *very* slow stream of gas passing through.
5. The reactor is raised to reaction temperature and the contents are stirred briskly.
6. The progress of the reaction can be checked, once the temperature is steady, by periodically shutting off the carbon monoxide supply at the needle valve. If gas is being consumed, the oil level will rise in the bubbler, closing the float valve. (There may be an induction period while the active form of the catalyst is generated.)
7. When the oil no longer rises after shutting off the gas supply the reaction is finished, and the system should be flushed with nitrogen.

In order to minimize back-diffusion of vapors into the system, and to maximize the concentration of carbon monoxide in solution (see Sec. 3.24), it is best to choose relatively involatile solvents (for example, xylene or *n*-butanol rather than benzene or methanol) and high-boiling-point coreactants such as tributylamine, rather than, say, triethylamine. Mechanical stirring via an oil-sealed gland may be necessary for larger-scale reactions, or where insoluble products are formed, but magnetic stirring is usually adequate for small-scale exploratory work. Where quantitative measurements of gas uptake or kinetic data are required, a reservoir and volumetric device such as a gas burette (Fig. 3.1) can be readily incorporated into the system, as discussed in Sec. 3.2.4.

3.5.2. Balloons—A Cautionary Note

Once equipment has been flushed with inert gas, a commonly used method of maintaining a slightly positive pressure in the system is to attach a thick-walled balloon or bladder filled with the gas, usually to the top of a reflux condenser. The balloon acts as a gas reservoir, ensuring a flow of gas out of, rather than air into, the reaction vessel in the event of a slight leak, and has the advantage of using less gas than when the apparatus is continuously purged. The thick-walled latex balloons supplied for this purpose are inexpensive and are available in a range of sizes.

However, while balloons may offer a convenient method of maintaining an inert atmosphere, they have also on occasion been mentioned in the literature as carbon monoxide reservoirs. On safety grounds it is clearly undesirable to fill balloons with carbon monoxide, in view of their obvious potential for exploding or otherwise discharging their contents unexpectedly. The use of balloons in carbonylation chemistry cannot therefore be recommended.

3.6. High-Pressure Operation

Throughout this book it is stressed that many extremely useful carbonylation reactions can be carried out in conventional glassware at ambient pressure. However, a number of valuable syntheses do require elevated pressures, so that the autoclave still has an important role to play in laboratory-scale carbonylation. Many well-designed, small-scale units are now commercially available, enabling more forcing conditions of pressure (and, if required, temperature) to be used. Some reactions (the synthesis of keto-esters from aryl halides, for example) proceed smoothly at 10–20 bar pressure of carbon monoxide, whereas at atmospheric pressure the reaction rates may be impractically slow or the desired product may not even be obtained at all.

Small units such as the Parr "Bench Top Reactors" shown in Fig. 3.4 provide all the facilities necessary for stirred operation at moderate pressures and elevated temperatures, yet they can be operated inside a fume cupboard using procedures similar to those described above for glass apparatus. Reactor volumes as small as 50 cm^3 are available, and larger units of this type are also supplied.

Recent advances in reactor engineering have resulted in autoclaves for larger-scale work that are much more convenient to use than earlier designs.[76] These are typified by the self-contained "Zipperclave" units (Fig. 3.5) supplied by Autoclave Engineers. This range provides standard

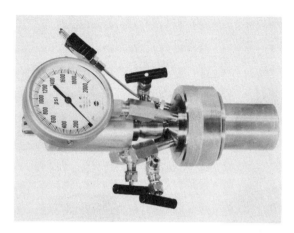

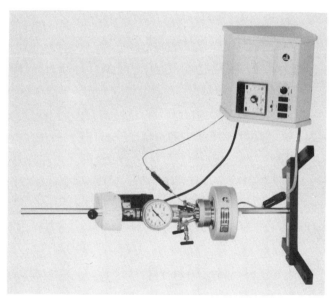

Figure 3.4. Parr Bench-top microreactors (25 or 50 cm^3) enable small-scale carbonylations to be carried out at elevated pressures and temperatures. A single vertical support carries the stirrer-motor and reactor, and the whole assembly can be set up in a conventional fume cupboard.

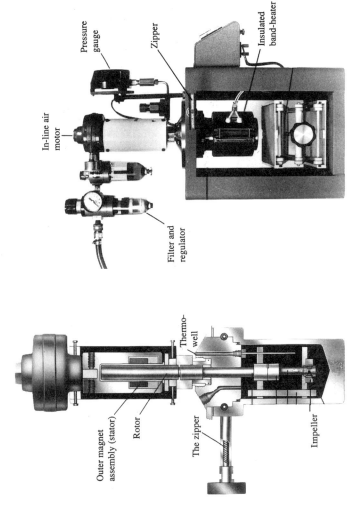

500 cm³ 'Zipperclave' System   Complete 'Zipperclave' Assembly

Figure 3.5.   Rapid-opening devices such as that used on the Autoclave Engineers' Zipperclave enable larger-scale carbonylations to be readily carried out, typically at up to 100 bar and 200°C. Reactor volumes of up to 4 liters are available.

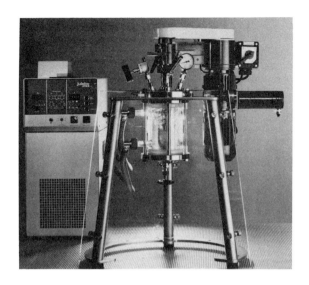

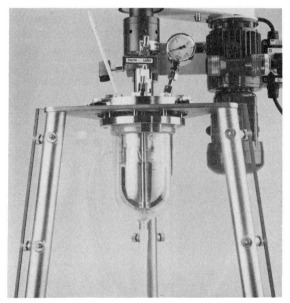

Figure 3.6. The Buchi laboratory autoclave, shown with a jacketed glass reactor for operation at up to 200°C and 12 bar. The stirrer incorporates a magnetic coupling drive. Reactors are available in a range of sizes, and also in metal for operation at higher pressures.

reactors with volumes of 500–4000 cm^3, and uses a pressure-sealing system that requires no bolts or screw threading. All that is necessary to secure the body to the cover, which carries the stirrer and associated fixtures, is to push a spring section. After reaction, the body containing the reaction mixture is removed simply by pulling the spring and lowering the body. A few years ago, Buchi AG introduced a stirred glass autoclave system with a maximum operating pressure of 12 bar, enabling the course of the reaction to be monitored visually. This unit is gaining in popularity with many chemists, and stainless steel reactors (250–2000 cm^3) for the system are available if higher pressures are required. A unit equipped with a glass reactor is shown in Fig. 3.6.

3.7.   Product Work-Up

Work-up of carbonylation reaction mixtures, especially those from high-pressure reactions, can lead to outgassing of dissolved carbon monoxide so that product work-up should always be carried out in a fume cupboard. Transition metal residues can often be removed by merely passing a solution of the reaction mixture through a short bed of filter-aid, and this has the advantage of retaining the metal in a small volume for subsequent recovery. Distillation of the product, if sufficiently volatile, is perhaps the best method of obtaining metal-free material, and this approach also concentrates the metal residue. When metal recovery is less important, column chromatography is probably the most widely used method of isolating pure product.

Chapter 4
Synthesis of Aldehydes

4.1. Introduction

The synthetic routes to aldehydes are probably more diverse and numerous than those to any other class of carbonyl compound, reflecting not only the intrinsic value of aldehydes as synthetic intermediates but also the scarcity of truly general routes to aldehydes themselves. As a result, a great number of relatively specific syntheses have been devised, often relying on partial oxidation or reduction procedures in which the danger of over-reaction is always present. In contrast to such redox-based procedures, carbonylation chemistry offers a limited number of general synthetic routes to aldehydes which are both selective and high yielding, and often use readily accessible starting materials. The two must generally useful substrates for carbonylative aldehyde synthesis are halocarbons and alkenes, and a number of catalytic or stoichiometric reactions will be described that allow the synthesis of a wide variety of functionalized aldehydes under relatively mild conditions.

Although this chapter concentrates on *transition metal* mediated carbonylations, other types of carbonylation reagent can also be useful. The long-established Gatterman–Koch reaction, for example, affords aromatic aldehydes by carbonylation of arenes in the presence of Lewis acids such as aluminum chloride or boron trifluoride. Even with $AlCl_3$, however, addition of a catalytic quantity of transition metal compound [copper(I) chloride] is necessary for optimum activity.

4.2.   Aldehydes: From Halocarbons

Carbonylferrate salts such as $K^+[HFe(CO)_4]^-$, which is readily prepared from $Fe(CO)_5$ and ethanolic potassium hydroxide, react with alkyl bromides and iodides to form alkyliron complexes. Under ambient conditions these complexes undergo insertion of carbon monoxide and, in the presence of excess CO, aldehyde is eliminated and $Fe(CO)_5$ is regenerated as shown in Scheme 4.1. Unfortunately the reaction is not catalytic and the hydridocarbonyl ferrate salt must be prepared in a separate step.[77]

Scheme 4.1

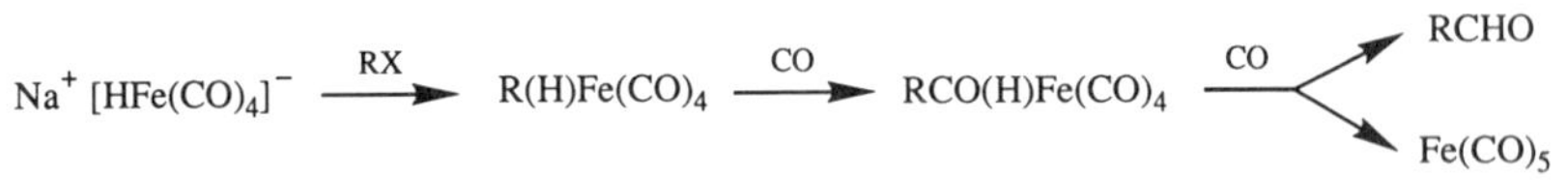

$$Na^+[HFe(CO)_4]^- \xrightarrow{RX} R(H)Fe(CO)_4 \xrightarrow{CO} RCO(H)Fe(CO)_4 \xrightarrow{CO} \begin{array}{c} RCHO \\ Fe(CO)_5 \end{array}$$

Using the commercially available salt $Na_2Fe(CO)_4$ (see also Chapter 12 for a synthesis of this very air-sensitive compound) a similar reaction takes place, but in this case the intermediate acyl complex is anionic and acid treatment is necessary to liberate the aldehyde (Scheme 4.2).[78]

Scheme 4.2

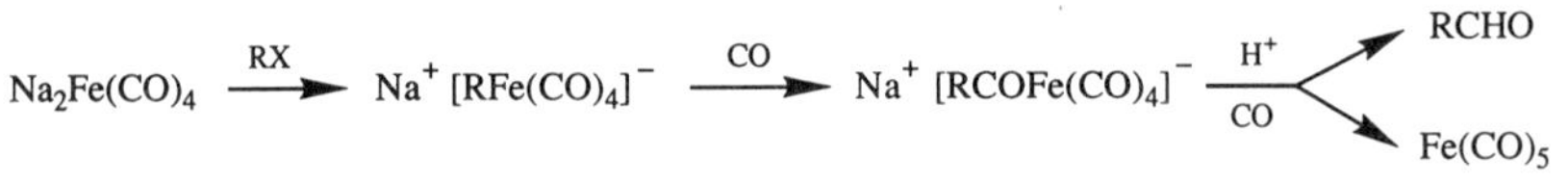

$$Na_2Fe(CO)_4 \xrightarrow{RX} Na^+[RFe(CO)_4]^- \xrightarrow{CO} Na^+[RCOFe(CO)_4]^- \xrightarrow[CO]{H^+} \begin{array}{c} RCHO \\ Fe(CO)_5 \end{array}$$

Aryl or heteroaryl halides are not normally reactive towards carbonyl-ferrate anions, but such halides can instead be converted to organolithium reagents, which react with iron pentacarbonyl to give acyl anions $[ArCOFe(CO)_4]^-$. On acidification, aldehydes are generated in moderate yield,[79] as shown in Scheme 4.3.

Scheme 4.3

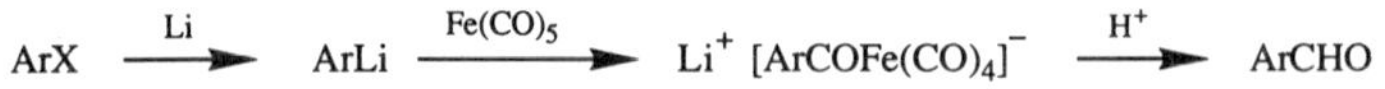

$$ArX \xrightarrow{Li} ArLi \xrightarrow{Fe(CO)_5} Li^+[ArCOFe(CO)_4]^- \xrightarrow{H^+} ArCHO$$

Aryl, heteroaryl and vinylic halides are *catalytically* converted to aldehydes on treatment with synthesis gas (CO/H_2) in the presence of palladium–phosphine complexes such as $Pd(PPh_3)_2Cl_2$. A stoichiometric amount of tertiary amine is required to remove the hydrogen halide formed in the reaction [Eq. (4.1)], and yields are generally good, though conditions are somewhat forcing (typically 100 bar and 100°C).[80] Yields are

$$R'_3N + RX + CO + H_2 \longrightarrow RCHO + R'_3NHX \qquad (4.1)$$

particularly high with iodo-compounds, which are carbonylated more rapidly than chlorides or bromides owing to the greater ease of oxidative addition of the $C-I$ bond to the catalyst. A possible mechanism for this reaction is shown in Scheme 4.4. The $C-Cl$ bond in a chloroarene may,

Scheme 4.4

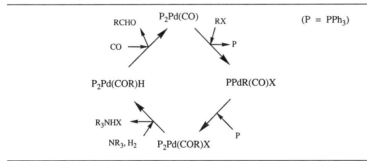

however, be activated to oxidative addition by formation of the arenetricarbonylchromium derivative,[81] and palladium-catalyzed formylation of the *complex* then proceeds in good yield at 30 bar total pressure (CO/H_2) and 130°C [Eq. (4.2)]. A recent communication indicates that, in the presence

$$\text{(complex)} - Cl + CO + H_2 \xrightarrow[\text{NEt}_3, 130°C]{Pd(PPh_3)_2Cl_2} \text{(complex)} - CHO \qquad (4.2)$$

of the bulky chelating ligand 1,3-bis(di-isopropylphosphino)propane (dippp), palladium catalysis can even allow low-pressure carbonylation of *free* chloroarenes to aldehydes.[82] By using formate ion as the hydrogen source, and $Pd(dippp)_2$ as catalyst [Eq. (4.3)], very high yields of aromatic aldehydes can be obtained from the corresponding chloroarenes under relatively mild conditions (6 bar, 150°C).

$$\text{ArCl} + \text{CO} + \text{HCO}_2\text{Na} \xrightarrow[\text{6 bar, 150°C}]{\text{Pd(dippp)}_2} \text{ArCHO} + \text{CO}_2 + \text{NaCl} \quad (4.3)$$

Alkyl halides are not usually good substrates for palladium-catalyzed carbonylation reactions, as such halocarbons tend to undergo catalytic dehydrohalogenation to form alkenes. By using the *platinum*-based catalyst precursor Pt(PPh$_3$)$_2$Cl$_2$, however, aldehydes can be obtained from primary alkyl iodides in good yield under mild conditions.[83]

Preparation of Pyridine-3-carboxaldehyde[80]
To a 50-cm^3 autoclave are added 3-bromopyridine (3.95 g, 25 mmol), triethylamine (10 cm^3), and Pd(PPh$_3$)$_2$Br$_2$ (0.30 g). The vessel is flushed with argon and then carbon monoxide, pressurized to 80 bar with synthesis gas (the commercially available CO/H$_2$ mixture; mole ratio 1:1), and heated to 145°C with magnetic stirring. After 24 h gas absorption ceases, and the vessel is cooled and the gases are vented. The reaction mixture is diluted with diethyl ether, filtered, evaporated under reduced pressure, and the residue finally distilled to give 2.15 g (80% yield) of pyridine-3-carboxaldehyde (bp 80°C at 16 mbar).

More recently, palladium-catalyzed chemistry has been extended to the synthesis of β, γ-unsaturated aldehydes, which can be obtained in moderate yield from the corresponding allylic halides,[84] as shown in Eq. (4.4) (X = Cl or Br).

$$\text{CH}_3(\text{R})\text{C}{=}\text{CHCH}_2\text{X} + \text{CO} + \text{H}_2 \xrightarrow{\text{[Pd]}} \text{CH}_3(\text{R})\text{C}{=}\text{CHCH}_2\text{CHO} \quad (4.4)$$

A closely related but lower-pressure formylation reaction has also been developed for conversion of aryl and benzyl halides to aldehydes.[85] This

$$\text{ArI} + \text{CO} + \text{PHMS} \xrightarrow{\text{Pd(PPh}_3)_4} \text{ArCHO}$$

[PHMS = poly(methylhydrosiloxane)]

$$(4.5)$$

approach uses poly(methylhydrosiloxane) as the hydrogen donor in place of molecular hydrogen, and the reaction is carried out in the presence of Pd(PPh$_3$)$_4$ at 80°C and 3 bar pressure of carbon monoxide, as shown in Eq. (4.5). Results obtained with a range of halides are given in Table 4.1.
One of the most versatile reactions of this type involves the use of tributyl tin hydride as hydrogen donor and Pd(PPh$_3$)$_4$ as catalyst [Eq.

Table 4.1. Formylation of Halides in the
Presence of $Pd(PPh_3)_4$ and
Poly(methylhydrosiloxane)[a]

Halide	Product (% yield)	
C_6H_5I	C_6H_5CHO	(96)
$4\text{-}BrC_6H_4I$	$4\text{-}BrC_6H_4CHO$	(95)
C_6H_5Br	C_6H_5CHO	(59)
$4\text{-}CH_3C_6H_5Br$	$4\text{-}CH_3C_6H_5CHO$	(48)
$4\text{-}NCC_6H_5Br$	$4\text{-}NCC_6H_5CHO$	(48)
		(82)
		(61)

[a] Reference 85.

(4.6)].[86] As shown in Table 4.2, a wide variety of substrates have been successfully carbonylated using this system, including aryl and vinyl iodides, benzylic halides, vinyl triflates, and allylic halides. Reaction condi-

$$RI + CO + Bu_3SnH \xrightarrow{\text{Pd(PPh}_3)_4} RCHO \qquad (4.6)$$

tions are mild (1–3 bar of carbon monoxide at around 50°C) and a range of functional groups can be tolerated. Regioselective formylation of allyl halides occurs at the less substituted allylic position and retention of geometry at the allylic double bond is normally observed.

Preparation of 4-Methoxybenzaldehyde[86]

Under 1 bar pressure of carbon monoxide, a solution of Bu_3SnH (0.646 g, 2.22 mmol) in toluene (10 cm^3) is added over $2\frac{1}{2}$ h to a stirred solution of 4-iodoanisole (0.467 g, 2.0 mmol), $Pd(PPh_3)_4$ (0.088 g, 0.076 mmol), and ethylbenzene (internal G.C. standard, 0.103 g), in toluene (4 cm^3) at 50°C. Analysis of the reaction mixture by gas phase chromatography indicates 100% conversion to 4-methoxybenzaldehyde, which is isolated in 72% yield by ether extraction and flash column chromatography.

Arenediazonium tetrafluoroborates, like iodoarenes, are readily converted to aromatic aldehydes in good yield by palladium-catalyzed carbonylation in the presence of poly(methylhydrosiloxane). Reactions are rather slow (up to 12 h), but a marked improvement in rate is obtained if

Table 4.2. Formylation of Halides and Triflates in the Presence of Bu_3SnH and $Pd(PPh_3)_4{}^a$

Substrate	Product (% yield)
C_6H_5I	C_6H_5CHO (93)
$4\text{-}CH_3COC_6H_4I$	$4\text{-}CH_3COC_6H_4CHO$ (100)
$4\text{-}NO_2C_6H_4Br$	$4\text{-}NO_2C_6H_4CHO$ (69)

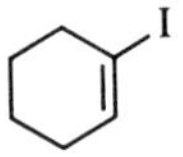

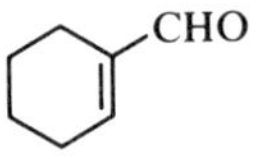

(89)

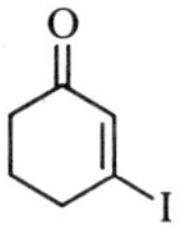

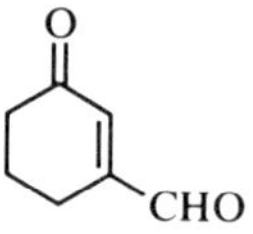

(84)

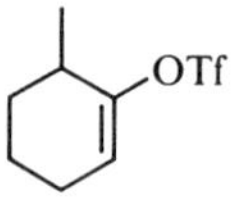

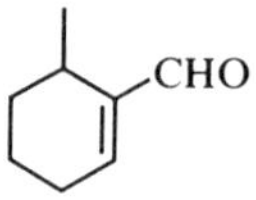

(96)

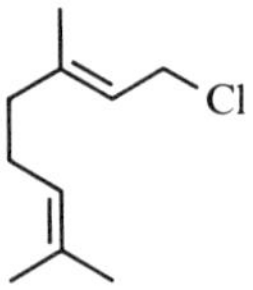

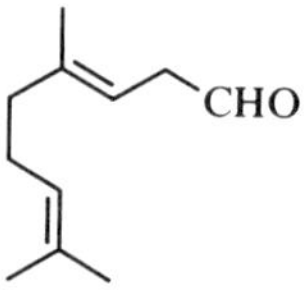

(54)

^{a} Reference 86.

the polymeric silane is replaced by Et_3SiH, when reaction times as short as two minutes can be achieved [Eq. (4.7)].[87] Functional group tolerance is high, with only the 2-nitro substituent producing a significant drop in yield.

$$ArN_2{}^+ \ + \ CO \ + \ Et_3SiH \ \xrightarrow{Pd(OAc)_2} \ ArCHO \qquad (4.7)$$

4.3. Aldehydes: From Alkenes

The transition metal catalyzed reaction of carbon monoxide and hydrogen with an alkene is termed *hydroformylation*, since it results in the addition of a hydrogen atom and a formyl group to the double bond. A mixture of linear and branched aldehydes is usually obtained, as shown in Eq. (4.8). Since the discovery[88] of this reaction by Roelen in 1938, hydroformylation has developed into an extremely important industrial

$$R\diagdown\diagup + CO + H_2 \longrightarrow \underset{R}{\overset{CHO}{\diagup\diagdown}} + R\diagup\diagdown\diagup CHO \quad (4.8)$$

process, and consequently has been the most intensively studied of all carbonylation reactions.[89] Industrially, the aldehyde products only rarely find direct application and are more often catalytically hydrogenated to alcohols, either *in situ* or in a separate step. The two major commercial products, 1-butanol (solvent) and 2-ethyl hexanol (plasticiser intermediate) are both derived from propene as shown in Scheme 4.5, via hydroformylation to butanal, followed either by direct hydrogenation or by aldol condensation and subsequent hydrogenation, respectively.

Scheme 4.5

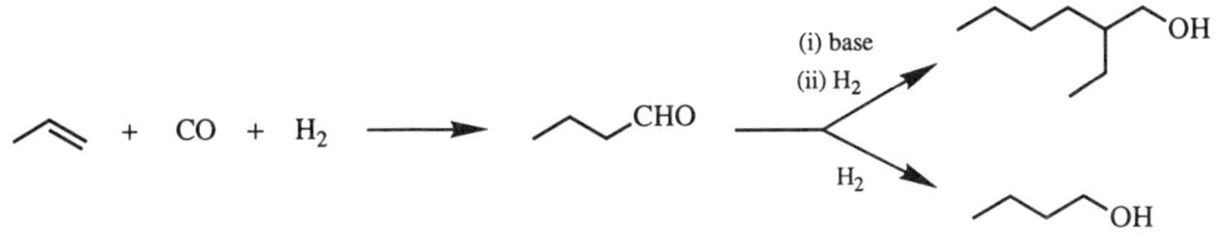

All the early work on hydroformylation was carried out using cobalt catalysts under relatively vigorous conditions (200–400 bar, 150–200°C) and gave mixtures of linear and branched aldehydes in a ratio of about 4:1. The accepted mechanism of the cobalt-catalyzed reaction (Scheme 4.6) involves generation of $HCo(CO)_4$ as the immediate catalyst precursor.[90]

Disadvantages of cobalt-catalyzed hydroformylation include only moderate selectivity to the generally more desirable linear isomers, low catalyst stability at the relatively high reaction temperatures, and substantial formation of by-products, including alkanes, ketones, and aldol condensation products. Some improvement can be made by the addition of donor ligands such as trialkylphosphines, since these increase both catalyst stability and selectivity to linear products. The linear-to-branched ratio typically improves to around 7:1, but aldehydes are no longer isolated since the ligand-substituted catalyst displays enhanced hydrogenation activity, and reduction of the aldehyde to alcohol occurs *in situ.*

Hydroformylation can be carried out under extremely mild conditions (1 bar of carbon monoxide/hydrogen, 25°C) in the presence of ligand-modified rhodium catalysts. The catalytic activity of these rhodium complexes is in fact some 10^3–10^4 times higher than that observed for the

Scheme 4.6

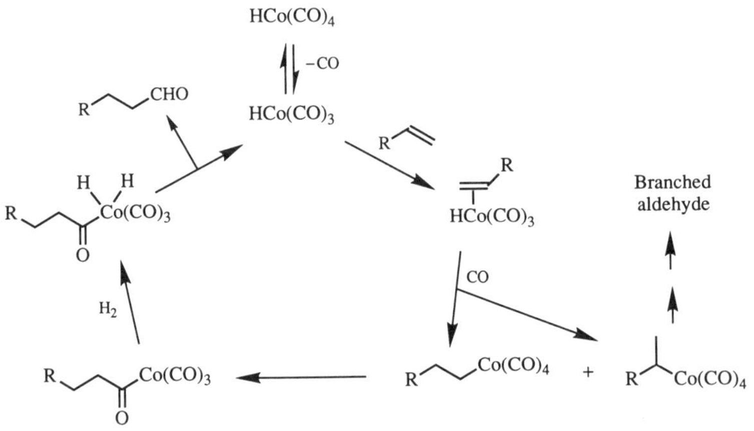

earlier cobalt catalysts, and side reactions such as hydrogenation are very significantly reduced. In the presence of bulky ligands such as triphenylphosphine, rhodium catalysis leads to very high selectivity for linear aldehyde, so that hydroformylation of 1-hexene to heptanal can be achieved under mild conditions with up to 94% selectivity, using $RhH(CO)(PPh_3)_3$ with an excess of triphenylphosphine.[91] Such selectivities are only achieved in the presence of high concentrations of PPh_3, and catalyst activity is significantly reduced under such conditions. Reasonable reaction rates *can* be achieved, however, at higher temperatures and pressures.[92] A novel catalyst consists of a binuclear rhodium-μ-thiolato complex bearing pendant amino groups.[93] In the presence of triphenylphosphine, this system offers both high selectivity for linear aldehyde and ease of catalyst recovery, since, even in the presence of a 40-fold molar excess of the phosphine, acidification of the reaction mixture leads to quantitative recovery of the original binuclear complex via its precipitation as an insoluble salt [Eq. (4.9)].

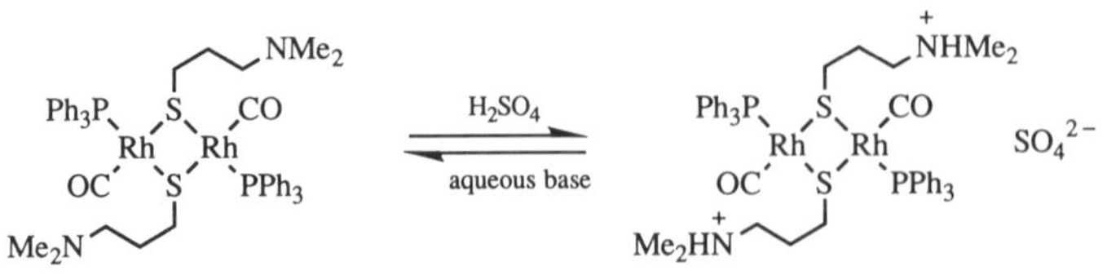

$$(4.9)$$

Although catalysts based on cobalt and rhodium have been the most intensively studied, other transition metals such as ruthenium, iron, palladium, and platinum also show activity. These metals have, in the past, offered few advantages over cobalt or rhodium, but some recent developments could change this situation. In particular, *exclusive* formation of linear aldehyde has been observed in the hydroformylation of 1-hexene using an aqueous ruthenium(III)/EDTA catalyst system,[94] and it has been shown that phosphine-modified platinum–tin complexes will catalyze hydroformylation of 1-alkenes with up to 99% selectivity for the linear aldehyde.[95] Unfortunately, competitive hydrogenation can be a problem with the latter system, especially when substituted alkenes are used.[96]

The reactivity of alkenes in hydroformylation follows a similar pattern to that observed in other carbonylation reactions—i.e., linear terminal alkenes react more readily than linear internal alkenes, which are in turn more reactive than branched alkenes. For example, using $RhH(CO)(PPh_3)_3$ as catalyst at 25°C and 1 bar total pressure, 1-heptene is hydroformylated 30 times faster than 2-heptene, and some 60 times faster than 2-methyl-1-pentene.[92] Relative reactivities are, however, highly dependent on reaction conditions and on any modification to the catalyst, so that at 70°C and 130 bar in the presence of an unliganded catalyst the rates of hydroformylation of terminal and internal alkenes differ by less than a factor of 2.[97]

By using organo*phosphite*-modified rhodium catalysts, even relatively unreactive alkenes such as 2-methyl-1-hexene, limonene, and cyclohexene can be hydroformylated under mild conditions.[98] High selectivities can also be achieved with such substrates (Scheme 4.7), leading to almost exclusive formation of 3-methylheptanal from 2-methyl-1-hexene. This result exemplifies an experimental "rule" due to Keulemans[99], which states that "in hydroformylation, formyl groups are not produced at quaternary carbon atoms."

Scheme 4.7

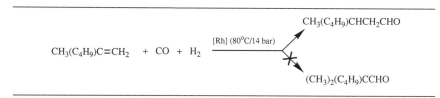

Rhodium complexes formed *in situ* with the ligand α,α-TREDIP (**1**) give very high regioselectivity for the branched isomer in the hydroformylation of styrene under mild conditions, and this type of catalyst has been used in the preparation of 2'-(2-methoxy-6-naphthyl)propanal (2), a precursor to the anti-inflammatory drug naproxen.[100]

1   2

Preparation of 2'-(2-Methoxy-6-naphthyl)propanal (2)[100]

6-Ethenyl-2-methoxynaphthalene (100 mg, 0.543 mmol) is added to a degassed solution of the diphosphine α, α-TREDIP (28 mg, 0.037 mmol) and tris(π-allyl)rhodium (8 mg, 0.035 mmol) in dichloromethane (2 cm³). The solution is stirred at room temperature under CO/H_2 (1:1 ratio; 1 bar total pressure) for 48 h and then passed through a short column of silica gel to remove rhodium salts. The solvent is removed under vacuum and examination of the crude product by ¹H NMR indicates the presence of two isomers, *2'-(2-methoxy-6-naphthyl)propanal* (2) (95%) and *3-(6-methoxy-2-naphthyl)propanal* (5%). Sublimation of this material gives a white solid product (110 mg, 95% yield) containing the same ratio of the two aldehydes.

Hydroformylation of dienes has been investigated, but in general mixtures of products are obtained. Octa-1,6-dienes, however, can be hydroformylated specifically at the terminal double bond,[101] and high selectivity for the linear aldehyde is obtained using a 2:1 mixture of carbon monoxide and hydrogen in the presence of $RhH(CO)(PPh_3)_3$ as catalyst [Eq. (4.10)]. By using a zero-valent rhodium complex (obtained by co-

$$(4.10)$$

condensation of rhodium vapor and a cycloalkadiene) as catalyst, cycloocta-1,5-diene may be selectively hydroformylated at 20°C and 70 bar to the corresponding cycloalkenecarboxaldehyde in good yield [Eq. (4.11)].[102]

$$(4.11)$$

The effects of functional groups on hydroformylation have been studied in some detail and can be broadly summarized as follows:

(i)   Electron-withdrawing substituents (carbonyl or nitrile groups, for example) which are conjugated with the double bond promote hydrogenation rather than hydroformylation, though this effect is less marked with rhodium than with cobalt. Where hydroformylation does occur, the formyl group tends to add at the carbon atom β to the substituent,[103] as in Eq. (4.12).

$$\text{CN} \quad + \quad CO \quad + \quad H_2 \quad \longrightarrow \quad \text{H} \overset{}{\underset{O}{\text{}}} \text{CN} \qquad (4.12)$$

(ii)   Electron-donor substituents conjugated with the double bond ($-OR$, $-NR_2$, $-Cl$) do not inhibit hydroformylation but favor addition of the formyl group at the carbon α to the donor group [Eq. (4.13)].[104]

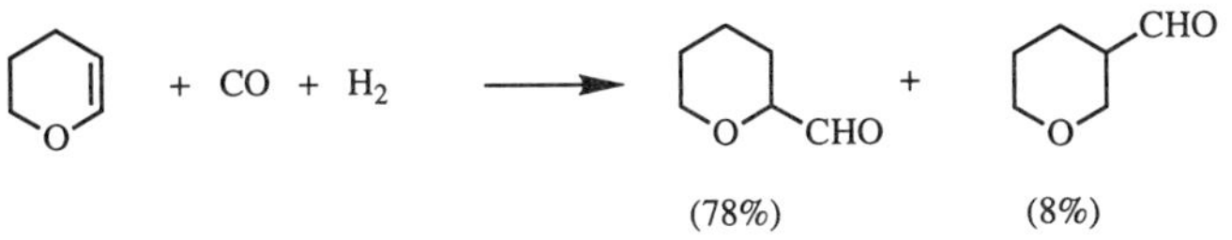

$$\begin{array}{c}\text{O} \end{array} \quad + \quad CO \quad + \quad H_2 \quad \longrightarrow \quad \underset{\text{CHO}}{\text{O}} \quad + \quad \underset{\text{O}}{\overset{\text{CHO}}{\text{}}} \qquad (4.13)$$

$$\qquad\qquad\qquad\qquad\qquad (78\%) \qquad\qquad (8\%)$$

(iii)   Substituents not conjugated with the double bond have less influence than conjugated substituents, though directing effects tend to be similar. Moreover, catalytic isomerization will often bring the substituent into conjugation with the double bond and may even result in loss of the carbon–carbon double bond altogether. Allylic alcohols, for example, isomerize to vinylic alcohols and thence to aldehydes[105] in the presence of $HCo(CO)_4$. Phosphine-containing rhodium catalysts, however, allow hydroformylation of most allylic alcohols since isomerization promoted by such complexes is slow, especially at high phosphine concentrations. Under such conditions 2-buten-1-ol affords the cyclic hemi-acetal (**3**) in 85% yield,[106] as shown in Scheme 4.8.

Scheme 4.8

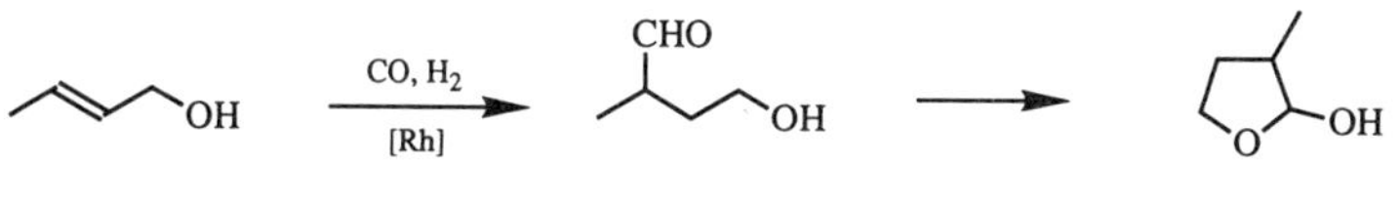

$$\underset{\text{OH}}{\text{}} \quad \xrightarrow[\text{[Rh]}]{CO, H_2} \quad \overset{\text{CHO}}{\underset{\text{OH}}{\text{}}} \quad \longrightarrow \quad \underset{\text{O}}{\overset{}{\text{}}}\text{OH}$$

3

The effects of *fluorinated* substituents on regioselectivity in hydroformylation have been studied in considerable detail, and a comprehensive review of this work is now available.[107] Dramatic differences are observed between different catalyst systems, so that, for example, the cobalt-catalyzed hydroformylation of 3,3,3-trifluoropropene affords 93% selectivity for linear aldehyde, whereas a phosphine-promoted ruthenium catalyst gives 92% selectivity for the branched isomer under otherwise similar conditions.

Although unsaturated aldehydes can be difficult to hydroformylate directly, protection via acetal formation often allows reaction with good yield and high selectivity, especially when using rhodium catalysts.[108]

An interesting review describes the application of hydroformylation to the synthesis of a range of vitamins, terpenes, and pharmacologically active compounds.[109] Hydroxycitronellal (**4**), for example, can be obtained by hydroformylation of 2,6-dimethyl-6-hepten-2-ol, which is in turn produced from 6-methyl-6-hepten-2-one via a Grignard reaction as shown in Scheme 4.9.

Scheme 4.9

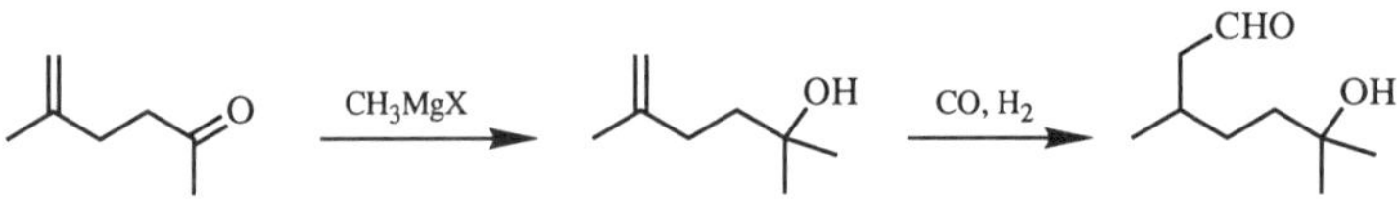

4.4. Asymmetric Hydroformylation

The synthesis of optically active compounds by asymmetric catalysis has progressed significantly in recent years, particularly in the area of asymmetric hydrogenation of prochiral alkenes. Optically active aldehydes *can* be obtained by hydroformylation of similar substrates, but only very recently have high enantiomeric excesses been reported. Early work concentrated on rhodium-based catalysts, since these operate under mild conditions and might therefore be expected to minimize racemization. Some progress was made with such systems, but it is in fact platinum-catalyzed reactions that have shown the most promise. Hydroformylation of styrene, for example, using PtCl₂/SnCl₂ in the presence of the chiral ligands (−)DIPHOL (**5**) or (−)DIOP (**6**) gives (+)-2-phenyl-propanal with

80%–90% enantiomeric excess.[110, 111] The same type of catalyst system allows hydroformylation of dimethyl itaconate in up to 82% enantiomeric excess (e.e.), though competitive hydrogenation is a problem here.[112] A number of other chelating phosphines, including Chiraphos (7), have been

used with rhodium and platinum systems,[113] but enantiomeric excesses rarely exceed 50%.

Asymmetric hydroformylation has been carried out using platinum catalysts bound to a polymer-supported chiral ligand. Optical yields are similar to those obtained in the corresponding homogeneous system (up to 70%), and the catalyst can be recycled with no apparent loss of activity.[114]

A major problem in asymmetric hydroformylation appears to be that the product aldehydes are susceptible to racemization via enol formation, so that the enantiomeric excess deteriorates as the reaction proceeds. The breakthrough in this field came with the development by Stille and co-workers of a system in which the aldehyde is converted to a nonenolizable derivative as it is formed.[115] In the presence of triethyl orthoformate, which converts the product aldehyde to its diethyl acetal, a catalyst formed from $PtCl_2$, $SnCl_2$, and the chelating chiral phosphine (−)BPPM (8) allows hydroformylation of a range of prochiral alkenes in essentially 100% enantiomeric excess. The product acetals may subsequently be hydrolyzed to aldehydes without loss of enantiomeric purity.[115]

4.5. Silylformylation

In view of the parallel behavior often observed for $H-H$ and $Si-H$ bonds, a number of workers have investigated carbonylation reactions in which dihydrogen, H_2, is replaced by a trialkylsilane, R_3SiH. Although no simple analogue of alkene hydroformylation has yet been found, alk*ynes* have very recently been shown to undergo catalytic "silylformylation" in the presence of the rhodium cluster carbonyl $Rh_4(CO)_{12}$, as shown in Eq. (4.14). Yields and geometrical selectivities (for the Z isomer) are both very high (often greater than 90%), as is the regioselectivity for silylation at the less hindered carbon.[116]

$$
\begin{array}{c}
Me_2PhSiH + CO \\
+ \\
R^1-\!\!\equiv\!\!-R^2
\end{array}
\quad
\xrightarrow[\text{Et}_3\text{N, C}_6\text{H}_6]{\text{Rh}_4(\text{CO})_{12}}
\quad
\begin{array}{c}
R^1 \quad\quad R^2 \\
\diagup\!\!=\!\!\diagdown \\
OHC \quad\quad SiMe_2Ph
\end{array}
+
\begin{array}{c}
R^1 \quad\quad R^2 \\
\diagup\!\!=\!\!\diagdown \\
PhMe_2Si \quad\quad CHO
\end{array}
\quad (4.14)
$$

Aldehydes may be converted to higher α-siloxyaldehydes by cobalt-catalyzed carbonylation in the presence of a trialkylsilane, as in Eq. (4.15). Silylformylation of the carbonyl group is favored over simple hydrosilylation by addition of triphenylphosphine, and to avoid formation of 1,2-bis(siloxy)alkene it is necessary to use the starting aldehyde in considerable excess.[117]

$$
RCHO + R^1_3SiH + CO \quad \xrightarrow{\text{Co}_2(\text{CO})_8} \quad
\begin{array}{c}
OSiR^1_3 \\
| \\
R \diagup\!\diagdown CHO
\end{array}
\quad (4.15)
$$

In view of the synthetic versatility of a strategically placed silyl group, it seems likely that silylformylation, and related reactions such as silylative decarbonylation (see Chapter 11), will see considerable development in the near future.

Chapter 5
Synthesis of Ketones

5.1. Introduction

Many of the carbonylative routes to ketones are closely related to those described in the previous chapter for the synthesis of aldehydes, with the obvious difference that a carbanion or carbonium ion source is now required in place of a hydride or proton donor. Favored starting materials are again halocarbons, alkenes, or alkynes, which are readily carbonylated to a range of symmetrical, unsymmetrical, or cyclic ketones. For example, one of the most intensively studied reactions of this type has been the catalytic coupling of an alkene with carbon monoxide and an alkyne. This leads to compounds containing the α,β-unsaturated cyclopentenone skeleton, and the reaction has been exploited in the synthesis of a series of natural products. In view of the importance of *cyclo*carbonylation, a separate section of this chapter is devoted to the synthesis of cyclic ketones.

5.2. Ketones: From Halocarbons and Related Compounds

As noted in the previous chapter, alkyl halides or tosylates react with the commercially available reagent disodium tetracarbonylferrate($-$II), $Na_2Fe(CO)_4$, to give anionic alkyliron complexes $[RFe(CO)_4]^-$. These in turn react further with active alkylating agents, as shown in Scheme 5.1, to give ketones in good yield.[118] Although the initial alkylation can be carried out with primary alkyl bromides and iodides, and primary and

Scheme 5.1

$$[Fe(CO)_4]^{2-} \; + \; RX \; \longrightarrow \; [RFe(CO)_4]^{-} \; \rightleftharpoons \; [RCOFe(CO)_3]^{-} \; \xrightarrow{R'Y} \; RCOR'$$

secondary alkyl tosylates, the reduced nucleophilicity of the monoanionic intermediate $[RFe(CO)_4]^{-}$ compared with its dianionic precursor means that only highly reactive alkylating agents (benzylic halides, primary iodides, or strongly activated alkenes such as acrylonitrile[119] are effective in the second step (Scheme 5.2).

Scheme 5.2

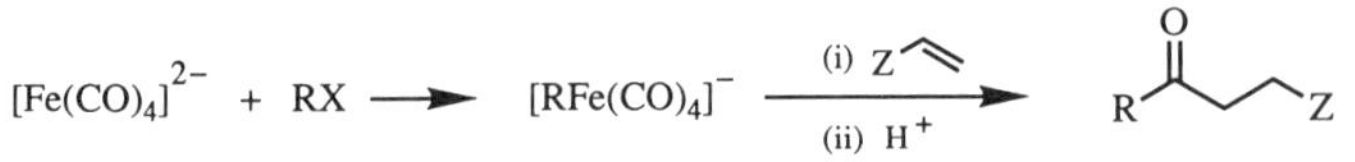

Diethyl 1,1-cyclopropanedicarboxylate undergoes nucleophilic ring opening with $Na_2Fe(CO)_4$, forming a dianionic alkyliron complex, which, under an atmosphere of carbon monoxide, rapidly undergoes insertion of carbon monoxide to give the corresponding acyl complex. *Double* alkylation (Scheme 5.3) then leads to a keto-diester, which may be formally regarded as the product of insertion of a cyclopropane into the $R-COR$ bond of a ketone.[120]

Scheme 5.3

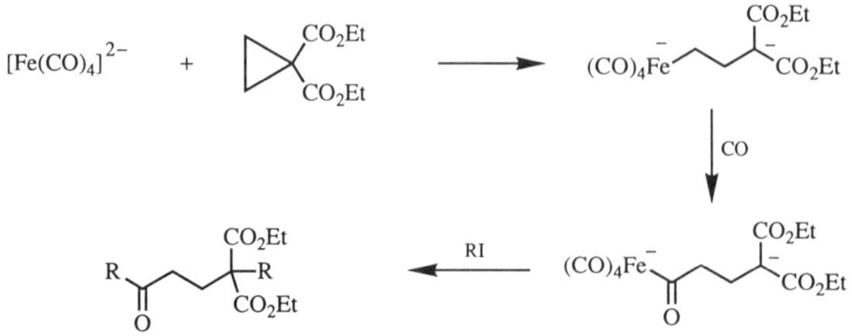

Under phase-transfer conditions $[Fe(CO)_4]^{2-}$, or its synthetic equivalent, can be generated *in situ* from iron pentacarbonyl and aqueous alkali, and on treatment with reactive halocarbons this forms symmetrical ketones.[121] An alternative approach involves the reaction of a carbanion such as an organolithium[122] or Grignard[123] reagent with $Fe(CO)_5$ itself to generate the anionic acyl complex $[RCOFe(CO)_4]^-$, which may be alkylated, as noted above, to give modest yields of ketones. For example, the benzoyl complex $[PhCOFe(CO)_4]^-$ (from phenyl lithium and $Fe(CO)_5$) reacts with benzyl bromide at $-40°C$ to give benzyl phenyl ketone in 57% isolated yield. Acyl halides can also be used in the second step since decarbonylation of the intermediate ketoacyl complex is rapid, and monoketones are therefore the major products (Scheme 5.4).[122] Aryl

Scheme 5.4

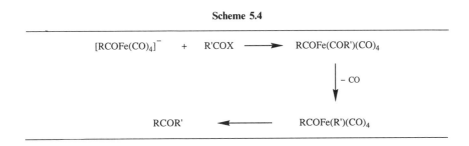

halides are not sufficiently reactive towards nucleophilic attack to take part in either the first- or second-stage reaction, but diaryliodonium salts have been successfully used in the second stage, as shown in Eq. (5.1), and give good yields of aromatic ketones.[124] Moreover, a recent communication indicates that aryl halides *can* be used in this type of reaction provided a catalytic quantity of $Pd(PPh_3)_4$ is added to the reaction mixture.[125] A probable mechanism involves oxidative addition of the aryl halide to palladium followed by acyl transfer from iron and finally reductive elimination of the product ketone with regeneration of $Pd(0)$.

$$[RCOFe(CO)_4]^- \ + \ [Ar_2I]^+ \ \longrightarrow \ RCOAr \ + \ ArI \qquad (5.1)$$

Synthesis of Ethyl 6-Keto-octanoate [Eq. (5.2)][118]

Ethyl-5-bromovalerate (2.16 g, 10.3 mmol) is added under nitrogen to a solution of $Na_2Fe(CO)_4$ (11.1 mmol) in 30 cm³ of *N*-methyl-2-pyrrolidone (distilled from calcium hydride), and the mixture is stirred for 1 hour before iodoethane (2.4 cm³) is added. After 24 h the solution is diluted with diethyl ether, washed with brine, dried, filtered, and the ether removed on the rotary evaporator. The residue is placed on a short (130-g) column of silica-gel, washed free of colored products with hexane, and then eluted with ether/hexane (1:1). Fractional distillation gives ethyl 6-keto-octanoate (1.4 g, 74%).

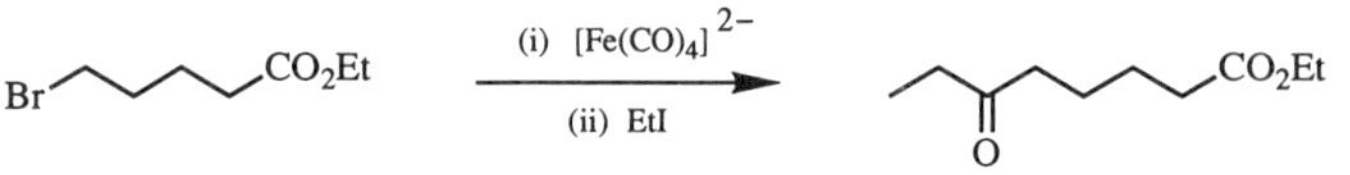

$$
\text{Br}\diagdown\diagup\diagdown\diagup\text{CO}_2\text{Et} \quad \xrightarrow[\text{(ii) EtI}]{\text{(i) } [\text{Fe(CO)}_4]^{2-}} \quad \diagup\diagdown\underset{\underset{\text{O}}{\|}}{\diagup}\diagdown\diagup\diagdown\diagup\text{CO}_2\text{Et} \qquad (5.2)
$$

The carbanion-based ketone synthesis,[122] has recently been extended to stabilized α-thio-derivatives $[\text{R}'\text{S}(\text{R}'')\text{HC}]^-$, which react with Fe(CO)_5 to give, after alkylation with iodomethane, the corresponding α-keto-sulfides [Eq. (5.3)].[126]

$$
[\text{R}'\text{S}(\text{R}'')\text{HC}]^- \xrightarrow{\text{Fe(CO)}_5} [\text{R}'\text{SCH}(\text{R}'')\text{COFe(CO)}_4]^- \xrightarrow{\text{MeI}} \text{R}'\text{SCH}(\text{R}'')\text{COMe} \qquad (5.3)
$$

A rather different route to an anionic acyl-iron intermediate for ketone synthesis involves attack of a carbanion on the 1,3-diene ligand of $[(\eta^4\text{-diene})\text{Fe(CO)}_3]$.[127] Under an atmosphere of carbon monoxide the first-formed alkyl complex inserts CO to give an acyl complex containing an intramolecularly coordinated C=C double bond, and alkylation with iodomethane affords the ketone (1) in good yield (Scheme 5.5).

Scheme 5.5

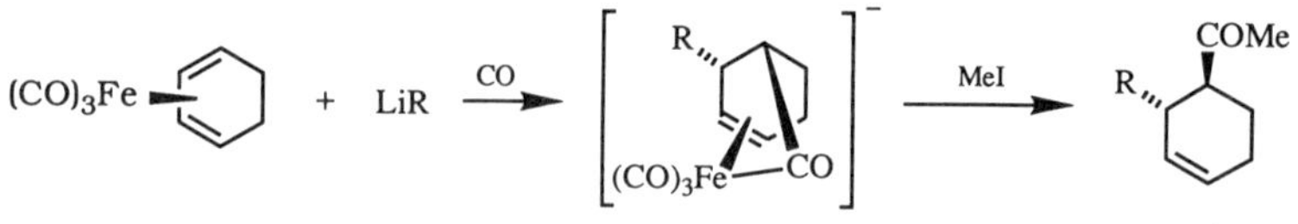

(R = CH_2CN; $\text{C(CH}_3)_2\text{CN}$; $\text{C(CH}_3)_2\text{CO}_2\text{Et}$)

Aryl iodides can be coupled with carbon monoxide and iodoalkanes in the presence of a stoichiometric amount of zinc–copper couple as halogen acceptor and $\text{Pd(PPh}_3)_4$ as catalyst.[128] Unsymmetrical ketones are thus produced, as shown in Eq. (5.4), under mild conditions (1 bar of carbon monoxide) and in good yield.

$$\text{ArI} + \text{RI} + \text{CO} \xrightarrow[\text{Zn-Cu}]{\text{Pd(PPh}_3)_4} \text{ArCOR} \qquad (5.4)$$

Preparation of n-Butyrophenone[128]

A flask containing a mixture of $Pd(PPh_3)_4$ (23.2 mg, 0.02 mmol) and zinc–copper couple (196 mg, 3.0 mmol) is purged with carbon monoxide, and a solution of iodobenzene (408 mg, 2.0 mmol) and 1-iodopropane (374 mg, 2.2 mmol) in dry THF (4 cm^3) is added under carbon monoxide. After stirring at 50°C for 24 h the mixture is worked up to give, after chromatography on silica-gel (hexane/ethyl acetate as eluent) a 90% yield of *n*-butyrophenone based on conversion of iodobenzene.

If the iodoalkane in Eq. (5.4) is replaced by a 1-alkyne, the reaction product is, somewhat surprisingly, a vinyl aryl ketone.[129] As shown in Scheme 5.6, this synthesis is thought to involve the formation of a vinyl zinc intermediate (**2**).

Scheme 5.6

$$\text{ArI} + \text{RC}\equiv\text{CH} + \text{CO} \xrightarrow[\text{Zn-Cu}]{[\text{Pd}]} \quad \underset{\mathbf{2}}{\text{Ar}\overset{\text{O}}{\diagup}\diagdown\overset{\text{ZnI}}{\diagup}\text{R}} \quad \longrightarrow \quad \text{Ar}\overset{\text{O}}{\diagup}\diagdown\overset{\text{H}}{\diagup}\text{R}$$

The palladium-catalyzed carbonylative coupling of halocarbons with organotin compounds is a potentially valuable route to unsymmetrical ketones.[130, 131] Benzylic bromides, for example, react with tetramethylstannane and carbon monoxide (20 bar), in the presence of $Pd(AsPh_3)_2Cl_2$, to give methyl benzyl ketones in fair yield [Eq. (5.5)].[131]

$$\text{ArCH(R)Br} + \text{Me}_4\text{Sn} + \text{CO} \xrightarrow{[\text{Pd}]} \text{ArCH(R)COCH}_3 \qquad (5.5)$$

Bis-allylic ketones may be synthesized under mild conditions as shown in Eq. (5.6), by reaction of allyl chlorides and allyltin reagents in the presence of carbon monoxide and a palladium catalyst.[132] This type of

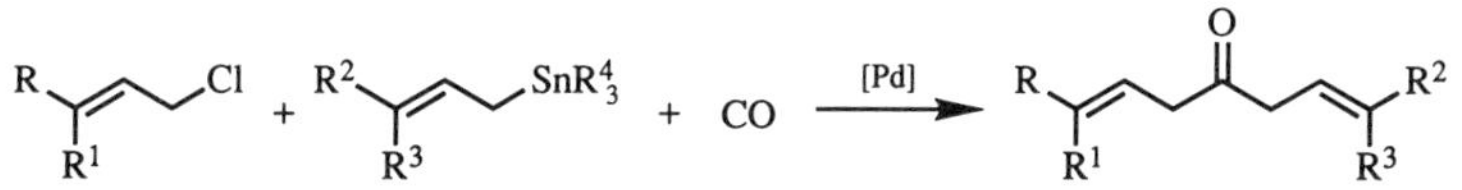

$$\underset{R^1}{R}\diagdown\diagup\diagdown\text{Cl} + \underset{R^3}{R^2}\diagdown\diagup\diagdown\text{SnR}_3^4 + \text{CO} \xrightarrow{[\text{Pd}]} \underset{R^1}{R}\diagdown\diagup\diagdown\overset{\text{O}}{\diagdown}\diagup\diagdown\underset{R^3}{R^2} \qquad (5.6)$$

reaction sequence has been extended to vinyl allyl ketones and provides, for example, a high-yield synthesis of egomaketone from prenyl chloride and 3-furanyltrimethyltin [Eq. (5.7)].[133] Closely related ketone syntheses

$$\text{(5.7)}$$

involve palladium-catalyzed reactions of vinyl iodides or triflates with vinyl-, alkenyl-, alkynyl-, or phenyl-tin compounds,[130] as shown, for example, in Eq. (5.8).

$$\text{(5.8)}$$

(X = I or CF$_3$SO$_3$)

For the specific reaction of bromo- or iodo-arenes with carbon monoxide and tetramethyl tin, giving acetophenones, a nickel-based catalyst system has proved satisfactory [Eq. (5.9)],[134] and Table 5.1 contains some representative results.

Table 5.1.   *Preparation of Aryl Methyl Ketones from Aryl Halides Using a Nickel Catalyst*[a]

Substrate	Product (yield)	Catalyst
⬡–I	⬡–COMe (28%)	Ni(CO)$_4$
⬡–I	⬡–COMe (78%)	Ni(CO)$_2$(PPh$_3$)$_2$
MeOC–⬡–I	MeOC–⬡–COMe (94%)	Ni(CO)$_2$(PPh$_3$)$_2$
NC–⬡–Br	NC–⬡–COMe (41%)	Ni(CO)$_2$(PPh$_3$)$_2$
thienyl–I	thienyl–COMe (51%)	Ni(CO)$_2$(PPh$_3$)$_2$

[a] Reference 134.

$$ArI + CO + SnMe_4 \xrightarrow[\text{HMPT}]{\text{[Ni]}} ArCOCH_3 + Me_3SnI \qquad (5.9)$$

Aryl triflates have recently been shown to be excellent substrates for palladium-catalyzed carbonylative cross coupling with organotin compounds.[135] Using a palladium(II) complex of the chelating ligand 1,1'-bis(diphenylphosphino)ferrocene (dppfc) as catalyst, and lithium chloride as promoter, high yields of ketone are obtained at 70°C and 1 bar of carbon monoxide [Eq. (5.10)]. The reaction is tolerant of alcohol, aldehyde, and ester functions, and allows transfer of most organic groups (alkyl, aryl, vinyl, or alkynyl) from the organostannane.

$$ArOSO_2CF_3 + CO + RSnR'_3 \xrightarrow[\text{LiCl}]{\text{Pd(dppfc)Cl}_2} ArCOR + R'_3SnOSO_2CF_3 \quad (5.10)$$

Organometallic compounds based on metals other than tin have also been used successfully as sources of anionic carbon in carbonylative ketone syntheses. Arylaluminum compounds, for example, react with aryl iodides and carbon monoxide in the presence of a palladium catalyst, as shown in Eq. (5.11), to give unsymmetrical diaryl ketones.[136] Choice of solvent is particularly important here (THF/HMPA mixtures are preferred) if formation of homo-coupling products is to be avoided. The tetraphenylborate ion [BPh$_4$]$^-$ can similarly be used in platinum-catalyzed cross-coupling reactions with carbon monoxide and iodoalkanes to give alkyl phenyl ketones.[137] Alkylmercury(II) iodides and aryl halides are carbonylated to the same type of product in the presence of an unliganded, iodide promoted, palladium catalyst.[138]

$$ArAlR_2 + CO + Ar'I \xrightarrow{\text{[Pd]}} ArCOAr' + R_2AlI \qquad (5.11)$$

Symmetrical diaryl ketones can be prepared from arylmercuric halides and carbon monoxide in the presence of palladium or rhodium catalysts [Eq. (5.12)], although yields are not particularly high.[139] A better-

$$2ArHgX + CO \longrightarrow Ar_2CO + Hg_2X_2 \qquad (5.12)$$

yielding, though more hazardous, synthesis of either symmetrical or unsymmetrical diaryl ketones involves the carbonylative coupling of arylmercuric halides and aryl iodides in the presence of nickel tetracarbonyl [Eq. (5.13)].[140] Some typical results are given in Table 5.2.

$$ArI + Ar'HgX + Ni(CO)_4 \longrightarrow ArCOAr' + NiXI + Hg \qquad (5.13)$$

Table 5.2. *Preparation of Unsymmetrical Diaryl Ketones Using*
Nickel Tetracarbonyl[a]

Aryl halide	Aryl mercuric halide	Product (yield)	
			(88%)
			(90%)
			(67%)
			(60%)

[a] Reference 140.

In contrast to the limited yields of symmetrical diaryl ketones available from direct carbonylation of arylmercuric halides [Eq. (5.12)], the corresponding rhodium-catalyzed carbonylative coupling of *vinyl*-mercuric halides shown in Eq. (5.14) gives excellent yields of divinyl ketones (**3**). The starting materials are readily accessible by hydroboration/ mercuration of alkynes, and carbonylation (which occurs under ambient conditions of temperature and pressure) proceeds with complete retention of geometry about the double bond.[141]

$$2 \quad \overset{R}{\underset{H}{\diagdown}}\!=\!\overset{H}{\underset{HgCl}{\diagup}} \quad + \quad CO \quad \xrightarrow{\ [Rh(CO)_2Cl]_2\ } \quad R\diagup\!\!\!\diagdown\!\!\!\diagup\!\!\overset{O}{\diagdown}\!\!\!\diagup\!\!\!\diagdown\!\!\!\diagup R \qquad (5.14)$$

3

Preparation of trans,trans-Trideca-5,8-dien-7-one (Structure 3;
$R = C_4H_9^-$)[141]

Lithium chloride (0.42 g) and *trans*-1-hexenylmercuric chloride (1.6 g) are weighed into a 250-cm^3 round-bottomed flask fitted with a serum-cap, gas inlet, and magnetic stirrer bar. The flask is flushed with carbon monoxide, and THF (40 cm^3) is then added by syringe. A solution of

$[Rh(CO)_2Cl]_2$ (9.7 mg) in THF (10 cm^3) is added, the flask is flushed with further CO, and the reaction mixture is then stirred under a carbon monoxide atmosphere for 24 h at room temperature. The reaction mixture is filtered to remove elemental mercury, diluted with pentane (80 cm^3), and washed three times with saturated brine. The organic phase is dried over sodium sulfate, concentrated, and finally purified by bulb-to-bulb distillation (bp 75–80°C at 0.35 mm Hg), to give the product (0.38 g, 78% yield) as a colorless oil.

Arenediazonium ions (as their BF_4^- and PF_6^- salts) can also be used as precursors to aryl ketones since, in the presence of palladium acetate, they couple readily with carbon monoxide and tetra-alkyl or tetra-aryl tin compounds.[142] High yields of unsymmetrical products are obtained, according to Eq. (5.15).

$$ArN_2^+X^- + CO + R_4Sn \xrightarrow{Pd(OAc)_2} ArCOR + N_2 + R_3SnX \quad (5.15)$$

A potentially useful route to α-keto carboxylic acids involves the *double*-carbonylation of iodoarenes in the presence of water, triethylamine, and $Pd(PPh_3)_2Cl_2$ as catalyst [Eq. (5.16)].[143] This, and related syntheses of α-keto acids, are discussed in more detail in Chapter 6 (Sec. 6.3).

$$ArI + CO + H_2O + NEt_3 \xrightarrow{[Pd]} ArCOCO_2H + Et_3NHI \quad (5.16)$$

Palladium-catalyzed carbonylation of siloxycyclopropanes leads to formation of 4-keto pimelate esters in good yield under very mild conditions (60°C, 1 bar).[144] In this reaction, which is shown in Eq. (5.17),

$$2 \underset{OR^2}{\overset{R^1\diagdown\text{OSiR}^3_3}{\triangle}} + CO + [PdCl_2] \longrightarrow R^2OOC\overset{R^1\ O\ R^1}{\diagup\diagdown\diagup\diagdown}COOR^2 \quad (5.17)$$
$$+$$
$$R^3_3SiCl + [Pd]$$

the siloxycyclopropane behaves formally as a nucleophilic homoenolate and carbon monoxide as a carbonyl dication (Scheme 5.7). Although Eq. (5.17) shows this reaction to be stoichiometric in palladium, only catalytic quantities of the transition metal are in fact required when the reaction is carried out in chloroform, apparently because this solvent is able to reoxidize palladium(0) back to palladium(II).

Scheme 5.7

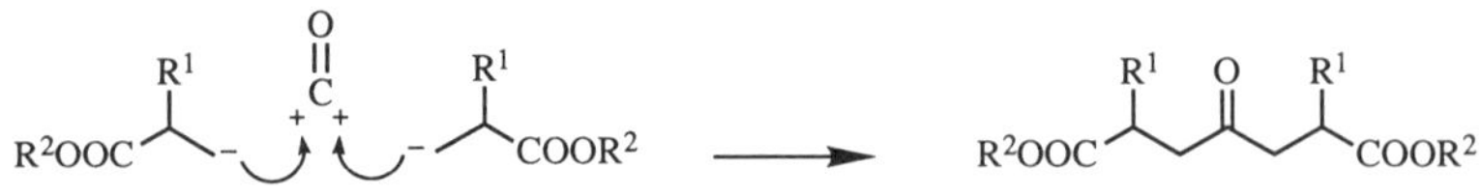

5.3. Ketones: From Alkenes and Alkynes

Alkyl ketones are sometimes produced as hydroformylation by-products, particularly at high concentrations of alkene and low pressures of carbon monoxide and hydrogen. Diethyl ketone, for example, can be obtained from ethylene in over 50% yield under such conditions, but this type of reaction is much less favorable with higher alkenes.[145]

Acyl dienes are accessible in good yield from reactions of alkyl or acyl halides with conjugated dienes and carbon monoxide in the presence of catalytic quantities of the $[Co(CO)_4]^-$ ion and a stoichiometric amount of base [Eq. (5.18)].[146] The probable reaction pathway has been established

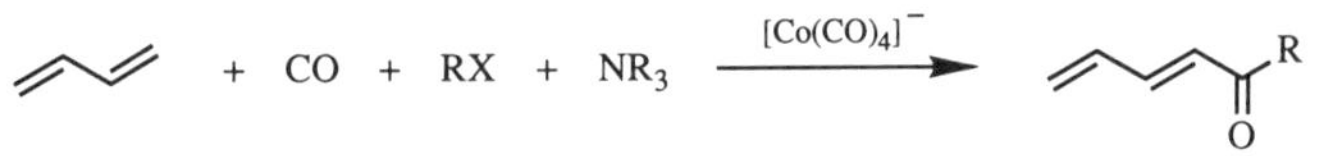

$$\diagup\hspace{-0.3em}\diagdown\hspace{-0.3em}\diagup \quad + \quad CO \quad + \quad RX \quad + \quad NR_3 \quad \xrightarrow{\;[Co(CO)_4]^-\;} \quad \diagup\hspace{-0.3em}\diagdown\hspace{-0.3em}\diagup\hspace{-0.3em}\diagdown\underset{O}{\overset{}{\diagup}}R \qquad (5.18)$$

by isolation of a series of intermediates, and a catalytic cycle based on interconversion of these species is shown in Scheme 5.8. Increased yields of acyl diene can be obtained under milder conditions (25°C, 1 bar) by

Scheme 5.8

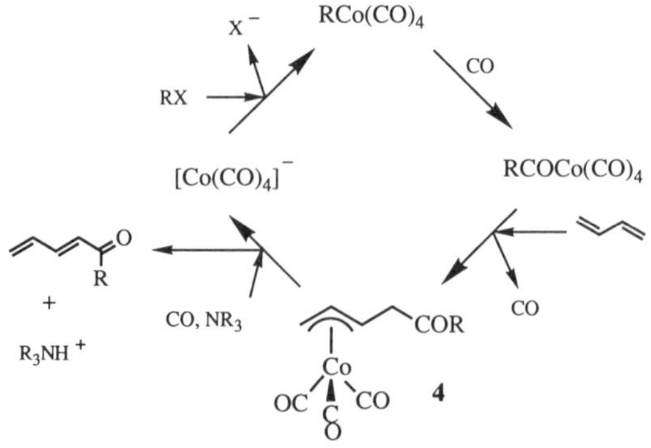

use of a phase-transfer catalyst.[147] The reaction is highly regio- and stereospecific, the acyl group normally being added to the least substituted carbon of the least substituted double bond with exclusive formation of the *trans*-acyl diene. The π-allylcobalt complex (**4**), formed as an intermediate in the acyl diene synthesis (Scheme 5.6), can alternatively be isolated and reacted with a stabilized carbanion to give the β,γ-unsaturated ketone (**5**), resulting from alkylation at the unsubstituted position of the π-allyl ligand [Eq. (5.19)].[148]

$$\text{(5.19)}$$

4   (W, Y = CO_2Me, $COMe$; Z = H, Me)   **5**

Under hydroformylation conditions, and in the presence of $Rh_4(CO)_{12}$ as catalyst, α,β-unsaturated ketones can be obtained from mixtures of ethene and substituted alkynes,[149] as in Eq. (5.20). Terminal alkynes in particular react with high regio- and stereospecificity to give *trans*-α,β-unsaturated ketones.

$$R^1C{\equiv}CR^2 \ + \ H_2C{=}CH_2 \ + \ CO \ + \ H_2 \ \xrightarrow{Rh_4(CO)_{12}} \ \begin{array}{c} R^1 \quad\quad R^2 \\ C{=}C \\ H \quad\quad C{-}Et \\ \quad\quad\ \ O \end{array} \quad \text{(5.20)}$$

Carbonylation of the organocuprate reagent $Li_2[CuR_2(CN)]$, obtained by reaction of an alkyl lithium with copper(I) cyanide, gives a product that can be used *in situ* for direct, nucleophilic, 1,4-acylation of α,β-unsaturated aldehydes and ketones [Eq. (5.21)].[150] Although this type of reaction is especially successful with cyclic α,β-unsaturated ketones, good yields of 1,4-addition products can also be obtained from α,β-unsaturated aldehydes.

$$R_2(CN)CuLi_2 \ + \ CO \ + \ \begin{array}{c} R^1 \quad\quad COR^4 \\ C{=}C \\ R^2 \quad\quad R^3 \end{array} \longrightarrow \quad \text{(5.21)}$$

Copolymerization of ethene with carbon monoxide at 25°C and 20–60 bar total pressure, in the presence of a divalent palladium catalyst such as $[Pd(MeCN)_3(PPh_3)]^{2+}$, leads to formation of the regularly alternating polyketone $(CH_2CH_2CO)_n$. The proposed catalytic mechanism involves initial formation of a palladium(II) hydride followed by alternate "insertions" of alkene and carbon monoxide and finally chain termination, via β-hydride elimination, to regenerate the starting hydride complex.[151]

5.4. Synthesis of Cyclic Ketones

Strained alkenes such as norbornadiene and its derivatives react photochemically with iron carbonyls [$Fe(CO)_5$ or $Fe_2(CO)_9$] to give fused-ring cyclopentanones in up to 70% yield [Eq. (5.22)].[152] Product stereochemistry is invariably *exo–trans–exo*, and unsymmetrical ketones can be prepared by the sequential addition of two different alkenes to the system. The reaction is thought to proceed via a bis-alkene complex of

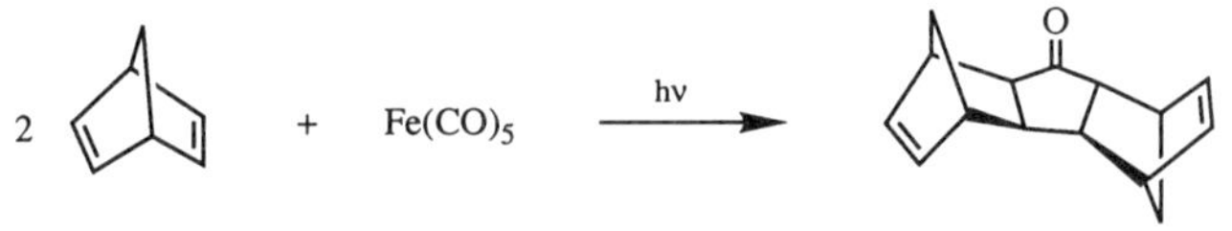

$$2 \quad \text{(norbornadiene)} \quad + \quad Fe(CO)_5 \quad \xrightarrow{h\nu} \quad \text{(product)} \qquad (5.22)$$

iron(0), which undergoes internal coupling to a metallacyclopentane, followed by carbon monoxide insertion and reductive elimination of cyclic ketone (Scheme 5.9).[153]

Scheme 5.9

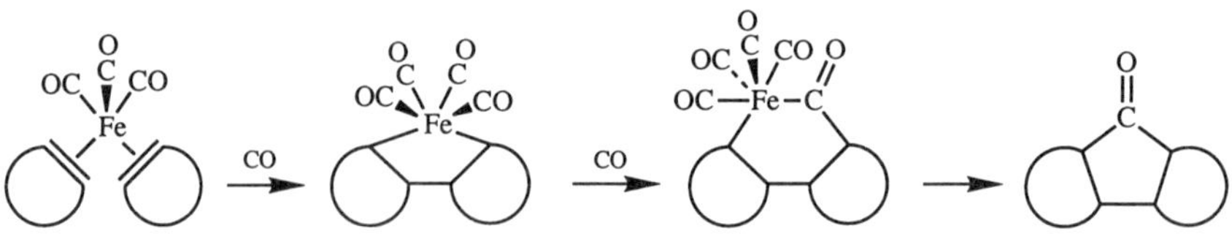

Tosylate esters containing a suitably positioned C=C double bond undergo carbonylative cyclization when treated with disodium tetracarbonylferrate(−II),[154] as shown in Eq. (5.23). The reaction is particularly effective for the synthesis of five- and six-membered rings, and although seven-membered ketones *can* be obtained, the latter synthesis is not regioselective and a mixture of isomers is formed. The reaction appears limited to monosubstituted alkenes, since the presence of a second substituent at either end of the double bond prevents cyclization.

$$ \text{(5.23)} $$

A high-yield synthesis of unsymmetrical cyclopentanones is provided by the reductive carbonylation of 3,3-dialkyl-1,4-pentadienes [Eq. (5.24)]. The reaction uses $Co_2(CO)_8$ as catalyst, and is particularly valuable as a route to *spiro*cyclopentanones.[155]

$$ \text{(5.24)} $$

$$ (R^1, R^2 = Me, -(CH_2)_4-, -(CH_2)_5-) $$

A related synthesis (sometimes known, after its discoverers, as the Pauson–Khand reaction),[156] involves the cobalt-promoted coupling of an alkene, an alkyne, and carbon monoxide to give an α,β-unsaturated cyclopentenone [Eq. (5.25)]. This reaction involves an intermediate

$$ \text{(5.25)} $$

binuclear alkyne-cobalt complex (6), and when using relatively unstrained and therefore less reactive alkenes it is preferable to synthesize this complex stoichiometrically first in a separate step.[157] It has recently been found that yields of cyclopentenone in the "stoichiometric" Pauson–Khand reaction can be significantly enhanced by addition of tributylphosphine oxide, and reaction rates can be improved by ultrasonic irradiation.[158] With

6

highly strained alkenes such as norbornadiene, however, only catalytic quantities of $Co_2(CO)_8$ need to be added to the reaction mixture.[159] Reactions involving very volatile or gaseous alkenes are assisted by the use of an autoclave, but since the reaction proceeds at only 1 bar total pressure of carbon monoxide and alkyne, many alkenes may be used at ambient pressure in hydrocarbon solvents. Fused-ring systems are readily obtained by this route, norbornene and acetylene, for example, reacting to give *exo*-hexahydro-4,7-methanoinden-1-one (**7**) in 74% yield [Eq. (5.26)].[159a]

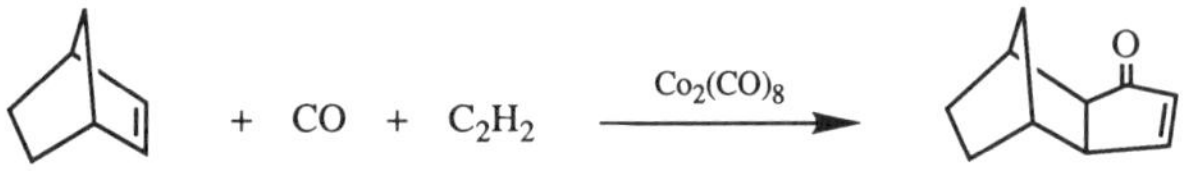

$$
\text{(structure)} + CO + C_2H_2 \xrightarrow{Co_2(CO)_8} \text{(structure)} \qquad (5.26)
$$

7

Synthesis of exo-Hexahydro-4,7-methanoinden-1-one (7)[159]
   A solution of dicobalt octacarbonyl (1 g, 3 mmol) and norbornene (3 g, 32 mmol) in iso-octane (100 cm³) is stirred at 60–70°C under acetylene (1 bar), and then under acetylene/CO (1:1, 1 bar) until gas absorption ceases (1550 cm³). The reaction solution is concentrated and then chromatographed on neutral alumina. Elution with a 1:1 mixture of petroleum ether (bp 40–60°C) and benzene gives the acetylene complex $Co_2(C_2H_2)(CO)_6$, and further elution with benzene–chloroform (1:1) gives a yellow oil, which, after distillation at 101–102°C (15 mm Hg), affords the product ketone (3.4 g, 74%), which slowly dimerizes on storage at room temperature (mp of dimer is 51°C).

   The Pauson–Khand reaction has been extended to the synthesis of natural products. Thus the complex $Co_2(2\text{-butyne})(CO)_6$ reacts regiospecifically with tetrahydro-2-(2-propenyloxy)pyran (**8**) to give a trisubstituted cyclopentenone (**9**) (32% yield), which, as shown in Scheme 5.10, can be converted to the antibiotic methylenomycin (**10**).[160]

Scheme 5.10

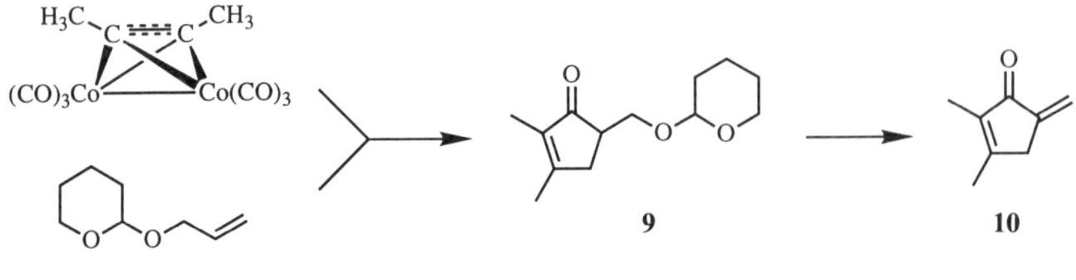

8   **9**   **10**

Synthesis of 2,3-Dimethyl(5-tetrahydropyran-2-yloxy)methylcyclopent-2-en-1-one[160]

(a) A solution of 2-butyne (2.8 cm³, 1.9 g, 35 mmol) in petroleum ether (50 cm³) is added over 0.5 h to a stirred solution of $Co_2(CO)_8$ (12 g, 35 mmol) in the same solvent (100 cm³) at 10°C under an atmosphere of nitrogen. The resulting red solution is stirred for 5 h at 10–15°C, and is then filtered through Celite filter-aid. After removal of the solvent under vacuum and chromatography on neutral alumina (petroleum ether as eluent), $Co_2(2\text{-butyne})(CO)_6$ is obtained as a dark red oil (11.3 g, 94% yield).

(b) A solution of this complex (2.5 g, 7.3 mmol) and the acetal (8) (3 g, 22 mmol) in dry toluene (100 cm³) is heated for 8 h under reflux in a nitrogen atmosphere. After filtering the dark blue solution through Celite filter-aid and removing the solvent under vacuum, the residue is extracted with chloroform and then with petroleum ether. Evaporation of the combined extracts and flash chromatography (diethyl ether/petroleum ether as eluent) gives the product (9) as a colorless oil (0.43 g, 32%).

Intramolecular cyclizations of cobalt complexes of substituted allyl-propargyl ethers lead to formation of 3-oxabicyclo[3.3.0]octen-6-en-7-ones in moderate yield [Eq. (5.27)].[161] A closely related *inter*molecular reac-

$$(5.27)$$

tion involves treating a solution of $Co_2(2\text{-butyne})(CO)_6$ in dry toluene with 2,5-dihydrofuran under an atmosphere of carbon monoxide and 2-butyne [Eq. (5.28)]. The resulting 3-oxabicyclo-octenone (11) is formed in 70% yield.[162]

$$(5.28)$$

11

A very short route to known prostaglandin synthons, shown in Eq. (5.29), involves reaction of the dicobalt complex of an appropriate alkyne with ethene in an autoclave at 160°C.[163]

$$(5.29)$$

The intramolecular version of the Pauson–Khand reaction has been extended as shown in Scheme 5.11 to give direct, high-yielding syntheses of bicyclo[3.3.0]enones which are suitable intermediates for further elaboration into coriolin and hirsutic acid precursors.[164]

Scheme 5.11

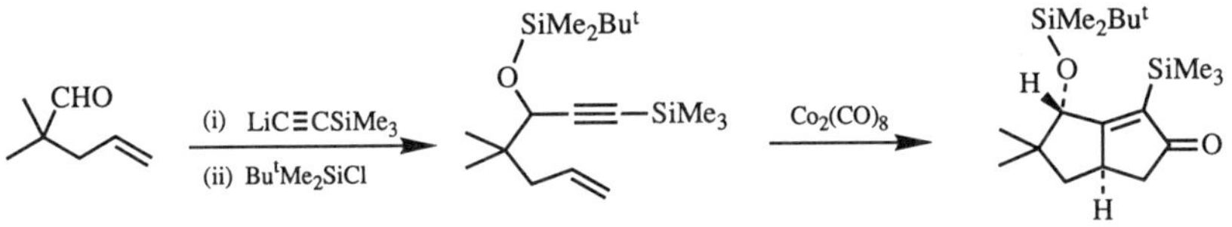

Attempted hydroformylation of enynes, with $Rh_4(CO)_{12}$ as catalyst, recently led to the discovery of a new cyclocarbonylation reaction [Eq. (5.30)],[165] closely related to the acyclic ketone synthesis shown in Eq. (5.20).

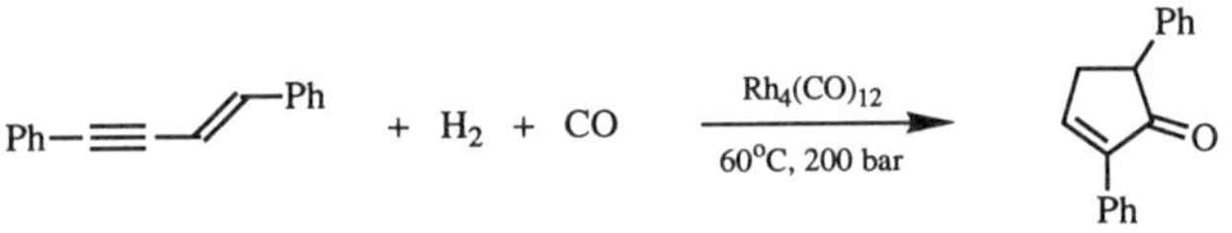

$$Ph-\!\!\equiv\!\!\diagup\!\!\diagup\!\!\diagdown\!\!^{Ph} \;+\; H_2 \;+\; CO \;\xrightarrow[60°C,\,200\,bar]{Rh_4(CO)_{12}}\qquad (5.30)$$

Carbonylation of 1-iodo-1,4-dienes in the presence of stoichiometric amounts of $Pd(PPh_3)_4$ leads to formation of α-methylene cyclopentenones [Eq. (5.31)].[166] The related reaction shown in Eq. (5.32) can be carried

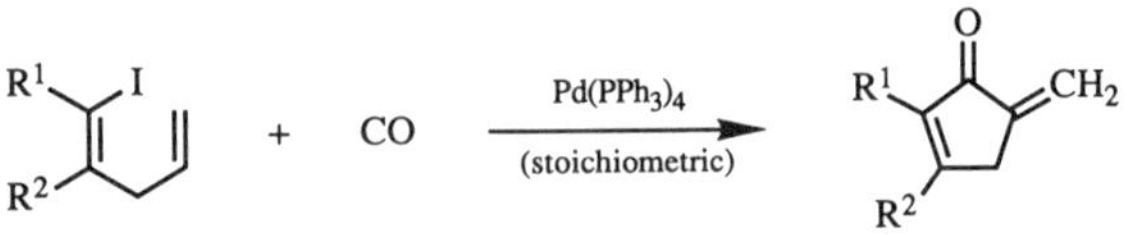

$$(5.31)$$

out using only catalytic quantities of palladium, but the presence of methanol results in hydroesterification of the exocyclic double bond with formation of the ester derivative (**12**).[167] The same reaction is applicable

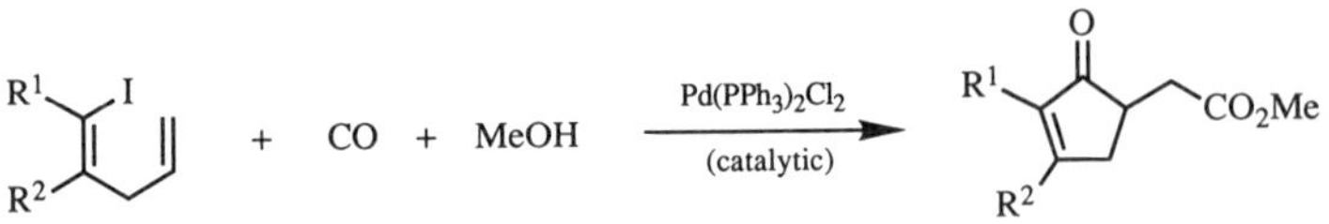

$$(5.32)$$

to the synthesis of cyclohexenones and spiro-compounds [Eq. (5.33)],[167] and all three variants of this cyclocarbonylation reaction [Eq. (5.31)–(5.33)] afford high yields of ketone.

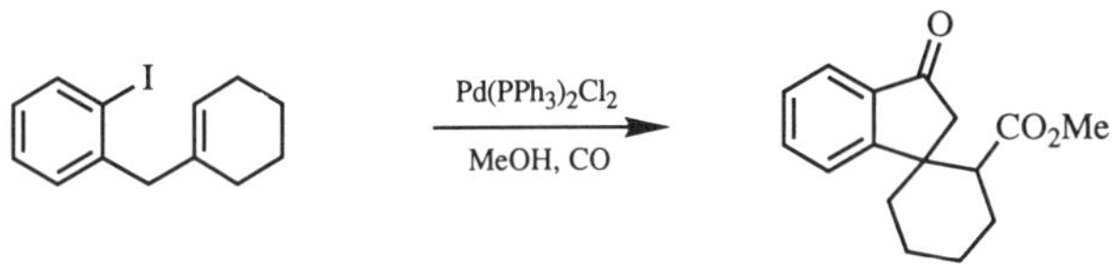

$$\text{(5.33)}$$

Addition of [(π-allyl)PdCl]₂ to norbornene affords a complex that, on carbonylation [Eq. (5.34)], gives the fused-ring cyclopentanones (**13**) and (**14**) as major products.[168] A related complex derived from norbornadiene

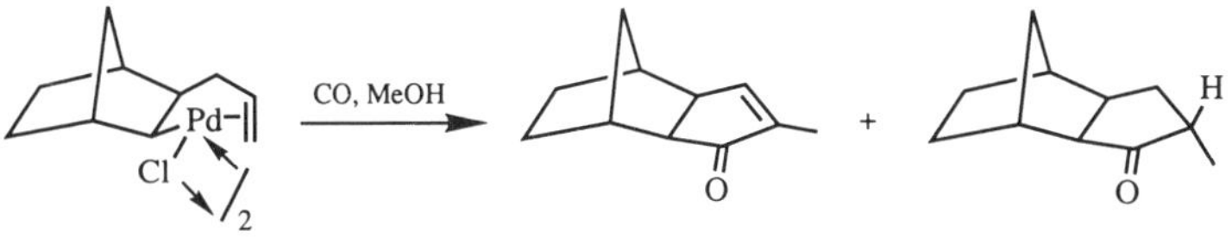

$$\text{(5.34)}$$

13   **14**

also gives cyclic ketones on reaction with carbon monoxide and methanol, but, as indicated in Scheme 5.12, the nature of the product is very dependent on the substitution pattern at the coordinated double bond.[168]

Scheme 5.12

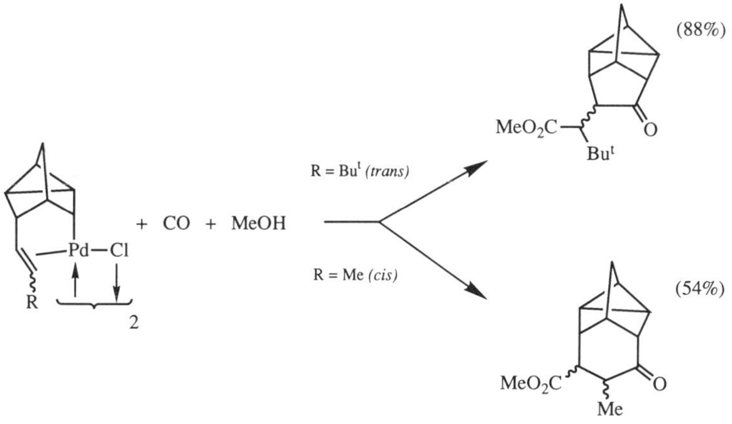

R = Bu^t *(trans)*   (88%)

R = Me *(cis)*   (54%)

A further variation on this theme is found in the catalytic reaction of allylic halides with carbon monoxide and either norbornadiene or norbornene [Eq. (5.35)].[169] In the presence of potassium acetate as acid acceptor, fused-ring methylenecyclopentanones are produced in reasonable yields.

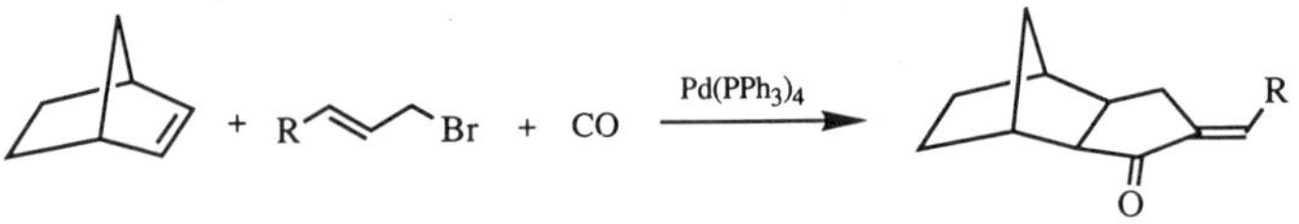

$$\text{(structure)} + R\diagup\hspace{-0.3em}\diagdown Br + CO \xrightarrow{Pd(PPh_3)_4} \text{(structure)} \qquad (5.35)$$

Stoichiometric reactions of nickel tetracarbonyl with alkenes, alkynes, or dihalo-compounds lead to a variety of cyclic ketones, and while use of the exceedingly toxic $Ni(CO)_4$ in the laboratory is not recommended, the following reactions serve to illustrate the range of products that can be obtained.

Substituted indenones are formed in good yield by reaction of 1.5 equivalents of $Ni(CO)_4$ with 1,2-di-iodobenzenes and alkynes (Scheme 5.13).[170] The same reaction, however, can be achieved *catalytically* in the presence of $Pd(PPh_3)_4$, using a stoichiometric amount of zinc as the halogen acceptor.[170] Carbonylation of 1,8-bis(bromomethyl)naphthalene

Scheme 5.13

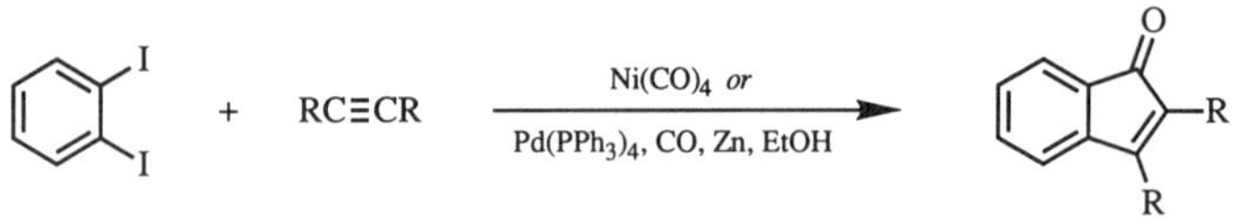

$$\text{(1,2-diiodobenzene)} + RC{\equiv}CR \xrightarrow[Pd(PPh_3)_4, CO, Zn, EtOH]{Ni(CO)_4 \ or} \text{(indenone product)}$$

(15) with $Ni(CO)_4$ gives 2,3-dihydrophenalen-2-one in 40% yield,[171] as shown in Eq. (5.36). Dialkyl- or diaryl-alkynes react stoichiometrically with

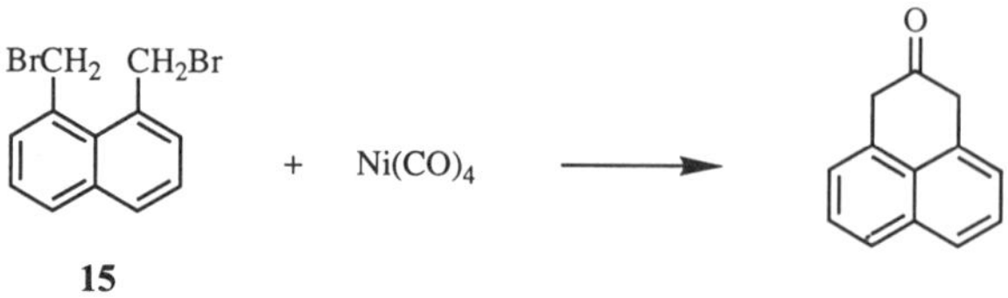

$$\underset{\textbf{15}}{\text{BrCH}_2 \ \text{CH}_2\text{Br (naphthalene structure)}} + \ Ni(CO)_4 \longrightarrow \text{(product structure)} \qquad (5.36)$$

$Ni(CO)_4$, in the presence of hydrochloric acid, to give tetrasubstituted cyclopentenones in up to 70% yield [Eq. (5.37)].[172]

$$2\ RC\!\equiv\!CR\ +\ Ni(CO)_4\ \xrightarrow{\ HCl\ }\ \underset{O}{\overset{\displaystyle R\underset{R}{\diagup}\ \diagdown R}{}}\quad(5.37)$$

A novel approach to functionalized cyclobutanones involves the photochemical reaction of "Fischer–carbene" complexes such as $(CO)_5Cr\!=\!C(OMe)Me$ with electron-rich alkenes [Eq. (5.38)].[173] The reaction is thought to involve *in situ* formation of methoxy(methyl)ketene, followed by a formal $[2+2]$ cycloaddition of the latter to the activated alkene.

$$(CO)_5Cr\!=\!\!\underset{Me}{\overset{OMe}{\diagup}}\ +\ \underset{R^2}{\overset{R^1}{\Big\|}}\ \xrightarrow[25^\circ C,\ 15\ hr]{hv,\ CH_3CN}\ \underset{O}{\overset{\displaystyle MeO\ \ H}{\ }}\quad(5.38)$$

$$(30 - 70\%)$$

The recently discovered carbonylative cyclization of 2-iodobenzyl-malonate esters and related compounds, shown in Eq. (5.39), is catalyzed by a number of late transition metal compounds, including Li_2CuCl_4, $Pd(PPh_3)_2Cl_2$, and $NiBr_2$. At 40 bar and 90–100°C, yields of ketone are well over 85%, and the reaction has been shown to be general for the formation of five-, six-, and seven-membered rings.[174]

$$\xrightarrow[\text{catalyst}]{CO,\ NEt_3}\quad(5.39)$$

$$(E = CN,\ SO_2Ph,\ or\ CO_2R;\ n = 1,\ 2,\ or\ 3)$$

Chapter 6

Synthesis of Carboxylic Acids

6.1. Introduction

From the standpoint of carbonylation chemistry, carboxylic acids and their derived esters, amides, anhydrides, and acyl halides are best regarded as a group of closely related and readily interconvertible compounds based on the formally disconnected structure $\{[E]^+...CO...[Nu]^-\}$. In most cases [E] represents a potentially electrophilic carbon-centered fragment (alkyl, vinyl, aryl, etc.), but can also be a hydrogen- or a heteroatom-centered group such as [RO] (in carbonate derivatives) or $[R_2N]$ (in carbamates). The group $[Nu]^-$ is represented by one of the more or less nucleophilic species $[HO]^-$, $[RO]^-$, $[R_2N]^-$, $[RCOO]^-$, or $[X]^-$ (where X = halogen).

In a number of cases this disconnection has much more than formal significance. The formation of tertiary amides, for example, by addition of nucleophilic $[R_2N]^-$ to free carbon monoxide followed by trapping of the resulting aminoacyl anion with carbon-centered electrophiles such as alkyl halides (Scheme 6.1a) was described in Chapter 2, as was the "Koch" synthesis of carboxylic acids (see also Sections 6.4 and 6.5). The latter involves addition of carbon monoxide to an electrophilic carbenium ion, followed by nucleophilic attack of water, as shown in Scheme 6.1b. Such noncatalytic reactions proceed only under very strongly acidic or basic conditions, where formation of powerful electrophiles or nucleophiles is not

Scheme 6.1

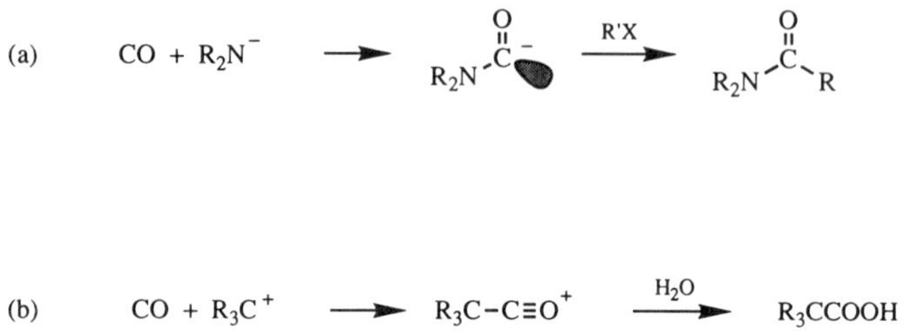

(a) $CO + R_2N^-$ $\longrightarrow$ $R_2N-\overset{\overset{\displaystyle O}{\|}}{C}$ $\xrightarrow{R'X}$ $R_2N-\overset{\overset{\displaystyle O}{\|}}{C}-R$

(b) $CO + R_3C^+$ $\longrightarrow$ $R_3C-C\equiv O^+$ $\xrightarrow{H_2O}$ R_3CCOOH

unexpected, but many carbonylation reactions involving transition metal catalysts occur under almost *neutral* conditions. Even here, however, the same general type of reactivity pattern, whereby carbon monoxide combines with both an electrophilic and a nucleophilic reactant, can often be discerned as indicated in Scheme 6.2.

Scheme 6.2

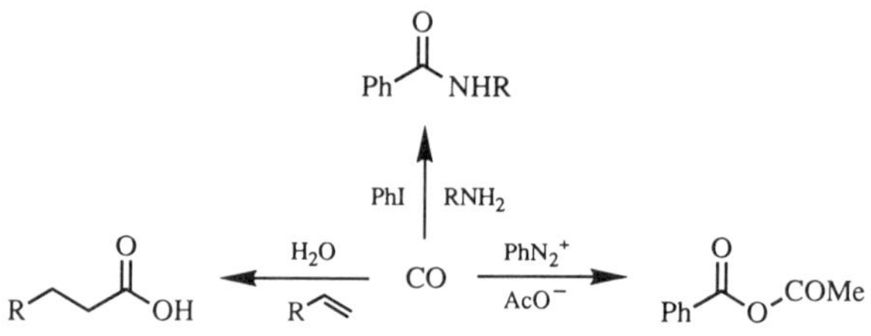

A consequence of this pattern, and in fact one of the most useful features of carbonylation chemistry, is that from a single starting material (e.g., a bromoarene) and a single catalyst precursor (e.g., $Pd(PPh_3)_2Cl_2$) is is often possible to synthesize a wide range of esters and amides, as well as the "parent" carboxylic acid, simply by varying the nucleophilic component (alcohol, amine, or water, respectively) present in the carbonylation reaction. Conversely, the ready interconvertability of carboxylic acid derivatives means that carbonylation may be used initially to generate a readily isolated and purified intermediate such as a volatile or soluble ester, or a crystalline amide, which can then be cleanly converted to a perhaps less tractable final product.

The following chapters are organized around the nature of the *isolated product* of a carbonylation reaction, even if this is not the direct product (e.g., an anhydride may be hydrolyzed to a carboxylic acid during workup) or is only an intermediate in a multistage synthesis. Reactions are further classified, where appropriate, on the basis of whichever reactant is involved in forming the C−CO bond. Reactions leading to *cyclization*, and thus affording lactones, lactams, or related compounds, are covered in detail in Chapters 9 and 10.

There are a number of valuable reviews[175] of carbonylative routes to carboxylic acids, esters, and amides, although early work in this field often tended to emphasize the exceedingly toxic nickel tetracarbonyl as reagent or catalyst. This compound is, however, being rapidly outmoded for almost every type of carbonylation by more active, selective, and far less hazardous catalysts based on complexes of cobalt, palladium, or rhodium. Indeed, most of the chemistry described in the present chapter depends on the use of these more recently developed homogeneous catalysts.

6.2. Carboxylic Acids: From Alkyl, Benzyl, and Related Halides

This group of halocarbons, having halogen attached at an sp^3 carbon, are susceptible to nucleophilic C−X bond cleavage by anionic transition metal complexes, and arylacetic acids in particular are readily obtained under mild conditions by catalytic carbonylation of benzylic halides in the presence of aqueous base [Eq. (6.1)], often under biphasic conditions. A great advantage of the biphasic system is that continuous extraction of the product acid into the aqueous phase (usually as a sodium salt) leaves the catalyst and any residual starting material or neutral by-products in the organic layer, allowing easy product separation and effectively heterogenizing the homogeneous catalyst.

$$ArCH_2Cl + CO + 2OH^- \xrightarrow{[Co(CO)_4]^-} ArCH_2COO^- + Cl^- + H_2O \qquad (6.1)$$

Catalysts based on nickel,[175] cobalt,[176–178] iron,[179, 180] ruthenium,[181] and palladium,[182, 183] are all effective for conversion of simple benzyl halides to arylacetic acids, but the cobalt-based system has been explored in much the greatest detail, and a wide range of substrates and procedures have been investigated. An alternative approach to heterogenizing the catalyst, for example, involves supporting $[Co(CO)_4]^-$ on an anion-exchange resin,[184] giving an active and readily separable solid catalyst, which is, however, rather less stable than that obtained in the biphasic homogeneous system.

Preparation of 2-Naphthylacetic Acid[177]

A mixture of 2-(bromomethyl)naphthalene (25 mmol), $Co_2(CO)_8$ (0.5 mmol), benzene (25 cm^3), $PhCH_2NEt_3Cl$ (1.1 mmol) and 5 M aqueous sodium hydroxide (25 cm^3), is stirred vigorously under carbon monoxide (1 bar) at ambient temperature for 12 h. After centrifuging the mixture, the aqueous layer is separated and acidified with dilute sulfuric acid to give 2-naphthylacetic acid in 64% yield.

The mechanism of such biphasic carbonylation reactions involving the $[Co(CO)_4]^-$ ion is shown in Scheme 6.3. Hydroxide ion is not transferred to the organic phase, but cleaves the acyl cobalt complex at the interface, so that the role of the "phase transfer" agent seems to be simply to provide a solubilizing counterion, $[R_4N]^+$, which retains $[Co(CO)_4]^-$ in the organic phase.[178] Since trialkyl*benzyl*ammonium ions can themselves be carbonylated to phenylacetic acid, tetra-*n*-alkylammonium salts are more generally useful,[185] although the benzylic derivatives *have* been used successfully under very mild conditions.[177]

Iron pentacarbonyl is also an effective catalyst precursor, allowing high-yield carbonylation of alkyl and aralkylhalides under relatively mild conditions,[179, 180] and this relatively recent discovery provides an

Scheme 6.3

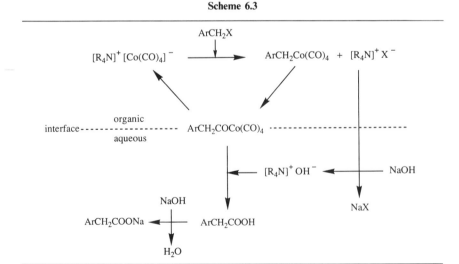

interesting example of the evolution that is sometimes possible from stoichiometric organometallic chemistry involving a preformed reagent [$Na_2Fe(CO)_4$, Scheme 6.4a][186] to a genuine high-yield, high-turnover, catalytic reaction (Scheme 6.4b).

Scheme 6.4

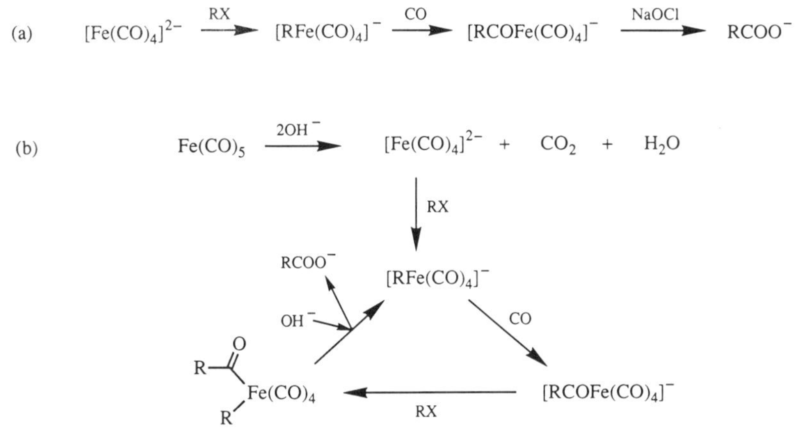
(a)
[Fe(CO)4]2−
RX
[RFe(CO)4]−
CO
[RCOFe(CO)4]−
NaOCl
RCOO−
(b)
Fe(CO)5
2OH−
[Fe(CO)4]2− + CO2 + H2O
RX
[RFe(CO)4]−
RCOO−
OH−
CO
Fe(CO)4
RX
[RCOFe(CO)4]−

The palladium-catalyzed carbonylation of benzyl halides[182] can likewise be carried out under biphasic conditions, but alkyl halides containing β-hydrogens are generally not suitable substrates owing to the strong tendency of alkyl–palladium complexes to undergo β-hydride elimination. This then results in catalytic dehydrohalogenation of the substrate, rather than carbonylation. Such benzylic carbonylations were originally believed to involve phase transfer of hydroxide ion into the organic layer, but later work[183, 187] suggests that, as in the corresponding cobalt- and iron-based reactions, the function of the quaternary salt may be to solubilize an anionic metal complex (in this case $[Pd(PPh_3)_nOH]^-$) in the organic phase (Scheme 6.5). An alternative two-phase system, requiring no phase transfer catalyst, involves the water soluble palladium complex $PdCl_2[Ph_2P(m\text{-}C_6H_4SO_3Na)]_2$ in aqueous base, with heptane or anisole as the organic phase. In this system the carbonylation of benzyl chloride to phenylacetic acid proceeds at 50°C and 1 bar, in over 90% yield.[188]

Scheme 6.5

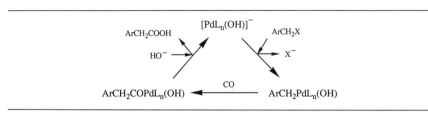

Secondary benzyl halides (alkyl/aryl or diaryl) are apparently much more difficult to carbonylate cleanly to carboxylic acids[189] than are the corresponding primary halides, and unusually, a biphasic system is not preferred. Selective, high-yield syntheses of diarylacetic and α-aryl propionic acids (including the important anti-inflammatory drug α-(4-isobutylphenyl)propionic acid, "Ibuprofen") can be achieved at low pressure (1 bar), using $[Co(CO)_4]^-$ as catalyst, ethanol as solvent, and sodium hydroxide as base.[190] This system apparently promotes rapid, irreversible nucleophilic cleavage of an intermediate acyl complex and thereby suppresses the side reactions (elimination, hydrogenolysis, isomerization, coupling, etherification, and *double carbonylation*) which occur under phase transfer conditions.[189]

Preparation of Bis(4-chlorophenyl)acetic Acid[190]

Ethanol (25 cm³), sodium hydroxide (50 mmol), NaCo(CO)₄ (1.6 mmol), and bis(4-chlorophenyl)chloromethane (25 mmol) are placed under carbon monoxide in a 100-cm³ flask fitted with a magnetic stirrer, thermometer, and gas burette filled with CO. The mixture is stirred at 16°C until gas absorption ceases, and is then acidified and worked up to give bis(4-chlorophenyl)acetic acid in 63% yield.

The realization that *doubly carbonylated* products can be formed in carbonylation reactions has emerged[177, 191] only over the past decade, but specific syntheses of many α-keto acids, amides, and esters are rapidly being developed (see also Chapters 7 and 8). Although, as might intuitively be expected, many such syntheses depend on the use of carbon monoxide pressures well above ambient, variation of other parameters such as solvent, catalyst, and base, can lead to efficient double carbonylation even at relatively low pressure. Selective, cobalt-catalyzed formation of phenylpyruvic acid (97% selectivity at 85% yield) from benzyl chloride [Eq. (6.2)] was thus first achieved[192] at 50 bar of carbon monoxide in propan-2-ol solution, but more recent work has shown that similar selectivity and an even higher yield can be obtained at less than 2 bar when 1,2-dimethoxyethane is used as solvent.[193]

$$PhCH_2Cl + 2CO + 2OH^- \xrightarrow{[Co(CO)_4]^-} PhCH_2COCOO^- + Cl^- + H_2O \quad (6.2)$$

The production of α-keto acids in this type of system is thought[194] to depend on enolization of the first-formed acyl ligand (Scheme 6.6), to give a vinyl group that migrates more readily from the metal to coordinated carbon monoxide. Strong evidence that enolization indeed plays a vital role is provided by the observation that double carbonylation is drastically inhibited when the benzylic hydrogen in α-phenyl bromoethane is replaced by the less-mobile deuterium.[195]

Scheme 6.6

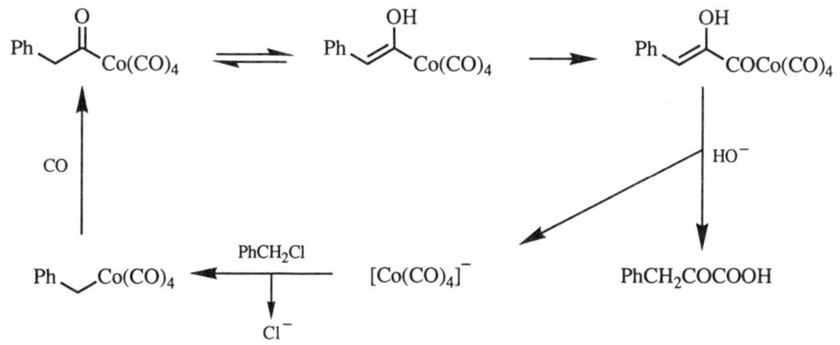

Attempts to achieve asymmetric catalytic carbonylation of racemic secondary benzyl halides [Eq. (6.3)] have recently met with some success,[196] though significant enantiomeric excesses ($>50\%$) are achievable only at low conversion.

Iron-based systems are not active for carbonylation of allylic halides, which tend to form kinetically inert π-allyliron complexes.[186] However,

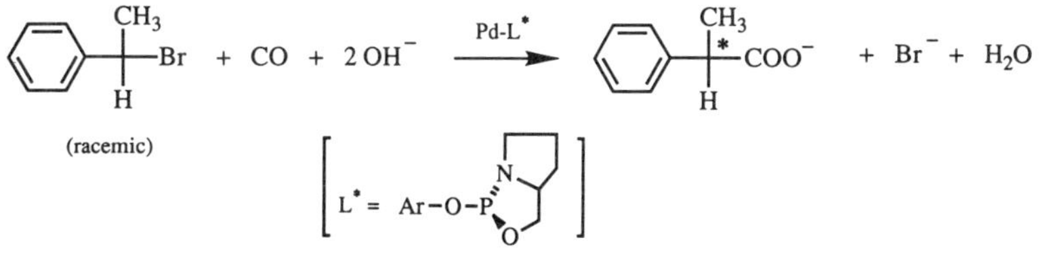

$$(6.3)$$

use of the palladium complex $PdCl_2[Ph_2P(m\text{-}C_6H_4SO_3Na)]_2$ or the anionic nickel complex $[Ni(CO)_3CN]^-$ (formed *in situ* from $Ni(CN)_2$ and carbon monoxide) allows catalytic conversion of allyl chlorides and bromides to carboxylic acids in good yield under mild two-phase conditions,[197] as shown in Eq. (6.4). When promoted by the addition of lanthanide salts, the latter catalyst also permits high-yield carbonylation of benzylic halides to arylacetic acids.[198]

$$R \diagup \diagdown CH_2Cl \quad + \quad CO \quad + \quad 2OH^- \xrightarrow{\quad [Ni(CO)_3CN]^- \quad} Ph \diagup \diagdown CH_2COO^- \qquad (6.4)$$

$$+ \quad Cl^- \quad + \quad H_2O$$

Preparation of (E)-4-Phenyl-3-butenoic acid[197]

Carbon monoxide is rapidly bubbled through a stirred solution of 5 M aqueous NaOH (25 cm^3), tetra-n-butylammonium hydrogen sulfate (0.2 mmol), and 4-methylpentan-2-one (25 cm^3). After a few minutes $Ni(CN)_2$ (1.0 mmol) is added and the mixture heated at 60°C for 3 h under carbon monoxide. After cooling to room temperature, a solution of $PhCH=CHCH_2Cl$ (11.6 mmol) in 4-methylpentan-2-one (20 cm^3) is added dropwise over a 3-h period to the stirred solution, and stirring is continued overnight. The aqueous phase is separated, washed with ether, acidified with 3 M H_2SO_4 (*caution*: HCN may be evolved), and extracted with ether (4 × 25 cm^3). The combined ether extracts are dried over MgSO$_4$ and evaporated to give (*E*)-4-phenyl-3-butenoic acid in 84% yield.

Stoichiometric carbonylation of α-chloroalkyl alkynes (propargyl halides) with nickel tetracarbonyl provides a general route to allenic acids [Eq. (6.5)].[199] Although yields are often only moderate, the selectivity achieved compares favorably with that obtained in more conventional syntheses such as the reaction of a propargyl Grignard reagent with carbon

Scheme 6.7

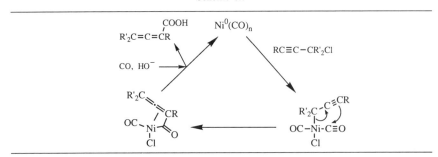

dioxide. A possible mechanism for the allenic acid synthesis is shown in Scheme 6.7, where the general resemblance (apart from the rather unusual insertion step) to benzylic carbonylation suggests that even here palladium *catalysis* might well provide an alternative method of conversion.

$$RC{\equiv}C-CR'_2Cl \ + \ CO \ + \ 2OH^- \ \xrightarrow{\ Ni(CO)_4\ } \ \underset{RC=C=CR'_2}{\overset{COO^-}{\underset{|}{}}} \ + \ Cl^- \ + \ H_2O \qquad (6.5)$$

6.3. Carboxylic Acids: From Aryl, Vinyl, and Heteroaryl Halides

Halocarbons in which the $C-X$ bond occurs at an sp^2 carbon are not, in general, susceptible to nucleophilic attack, so that the anionic, 18-electron, species $[Co(CO)_4]^-$ and $[Fe(CO)_4]^{2-}$, described in Section 6.2, *usually* fail to activate such halides towards carbonylation (but see below). With palladium- and some nickel-based systems, however, $C-X$ bond activation occurs by oxidative addition to the metal, a reaction well established for both sp^2 and sp^3 hybridized carbon atoms. Aryl halides such as chloronaphthalenes and substituted bromo- or iodobenzenes can thus be converted to carboxylic acids under mild conditions (100°C, 1 bar of carbon monoxide) in the presence of nickel tetracarbonyl,[200] but a less hazardous and more effective synthesis is achieved using the biphasic, palladium-catalyzed procedure described earlier for carbonylation of benzylic halides (Section 6.2).[182, 201] This approach allows selective mono-carbonylation of polyhalogenated aromatic substrates such as 1,4-dibromobenzene, since transformation of the first $C-X$ group to carboxyl enables the product to be rapidly removed (as its sodium salt) from the organic to the aqueous phase, where it is no longer in contact with the catalyst. In aqueous DMF as a single-phase reaction medium, even ligand-free palladium salts catalyze the carbonylation of a wide range of iodoarenes to aromatic carboxylic acids under ambient conditions of temperature and pressure.[202]

Preparation of 4-Bromobenzoic Acid[201]
Into a 100-cm³ flask fitted with a mechanical stirrer and thermometer are introduced 1,4-dibromobenzene (7.0 g), *p*-xylene (30 cm³), aqueous sodium hydroxide (70 cm³ of a 50% solution), triphenylphosphine (1.2 g), PdCl₂(PhCN)₂ (0.1 g), and tetra-*n*-butylammonium iodide (0.6 g). The temperature is raised to 90°C and the mixture is stirred vigorously under carbon monoxide (1 bar constant pressure) for 5 h. The phases are separated and the aqueous layer is acidified and then extracted with ether, to give an extract

that on drying, filtering, and evaporating to dryness affords 4-bromobenzoic acid in 77% yield.

At very much higher pressures (180 bar), homogeneous palladium catalysis allows double-carbonylation of iodobenzene to the keto acid PhCOCOOH in 53% yield, although significant quantities of benzoic acid and benzaldehyde are also formed.[203] The mechanism of this particular carbonylation has not been investigated, but it is certainly unrelated to the benzylic double-carbonylations mentioned earlier (enolization of a benzoyl complex being impossible), and probably resembles that established for the palladium-catalyzed keto ester synthesis described in Chapter 7.

In view of the relatively high cost of palladium, even in catalytic quantities, it is significant that $[Co(CO)_4]^-$, *under conditions of photostimulation*,[204] will catalyze carbonylation of aryl and vinyl halides (including the normally less-reactive aryl chlorides)[205] at low pressure, with high efficiency [Eq. (6.6)]. Under these conditions the catalyst can be generated

$$ArX + CO + 2OH^- \xrightarrow{h\nu, \ [Co(CO)_4]^-} ArCOO^- + X^- + H_2O \qquad (6.6)$$

in situ from simple Co(II) or Co(III) salts,[206] thus avoiding any need to handle the air-sensitive complex $Co_2(CO)_8$ or its derived anion. The mechanistic role of photolysis in these reactions has yet to be explored, but photochemical regeneration of the coordinatively unsaturated, 16-electron species $[Co(CO)_3]^-$ (to which oxidative addition of the $C-X$ bond could occur; Scheme 6.8) seems a strong possibility.

Scheme 6.8

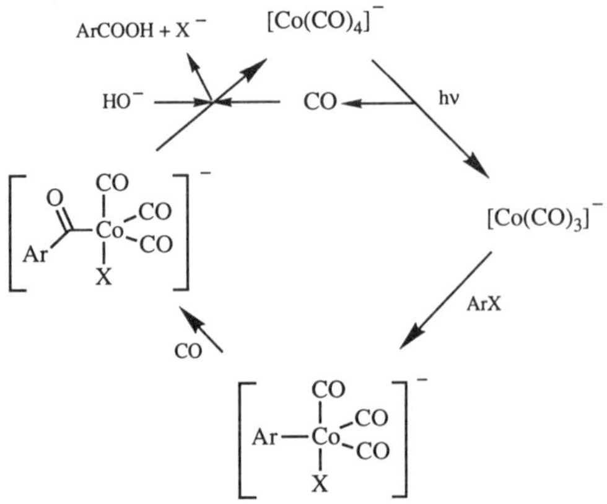

Even photolysis is unnecessary if an alkylating agent such as methyl iodide or dimethyl sulfate is added to the cobalt-catalyzed reaction.[193, 207] At 55°C and under 1 bar of carbon monoxide an essentially quantitative yield of benzoic acid is obtained from bromobenzene, and more remarkably still, doubly carbonylated products are formed at lower temperatures, though these must be liberated by hydrolysis from their condensation products with pyruvic acid (Scheme 6.9). The mechanism of this obviously highly complex reaction has not yet been established, although the stereochemistry of *vinylic* carbonylation with this catalyst has been investigated.[208] Very recently, the anionic cyanonickel complex

Scheme 6.9

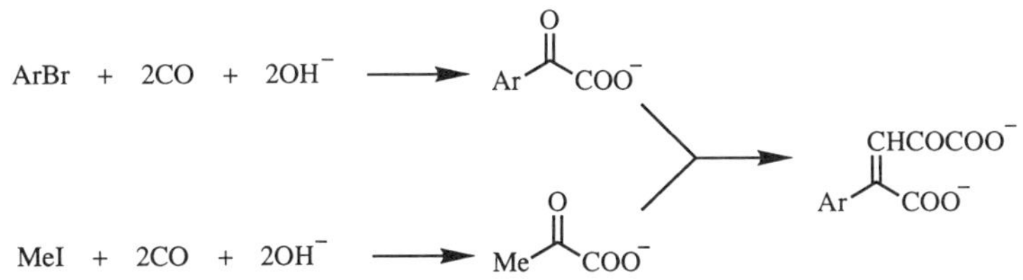

$[Ni(CO)_3CN]^-$, which was noted above as an effective catalyst for carbonylation of allylic and benzylic halides, has also been shown to catalyze the corresponding reactions of aryl[209] and vinyl[210] halides at 90–100°C and 1 bar pressure. Although these reactions appear to avoid the use of nickel tetracarbonyl, the ease with which this compound can be formed (see Chapter 3) suggests that extreme care should be taken when working up such reaction mixtures.

6.4. Carboxylic Acids: From Alcohols

The direct synthesis of acetic acid from methanol and carbon monoxide has been a major industrial process for well over 20 years. The original (BASF) process used a homogeneous cobalt catalyst, which required the use of high temperatures ($>200°C$) and pressures ($\simeq 700$ bar),[211] but in 1971 the low-pressure rhodium-catalyzed reaction shown in Eq. (6.7) was commercialized by Monsanto, and this is now the dominant technology in the field.

$$CH_3OH \ + \ CO \ \xrightarrow{\ [Rh(CO)_2I_2]^- \ } \ CH_3COOH \qquad (6.7)$$

The mechanism of the "Monsanto" process has been studied intensively for several years,[212] and it now seems clear that carbonylation depends on generation of iodomethane in a preequilibrium reaction between methanol and the hydrogen iodide added as promoter. The catalytic cycle which then converts iodomethane to acetyl iodide is shown in Scheme 6.10, and the acyl halide is finally hydrolyzed to acetic acid with regeneration of hydrogen iodide. The reaction rate is essentially independent of carbon monoxide pressure (so the process can be operated, if required, at atmospheric pressure) and the rate-determining step appears to be oxidative addition of iodomethane to the anionic square-planar complex $[Rh(CO)_2I_2]^-$. Very recently it has been suggested[213] that the increase in

Scheme 6.10

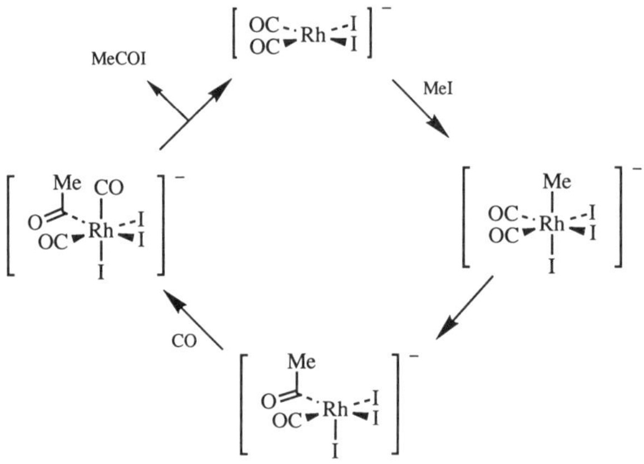

reaction rate that occurs with increasing iodide concentration may be due to an equilibrium involving the dianionic species $[Rh(CO)_2I_3]^{2-}$, which attacks iodomethane more rapidly (via the S_N2 mechanism shown in Scheme 6.11) than $[Rh(CO)_2I_2]^-$ itself.

Rhodium catalysis has been investigated for carbonylation of alcohols other than methanol, but rates diminish rapidly with increasing chain length, and mixtures of isomeric acids are obtained from C_3 and higher

Scheme 6.11

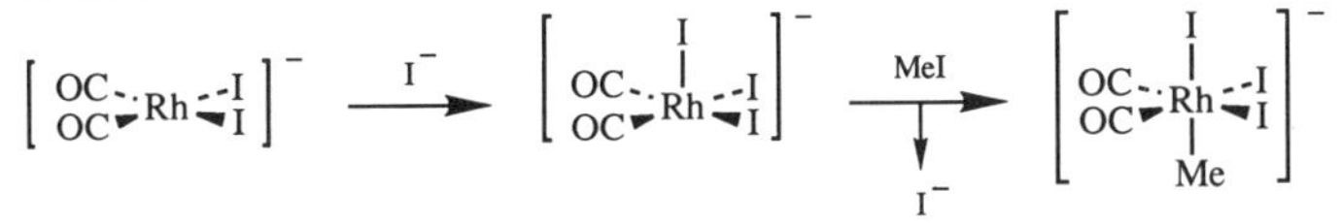

$$\left[\begin{matrix} OC\cdots_{,}Rh\llcorner_{I}^{-I} \\ OC\blacktriangleright \quad \blacktriangleleft_{I} \end{matrix}\right]^{-} \xrightarrow{I^-} \left[\begin{matrix} & \overset{I}{\underset{|}{}} & \\ OC\cdots_{,}Rh\llcorner_{I}^{-I} \\ OC\blacktriangleright \quad \blacktriangleleft_{I} \end{matrix}\right]^{-} \xrightarrow{MeI} \left[\begin{matrix} & \overset{I}{\underset{|}{}} & \\ OC\cdots_{,}Rh\llcorner_{I}^{-I} \\ OC\blacktriangleright \underset{Me}{|} \blacktriangleleft_{I} \end{matrix}\right]^{-}$$

alcohols.[214] The value of this process for the synthesis of more complex organic structures is probably therefore rather limited.

Strong acids such as boron trifluoride, sulfuric, or phosphoric acids can catalyze the carbonylation of methanol to acetic acid even in the *absence* of transition metals, but extreme conditions (300°C, 600 bar) are required.[215] This reaction (the Koch synthesis) depends on the interaction of carbon monoxide with a free carbenium ion followed by hydrolysis of the resulting acylonium ion (Scheme 6.12). Since the Koch reaction proceeds under much milder conditions (40°C, 70 bar) when alcohols that afford stable, tertiary carbenium ions are used,[216] the extreme conditions required for methanol carbonylation presumably reflect the high free energy of the methyl cation and its consequently very low equilibrium concentration.

Scheme 6.12

$$CH_3OH \underset{H_2O}{\overset{H^+}{\rightleftharpoons}} CH_3^+ \underset{}{\overset{CO}{\rightleftharpoons}} CH_3CO^+ \xrightarrow{H_2O} CH_3COOH + H^+$$

Carbonylation of the higher alcohols can even be achieved under *ambient* conditions of temperature and pressure if the effective concentration of carbon monoxide *in solution* is increased to a sufficiently high level. One approach is to use a sulfuric acid solution supersaturated with carbon monoxide (ca. 0.3 molar) generated *in situ* by dehydration of formic acid.[217] This method has recently been used to obtain diaryl-acetic and -propionic acids in high yield from the corresponding diarylcarbinols [Eq. (6.8)].[218] An alternative technique involves the addition of a copper(I) or silver(I) salt to 98% sulfuric acid under an atmosphere of

$$ArCH(R)OH + CO \xrightarrow[HCOOH]{H_2SO_4} ArCH(R)COOH \qquad (6.8)$$

carbon monoxide. Kinetically labile, soluble metal carbonyl complexes $[M(CO)_n]^+$ are readily formed under ambient conditions, and at a metal ion concentration of 0.2 M, the effective concentration of carbon monoxide approaches that achieved by the formic acid method. Whether the metal ion has any function other than simply increasing the carbon monoxide concentration in solution is not yet certain. An interesting feature of this latter synthesis is that, provided the starting alcohol is C_4 or higher, even primary and secondary alcohols give *tertiary* carboxylic acids, implying that extensive isomerization of the intermediate carbenium ions (as shown in Scheme 6.13) can occur.[219] Diols can be used as substrates in this type of reaction,[220] but mixtures of lactones, and ditertiary carboxylic acids are produced, in proportions that vary with chain length.

Scheme 6.13

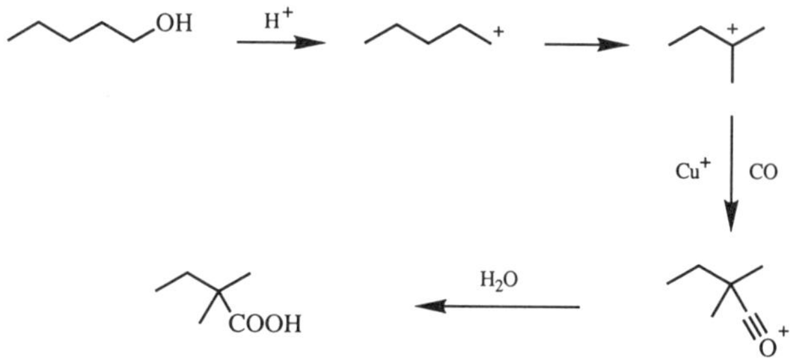

By working in media of *extremely* high acidity such as HF/SbF_5 or $SbCl_5/SO_2$, very high concentrations of carbenium ions can be obtained, allowing carbonylation of most alcohols other than methanol at ambient pressure and low temperature (typically $-78\,^\circ C$).[221] Isomerization of the intermediate carbenium ion remains a problem, however.

6.5. Carboxylic Acids: From Alkenes and Alkanes

Addition of water and carbon monoxide to an alkene to form a carboxylic acid, a reaction sometimes termed *hydrocarboxylation* [Eq. (6.9)],[222] is catalyzed by a variety of transition metal complexes including $Ni(CO)_4$, $Co_2(CO)_8$, and $HPtCl_6$, and by strong acids via carbenium ion chemistry (cf. Sec. 6.4). Stoichiometric carbonylations based on zirconium[223] and boron[224] reagents are also known.

$$R\diagdown\diagdown + CO + H_2O \longrightarrow R\diagup\diagdown\diagup COOH \quad \text{and/or} \quad \overset{\displaystyle COOH}{\underset{R}{\diagdown}\diagup} \qquad (6.9)$$

The transition metal catalyzed syntheses are all believed to involve initial generation of a hydrido complex, which then reacts by successive insertion of alkene and carbon monoxide to give an acyl complex. Finally, hydrolysis of the metal–carbon bond completes the catalytic cycle shown in Scheme 6.14 by formation of the product carboxylic acid and regeneration of the starting hydrido complex.

Scheme 6.14

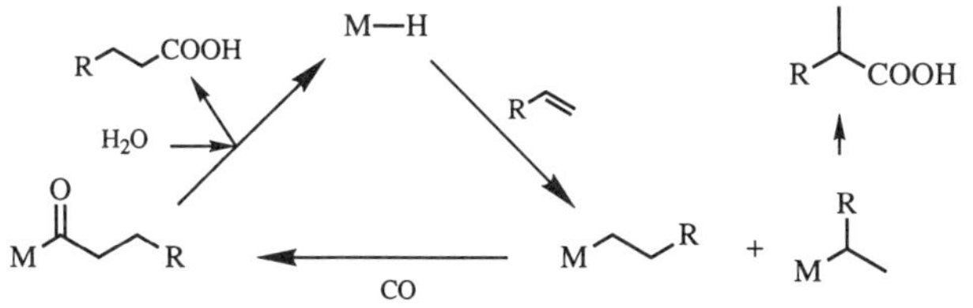

Hydrocarboxylation generally leads to mixtures of products, owing not only to the occurrence of both Markownikov and anti-Markownikov addition of the metal hydride to the alkene, but also to metal-catalyzed alkene isomerization. Cobalt-catalyzed hydrocarboxylation of 1-pentene (145°C, 180 bar) thus affords hexanoic acid (52%), 2-methyl pentanoic acid (17%), and 2-ethyl butanoic acid (5%), the latter presumably arising from carbonylation of 2-pentene.[225] Better selectivity toward linear acids can be achieved by using a platinum catalyst modified with tin(II) chloride, so that carbonylation of 1-dodecene with this catalyst, at 90°C/200 bar in aqueous acetone as solvent, gives nearly 70% of the linear product.[226] The selectivity enhancement is thought to be due to replacement of chloride ligands on platinum by the sterically more demanding $-SnCl_3$ groups, and indeed a number of stable trichlorotin complexes are known, including $[PtCl_2(SnCl_3)_2]^{2-}$ and $[Pt(SnCl_3)_5]^{3-}$. Although these catalytic reactions require relatively vigorous conditions of temperature and pressure, some commercially available zirconium and boron regents allow *stoichiometric* hydrocarboxylation under very much milder conditions (0–25°C, 1–2 bar). A remarkable feature of the zirconium hydride $(C_5H_5)_2Zr(H)Cl$ is that, presumably for steric reasons, its reactions with linear alkenes give exclusively *terminal* alkyl complexes, even if the alkene double bond is initially *internal*. Insertion of carbon monoxide followed by oxidative hydrolysis with hydrogen peroxide, as shown in Scheme 6.15, leads to linear carboxylic acids in very good yields.[223] In this case the previously unhelpful ability of metal hydride complexes to isomerize alkenes is turned into a positive advantage.

Scheme 6.15

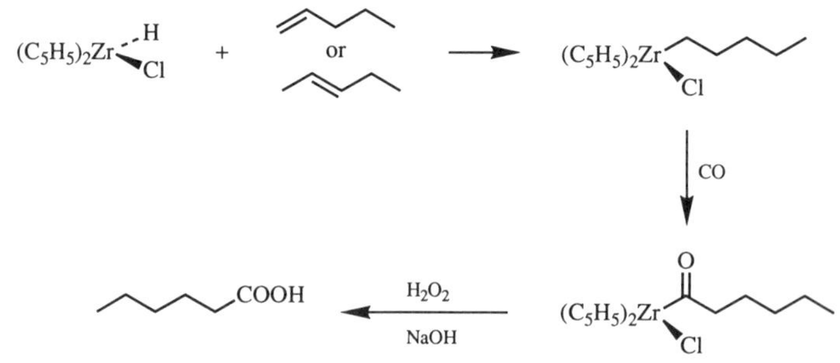

As noted in Chapter 1, carbonylation of an organoborane normally requires high pressures and results in transfer of all three alkyl groups to

Scheme 6.16

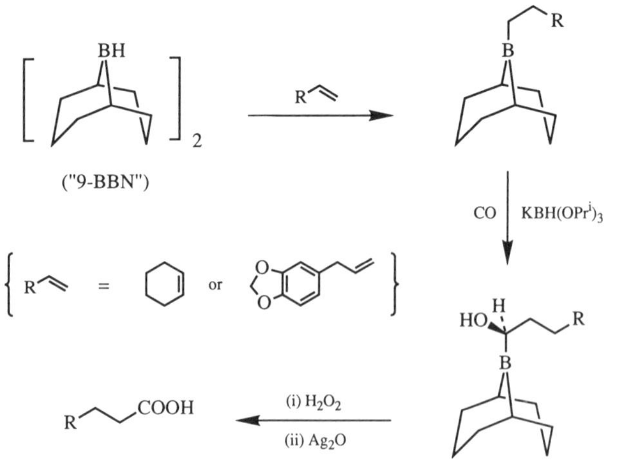

carbon monoxide, thus giving a tertiary alcohol rather than a carbonyl compound. However, if the reaction is carried out in the presence of a complex reducing agent such as an alkoxyaluminum hydride, carbonylation occurs under essentially ambient conditions, and only one alkyl group is transferred. The reduced species can then be oxidized *in situ* to an aldehyde and thence to a carboxylic acid.[227] Organoboranes which react by *selective* transfer are accessible via alkene hydroboration with the commercially available borane "9-BBN," and this approach has been developed as shown in Scheme 6.16, to provide a mild, high-yielding synthesis of labeled carboxylic acids $R^{13}COOH$ from terminal alkenes.[228]

The copper-promoted strong-acid (Koch) chemistry described in Sec. 6.4 for carbonylation of alcohols can also be applied to the hydrocarboxylation of alkenes. Low temperatures and pressures provide satisfactory reaction rates, and tertiary carboxylic acids are again formed from C_4 or higher alkenes, if necessary via extensive carbenium ion rearrangements.[229] Cyclohexene, for example, yields only 1-methylcyclopentanecarboxylic acid (Scheme 6.17), via ring contraction of the first-

Scheme 6.17

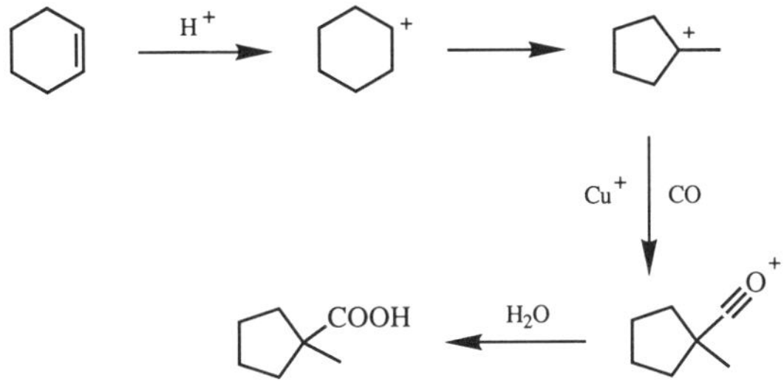

formed carbenium ion. The course of such reactions can be modified by including in the system an alkane containing a tertiary $C-H$ group. It is then possible for a tertiary carbenium ion to be formed by hydride transfer from the alkane to the secondary cation, rather than by rearrangement of the latter, and indeed good yields of *alkane* carbonylation products (carboxylic acids) can sometimes be obtained in this way (Scheme 6.18).[230]

In media of extremely high acidity such as HF/SbF_5, carbenium ions can be generated by protolysis of an alkane $C-H$ or $C-C$ bond (Scheme 6.19),[231] allowing direct carbonylation of alkanes to carboxylic acids. Selectivity for $C-H$ protolysis in the carboxylation of propane is markedly enhanced by the addition of bromide ion.[232]

Scheme 6.18

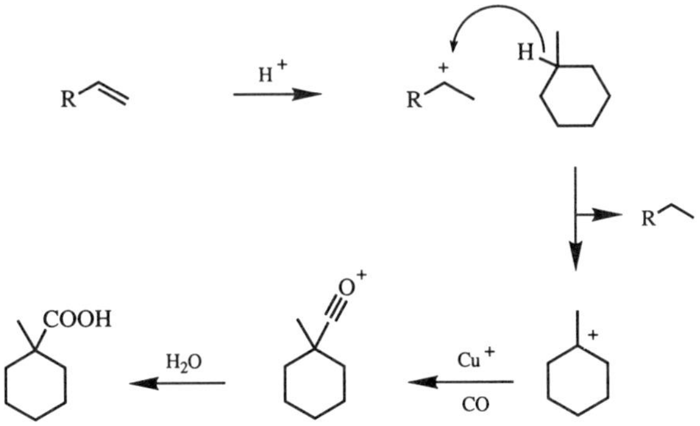

Scheme 6.19

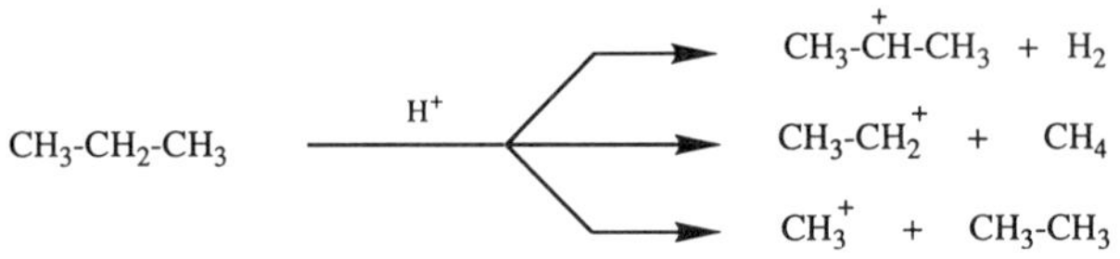

6.6.   Carboxylic Acids: From Alkynes

The low-pressure hydrocarboxylation of acetylene to acrylic acid, in the presence of stoichiometric or semicatalytic quantities of nickel tetracarbonyl [Eq. (6.10)], was one of the first large-scale processes to use carbonylation chemistry,[233] and although alternative routes to acrylic acid such as propene oxidation are now more competitive, plants based on this reaction still generate more than 150,000 tons of acrylic acid per annum. A genuinely catalytic process based on $Ni(CO)_4$ was also developed,[234] but the relatively vigorous conditions required (200°C, 60 bar) seem to have prevented its commercialization. The carbonylation process generally requires an acid promoter, apparently in order to form a hydrido–nickel complex, which is then able to initiate the reaction by "inserting" alkyne. As indicated in Chapter 2, such insertions occur in a *cis* manner, and indeed with substituted alkynes, $RC \equiv CR'$ the acid formed is generally the product of apparent *cis* addition of $H - COOH$ to the triple bond.[233] The reaction rate falls off as the steric bulk of the substituents increases, but the relative proportions of the isomeric acids formed by "Markownikov" or "anti-Markownikov" addition seem to depend on a more complex balance of steric and electronic factors.[235]

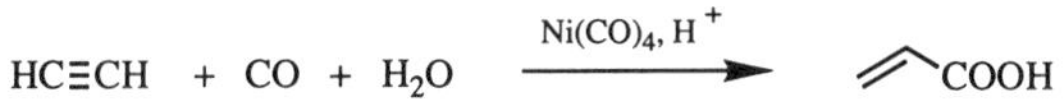

HC≡CH + CO + H2O
Ni(CO)4, H+
COOH

(6.10)

Cobalt catalysis of this type of reaction leads, at high temperatures (100°C) and pressures (100–200 bar), to an efficient second carbonylation, so that succinic acid, for example, can be obtained directly from acetylene in over 80% yield,[236] as shown in Eq. (6.11). At lower pressures, however, acrylic acid is again the major product.

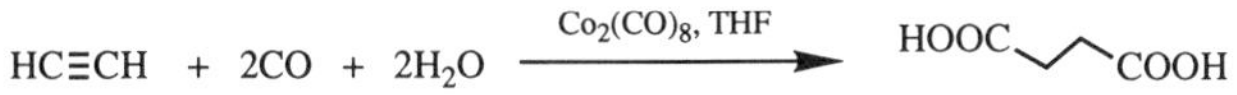

HC≡CH + 2CO + 2H2O
Co2(CO)8, THF
HOOC
COOH

(6.11)

Nickel-catalyzed hydrocarboxylation of alkynes in the presence of allylic derivatives (halides, ethers, alcohols, etc.) leads to formation of 2,5-dienoic acids in high yields under very mild conditions [Eq. (6.12)].[237]

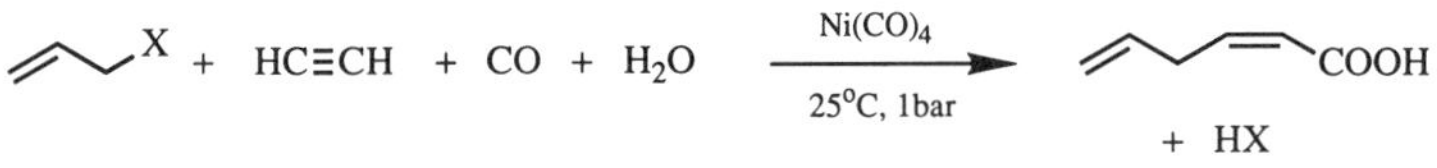

X
+ HC≡CH + CO + H2O
Ni(CO)4
25°C, 1bar
COOH
+ HX

(6.12)

This remarkable four-component reaction probably involves successive insertions of alkyne and carbon monoxide into a π-allyl nickel(II) complex, followed by hydrolysis and regeneration of a nickel(0) species which is required for reaction with the allylic substrate (Scheme 6.20).

Scheme 6.20

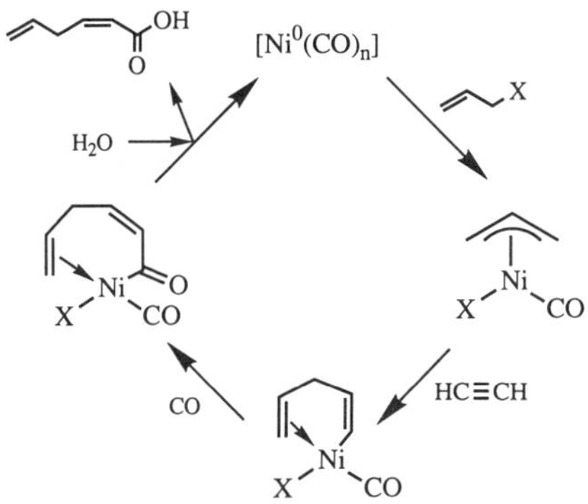

OH
O
[Ni0(CO)n]
X
H2O
Ni
X
CO
O
Ni
X
CO
CO
HC≡CH
Ni
X
CO

6.7. Carboxylic Acids: From Diazonium Ions

Although arenediazonium salts react stoichiometrically with nickel[238] and iron[239] carbonyls to give aromatic carboxylic acids in moderate yield, a more reliable procedure involves direct, catalytic carbonylation in the presence of palladium acetate.[240] The reaction proceeds at room temperature under carbon monoxide (ca. 9 bar), and gives a mixed anhydride as the primary product (Scheme 6.21). Hydrolytic workup affords the aromatic acid in 50%–90% yield.

Scheme 6.21

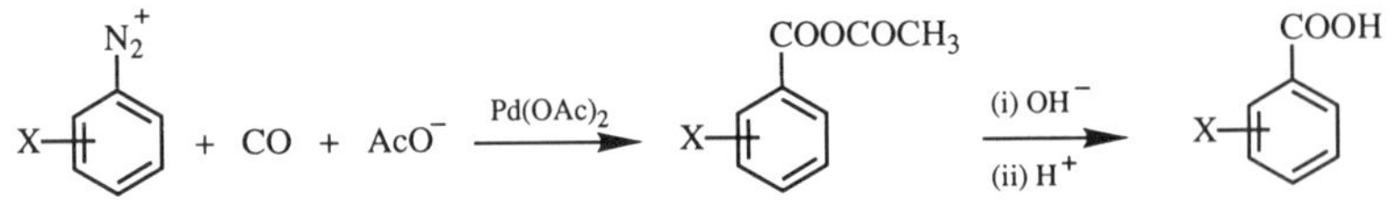

$$X\!-\!\!\bigcirc\!\!-\!\!N_2^+ \ +\ CO\ +\ AcO^- \ \xrightarrow{Pd(OAc)_2}\ X\!-\!\!\bigcirc\!\!-\!\!COOCOCH_3 \ \xrightarrow[\text{(ii) H}^+]{\text{(i) OH}^-}\ X\!-\!\!\bigcirc\!\!-\!\!COOH$$

Carboxylation of Diazonium Salts[240]

The diazonium fluoroborate (10 mmol), sodium acetate (30 mmol), palladium acetate (0.2 mmol), and acetonitrile (60 cm^3) are placed in a 300-cm^3 glass-lined autoclave cooled to 0°C. After flushing with nitrogen and then carbon monoxide, the CO pressure is increased to 9 bar and the mixture is stirred for 1 h at room temperature. After venting and flushing the autoclave with nitrogen, the solution is evaporated and the residue taken up in aqueous sodium hydroxide. This solution is washed with ether, treated with charcoal, filtered, and acidified with concentrated HCl. The acids formed are extracted into ether, and removal of the ether and acetic acid under reduced pressure affords the aromatic carboxylic acid in essentially pure form.

6.8. Carboxylic Acids: From Aldehydes—The Wakamatsu Reaction

A remarkable cobalt-catalyzed synthesis of N-acyl amino acids, based on the carbonylation of aldehydes and primary amides [Eq. (6.13)],[241] was reported by Wakamatsu and co-workers in 1971. More recent studies

$$RCHO\ +\ R'CONH_2\ +\ CO\ \xrightarrow[H_2]{Co_2(CO)_8}\ \underset{\displaystyle RCHCOOH}{\overset{\displaystyle NHCOR'}{|}} \qquad (6.13)$$

have shown that the reaction proceeds rapidly at 100°C under a relatively high pressure (140 bar) of synthesis gas (CO/H_2, 3:2).[242] The presence of hydrogen is necessary to stabilize $HCo(CO)_4$, which appears to be the active catalyst, and yields of over 90% of N-acyl amino acid are achievable in 15–20 min. The mechanism of the reaction is not completely certain, but the pathway shown in Scheme 6.22 is generally favored at present. In particular, a facile intramolecular cleavage of the acyl cobalt complex seems necessary to account for the absence of hydrogenolysis products (aminoaldehydes).

Scheme 6.22

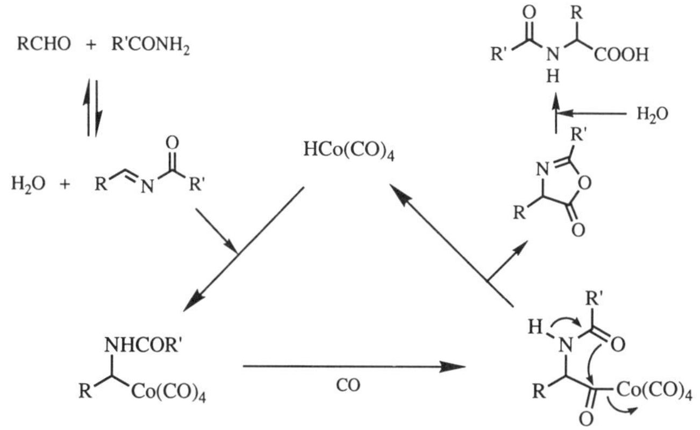

By combining the Wakamatsu reaction with known catalytic routes to aldehydes (e.g., isomerization of allylic alcohols or hydroformylation of alkenes) it has proved possible to achieve direct, one-step syntheses of N-acyl amino acids from precursors other than aldehydes themselves. Thus, trifluoropropene and acetamide are converted to N-acetyl-trifluoronorvaline in 80% yield with 96% selectivity using $Co_2(CO)_8$ as both hydroformylation and amidocarbonylation catalyst, although N-acetyltrifluorovaline is the major product (83% yield, 94% selectivity) when $Rh_6(CO)_{16}$ is used in combination with $Co_2(CO)_8$ (Scheme 6.23).[243] In similar fashion, the addition of an isomerization catalyst such as $RhH(CO)(PPh_3)_3$ or $Pd(PPh_3)_2Cl_2$ allows the direct amidocarbonylation of allylic alcohols, and mild Lewis acids such as titanium isopropoxide promote the *in situ* rearrangement of oxiranes to aldehydes, again permitting direct amidocarbonylation (Scheme 6.24).[243]

Scheme 6.23

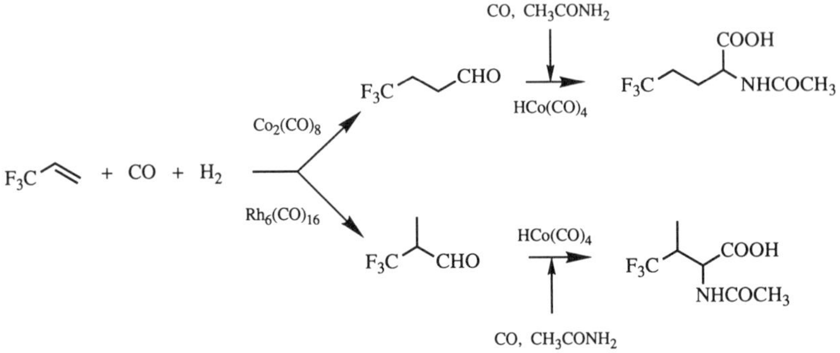

Scheme 6.24

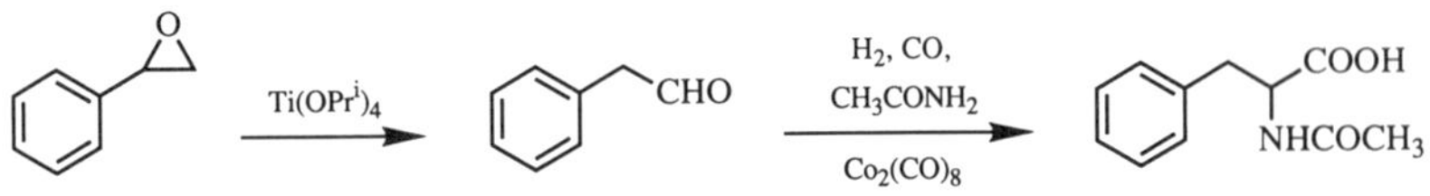

Preparation of N-Acetylphenylalanine[243]

A mixture of styrene oxide (2.40 g, 20 mmol), acetamide (1.18 g, 20 mmol), $Co_2(CO)_8$ (0.23 g, 0.67 mmol), $Ti(OPr^i)_4$ (0.19 g, 0.67 mmol), and THF (30 cm^3), are placed in a 50-cm^3 stainless steel autoclave, and the vessel is then pressurized with carbon monoxide (80 bar) and H_2 (20 bar) and heated to 110°C with stirring for 16 h. After cooling and venting, the reaction mixture is evaporated and the residue extracted with a mixture of ethyl acetate and 100 cm^3 of 5% aqueous sodium carbonate. The aqueous layer is separated, filtered, acidified to pH 1 with phosphoric acid, and extracted with ethyl acetate. The organic layer is dried over magnesium sulfate, and concentrated on the rotary evaporator to give colorless crystals of N-acetylphenylalanine (3.80 g, 92% yield).

Chapter 7
Synthesis of Esters

7.1. Introduction

Of all the carboxylic acid derivatives accessible by carbonylation chemistry, esters have been perhaps the most widely investigated. This has been for reasons of experimental convenience (the reactant alcohol can often be used as solvent), and because the ester linkage is sufficiently stable to withstand isolation and purification procedures, yet sufficiently reactive to make esters valuable intermediates for further synthetic transformation.

7.2. Esters: From Halocarbons

Halocarbons such as iodoalkanes, alkyl chloroacetates, and benzylic halides, which are all susceptible to attack by nucleophilic metal carbonyl anions, are readily converted to esters by reaction with carbon monoxide and an alcohol in the presence of base, using $NaCo(CO)_4$ as catalyst [Eq. (7.1)]. Reactions proceed at $25°C$, under 1 bar of carbon monoxide, and very good yields and selectivities can be achieved.[244] With less reac-

$$RX + CO + R'OH \xrightarrow{[Co(CO)_4]^-} RCOOR' + HX \qquad (7.1)$$

tive alkyl halides however, higher temperatures are required, leading to isomerization of the intermediate alkyl–cobalt complex (via a sequence of β-hydride eliminations and reinsertions) and hence to a mixture of carbonylated products.[244] Carbonylation of alkyl chloroacetates to malonate esters has also recently been achieved at low pressure in the vapor phase,[245, 246] using a heterogeneous rhodium-on-carbon catalyst at 175–300°C.

Esters may be obtained from alkyl and benzyl halides using the stoichiometric reagent $Na_2Fe(CO)_4$. The carboxylic acid synthesis discussed in the previous chapter (Sec. 6.2) can be modified by cleaving the intermediate iron acyl (or even its precursor alkyl complex) with *alcoholic*, rather than aqueous, bromine, leading to ester formation in good yield (Scheme 7.1).[247]

Scheme 7.1

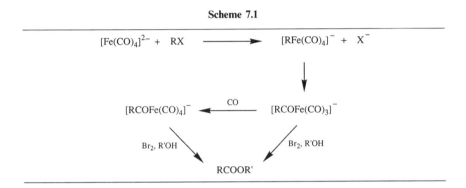

The cyclopentadienyl iron anion $[(\eta^5\text{-}C_5H_5)Fe(CO)_2]^-$ behaves similarly toward alkyl halides, and the resulting alkyliron complexes

Scheme 7.2

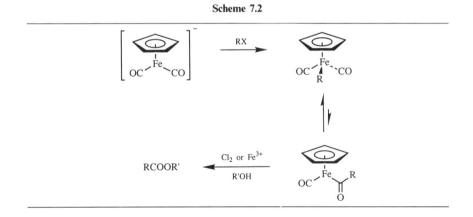

undergo spontaneous carbonyl insertion and cleavage to esters, on treatment with a wide range of oxidizing agents (Fe^{3+}, Ce^{4+}, Cl_2, etc.) in the presence of alcohols (Scheme 7.2).[248]

Alkyliron complexes $(\eta^5\text{-}C_5H_5)Fe(CO)_2R$ undergo carbonyl insertion on treatment with donor ligands such as triphenyl phosphine, to give iron acyls which are configurationally asymmetric. When resolved into their enantiomers such complexes allow extensive elaboration of the acyl ligand with essentially complete control of stereochemistry at any new chiral center that may be formed (Scheme 7.3). Moreover, chirality is almost invariably retained when the organic product (ester, acid, or amide) is liberated by oxidative cleavage of the iron–acyl bond. Both enantiomers of the acetyl complex $(\eta^5\text{-}C_5H_5)Fe(PPh_3)(CO)(COCH_3)$ are now commercially available.[249]

Scheme 7.3

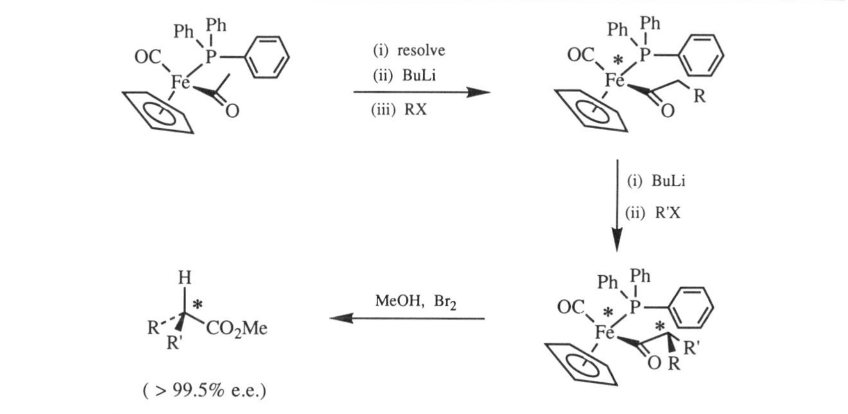

(> 99.5% e.e.)

Benzylic halides may be *catalytically* carbonylated to esters in the presence of $Pd(PPh_3)_2Cl_2$,[250] $[Rh(CO)_2Cl]_2$,[251, 252] $Ni(PPh_3)_2(CO)_2$,[250] or $Ni(CO)_4$.[253, 254] Iodoalkanes often tend to undergo dehydrohalogenation rather than carbonylation in the presence of the palladium-based catalyst, although this is not a problem with perfluoroalkyl iodides, which give excellent yields of esters.[255] The platinum-based catalyst $Pt(PPh_3)_2Cl_2$ does, however, allow ester formation even from conventional iodoalkanes, apparently as a consequence of the stability of platinum alkyls relative to the corresponding alkene–hydride complexes.[256] A recent report indicates that iodoalkanes may also be *photochemically* carbonylated to esters in up to 85% yield under ambient conditions, using a wide range of transition metal carbonyl compounds as catalysts; $Mn_2(CO)_{10}$, $Pt(PPh_3)_2(CO)_2$, and $Ru_3(CO)_{12}$, appear to be particularly effective.[257]

Esters are best obtained from aryl-, vinyl-, and heteroaryl bromides or iodides via the palladium-catalyzed synthesis developed by Heck and co-workers.[250] Reactions are carried out at 60–100°C and atmospheric pressure in the presence of 1–2 mol% of $Pd(PPh_3)_2Cl_2$, using the reactant alcohol as solvent and a tertiary amine (e.g., tri-*n*-butylamine) as the acid acceptor. With iodo-compounds, even palladium acetate without any added phosphine may be used as catalyst. Yields are generally high and are unaffected by most functional groups, including ester, ether, ketone, and nitrile. Vinylic halides may be carbonylated with almost complete retention of stereochemistry, so that, for example, in *n*-butanol at 60°C, *E*-3-iodo-3-hexene affords 74% of the corresponding *E*-carbobutoxylated product, with only 6% of the *Z*-isomer (Scheme 7.4). The scope of the reaction is

Scheme 7.4

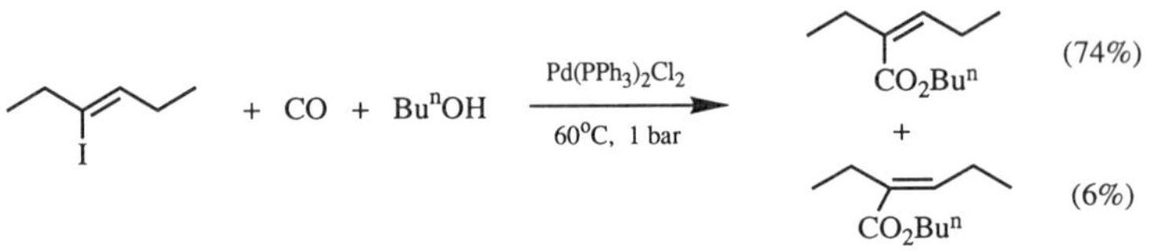

illustrated in Table 7.1, and a generalized procedure for palladium-catalyzed ester synthesis is given below.

Ester Synthesis from Aryl, Heteroaryl, or Vinyl Halides[250]

The catalyst (palladium acetate or $Pd(PPh_3)_2Cl_2$, 0.25 mmol) is placed in a 100-cm³ flask containing a magnetic stirring bar, and the flask is then attached to a thermostatted ambient-pressure carbonylation apparatus (see Chapter 3). The apparatus is flushed several times with carbon monoxide, and the reagents (17.2 mmol of halide, 21.2 mmol of alcohol, and 19.0 mmol of tri-*n*-butylamine) are injected into the flask through a rubber septum.

The flask is raised to the reaction temperature (typically 60–100°C) and the reaction mixture is stirred under a constant 1-bar pressure of carbon monoxide until gas absorption ceases. The mixture is cooled, extracted with ether, and the extract is washed several times with 20% HCl, then with saturated sodium bicarbonate solution, and finally with water. After drying and filtering the solution, ether is removed by evaporation, and the crude ester is purified by chromatography, distillation, or recrystallization.

Table 7.1. *Carbobutoxylation of Aryl, Vinyl, and Benzyl Halides*[a]

Substrate	Product (% yield)	Catalyst
	CO_2Bu^n (70%)	$Pd(OAc)_2$
	CO_2Bu^n (83%)	$Pd(OAc)_2$
	CO_2Bu^n (89%)	$Pd(PPh_3)_2Br_2$
	CO_2Bu^n (79%) / CO_2Bu^n (6%)	$Pd(PPh_3)_2I_2$
	CO_2Bu^n (69%) / CO_2Bu^n (11%)	$Pd(PPh_3)_2I_2$
	CO_2Bu^n (80%)	$Pd(PPh_3)_2Br_2$
	CO_2Bu^n (45%)	$Pd(PPh_3)_2Cl_2$

[a] Reference 250.

The mechanism of this type of carbonylation has recently been investigated by FT infrared observation of the various intermediates in the catalytic cycle.[258] Despite the occurrence of carboalkoxy-palladium species $L_2(X)PdCOOR$ in other contexts,[259] no evidence was found for their participation in the present synthesis, and the probable reaction mechanism (Scheme 7.5) involves oxidative addition of the halide to Pd(0) followed by carbon monoxide insertion, and finally nucleophilic cleavage of the acyl complex by alkoxide.[258] Alcoholate complexes of various elements (including boron,[260] aluminum,[261] tin,[262] and titanium[263]) can act as alkoxide sources in place of the conventional combination of an alcohol and a tertiary amine, although addition of a rhodium cocatalyst is then sometimes necessary—e.g., when trialkyl borates are used. Alternatively, sodium carbonate in the presence of a *catalytic* quantity of amine can replace the full stoichiometric amount of organic base.[264]

Scheme 7.5

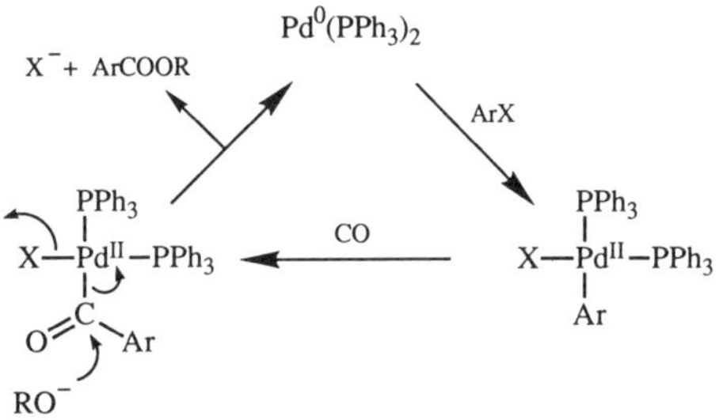

A number of ester-containing natural products including curvularin (**1**) and zearalenone (**2**) have been synthesized with the aid of palladium-catalyzed organohalide carbonylation,[265, 266] and although in these particular syntheses ester formation is *not* the ring-closing step, a wide range of lactones in which carbonylative ring closure *does* occur are accessible using intramolecular variants of this chemistry (see Chapter 9 for a detailed discussion).

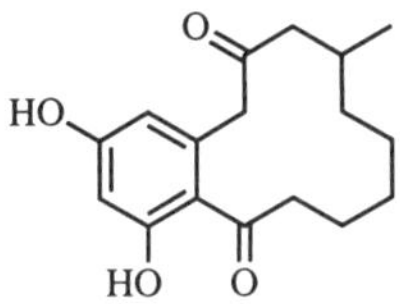

1 curvularin 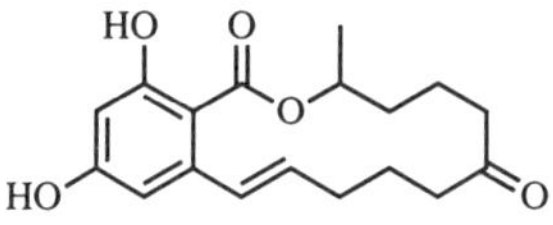   **2** zearalenone

As in the corresponding carboxylic acid synthesis, working at elevated pressure (ca. 90 bar of carbon monoxide) leads to double carbonylation, so that α-keto esters can be produced in moderate yield. Extensive mechanistic studies indicate that double insertion of carbon monoxide into the Pd–aryl bond does *not* occur, and that the key steps are attack by alcohol on coordinated carbon monoxide in a cationic acyl–carbonyl complex, followed by reductive elimination of the keto ester (Scheme 7.6).[267, 268] Selectivity for double carbonylation is maximized by the use of high pressures, sterically hindered alcohols, and solvents of low polarity.

Scheme 7.6

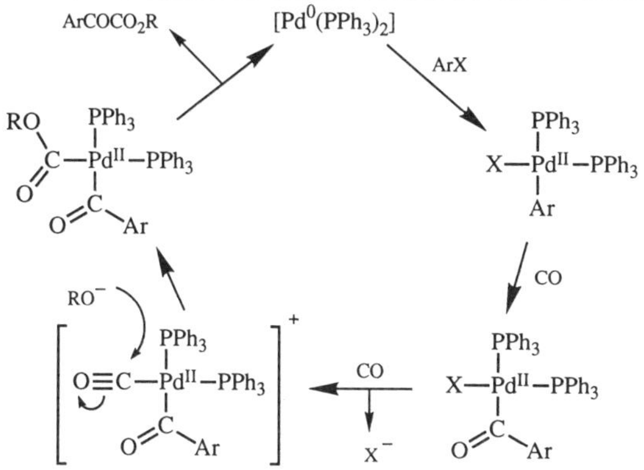

Palladium-catalyzed carbonylations of benzyl, aryl, or vinyl halides in the presence, not of alcohols, but of small-ring cyclic ethers lead to formation of halo-alkyl esters in moderate yields [Eq. (7.2)]. The reaction mechanism is thought to involve halide-promoted transfer of an acyl group from palladium to the ether oxygen, with consequent ring-opening.[269]

$$RX + CO + \begin{bmatrix} O \\ (CH_2)_n \end{bmatrix} \xrightarrow{[Pd^0]} RCOO-(CH_2)_nX \qquad (7.2)$$

(R = benzyl, aryl, or vinyl; n = 2, 3, or 4)

A remarkable new generation of cobalt-based catalysts for conversion of aryl halides to esters has recently been developed.[270] As noted earlier, $[Co(CO)_4]^-$ is normally ineffective for carbonylation of aryl halides, although reaction does occur under conditions of photostimulation.[271] However, Foà and co-workers have discovered an anionic cobalt catalyst, formed by addition of an alkoxy group to a carbonyl ligand of the alkyl cobalt complex $ECH_2Co(CO)_4$ (E is a strongly electron-withdrawing group such as $-C\equiv N$ or $-COOMe$), which promotes carbonylation of aryl and vinyl halides under mild, nonphotolytic conditions (1 bar, 60°C).[270] The role of the electron-withdrawing group is to inhibit CO insertion into the alkyl–cobalt bond, but full details of the mode of action of this catalyst are yet to be established. One possibility is that reversible dissociation of carbon monoxide from the anionic complex $[ECH_2Co(CO)_3(COOR)]^-$ allows oxidative addition of the organic halide to the resulting 16-electron species, followed by reductive elimination of the product ester, as in Scheme 7.7. This reaction is particularly attractive as the alkyl–cobalt catalyst can be produced *in situ* from $[Co(CO)_4]^-$ and an alkyl halide, and very high yields of aromatic carboxylate esters are obtained.

Scheme 7.7

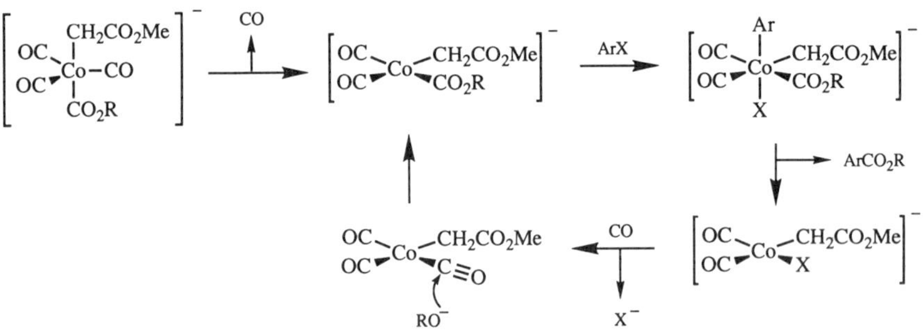

Preparation of Methyl 2-Naphthoate[270]

Methanol (25 cm³), $Co_2(CO)_8$ (1 mmol), methyl chloroacetate (3.5 mmol), 2-chloronaphthalene (30 mmol), and K_2CO_3 (40 mmol) are placed under carbon monoxide in a flask equipped with a magnetic stirrer, thermometer, and condenser connected to a carbon monoxide gas burette. The mixture is stirred at 60°C under a constant 1-bar pressure, and when gas absorption ceases the flask is cooled and the contents are acidified with dilute (1 M) HCl. The product is extracted with ether and purified by column chromatography to give methyl 2-naphthoate in 91% yield.

Even *chloro*-arenes are now viable substrates for carbonylative ester syntheses since the discovery by Milstein and co-workers of a high-activity palladium catalyst incorporating the bulky chelating phosphine 1,3-bis(di-isopropylphosphino)propane (dippp).[272] Relatively high temperatures and moderate pressures are needed (150°C, 5 bar of carbon monoxide), but with this catalyst carbonylation of 1,4-dichlorobenzene in 1-butanol, for example, gives the corresponding terephthalate ester in over 80% yield

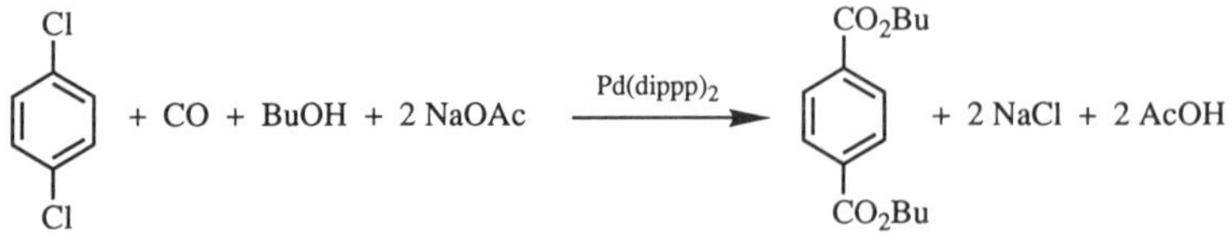

$$\text{(7.3)}$$

[Eq. (7.3)]. ^{31}P NMR data indicate that the sterically hindered, zerovalent complex Pd(dippp)$_2$ exists in a trigonal configuration, with one unused donor group.[272]

7.3. Esters: From Alkenes

The synthesis of an ester by addition of carbon monoxide and an alcohol to an alkene ["hydroesterification," Eq. (7.4)] has a fairly obvious relationship to the hydrocarboxylation reaction described in Sec. 6.5, where water replaces the alcohol and a carboxylic acid is formed. Not surprisingly, therefore, the same types of catalyst [Co$_2$(CO)$_8$, H$_2$PtCl$_6$, Pd(PPh$_3$)$_2$Cl$_2$] are effective for both reactions, and again, addition of SnCl$_2$ can markedly enhance selectivity for linear products.

$$R\diagdown + CO + R'OH \longrightarrow \underset{R}{\diagup}\overset{O\diagdown OR'}{} + R\diagup\diagdown\overset{O}{\diagdown}OR' \qquad (7.4)$$

Hydroesterification of 1-alkenes at 80°C and 200 bar in the presence of ligand-stabilized bimetallic catalysts based on tin(II) chloride and either platinum[273] or palladium[274] complexes affords linear esters with up to 98% selectivity. The highest yields of linear ester are achieved when the molar ratio of transition metal to tin is between 5 and 10. With platinum-based catalysts, selectivity for linear ester increases with alkene chain length, though the reaction rate reaches a maximum at around C$_7$ and such catalysts are relatively inactive for carbonylation of 2-substituted, internal, or cyclic alkenes.[273]

Slightly lower selectivity is exhibited by the palladium-based catalyst derived from Pd(PPh$_3$)$_2$Cl$_2$ and SnCl$_2$, but this type of system has the advantages of much greater flexibility with respect to the structure of the substrate, and significantly lower cost. Linear, branched, internal, and cyclic alkenes can be carbonylated to esters in good yield, and the presence of alkyl groups adjacent to the double bond in 1-alkenes can greatly enhance selectivity for linear products without diminishing activity too drastically. For example, at 30% conversion, the hydroesterification of 2-methyl-1-pentene gives methyl 3-methyl hexanoate with >99% selectivity.[274]

Hydroesterification reactions of unsaturated nitrogen heterocycles have recently been investigated as potential routes to piperidine- and tropane-based alkaloids. With Pd(PPh$_3$)$_2$Cl$_2$ as catalyst, strongly basic substrates such as 1,2,3,6-tetrahydropyridine failed to react, but quaternization or protonation allowed moderate or high conversion to mixtures of isomeric esters.[275]

A novel variant of palladium-catalyzed hydroesterification uses an *enolizable ketone*[276] in place of the alcohol, as shown in Scheme 7.8. The small, equilibrium concentration of enol is apparently trapped by reaction with an acylpalladium intermediate, thus driving the keto/enol equilibrium in the direction of enolization. The product vinylic ester is formed with >90% selectivity, although conversion seems limited to about 35%.

Scheme 7.8

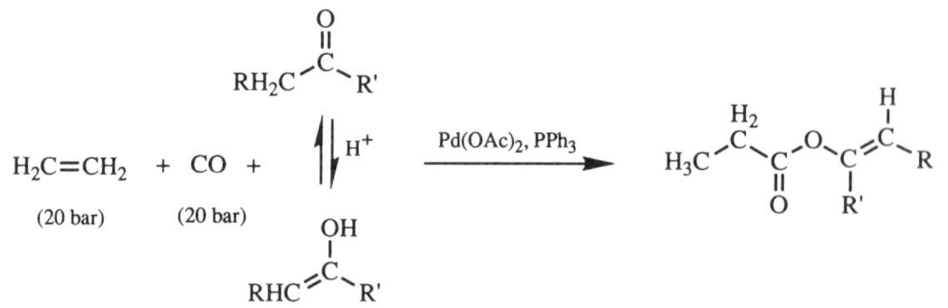

Hydroesterification of conjugated dienes can lead to a variety of products. With $Co_2(CO)_8$ or $PdCl_2$ as catalyst at 100°C and 200 bar pressure, monocarbonylation of butadiene occurs to give an alkyl-3-pentenoate ester,[277] but at higher temperatures and pressures the cobalt catalyst allows carbonylation of the second double bond, thus providing a direct route to dialkyl adipate esters (potential nylon intermediates) from butadiene.[278] The palladium-catalyzed ($PdCl_2$) *mono*carbonylation of 1,4-dienes seems to be a fairly general reaction, giving, for example, 4-methyl-3-pentenoates from isoprene or 4-chloro-3-pentenoates from chloroprene (i.e., with carbonylation occurring at the unsubstituted double bond).[277] Palladium catalysis *in the absence of halide ions*, however, leads to formation of 3,8-nonadienoate esters in high yields.[279] Such reactions are thought to involve diene coupling on the metal center to give a diallyl complex which subsequently undergoes carbonylation as shown in Scheme 7.9. The role of halide ions in preventing this type of reaction appears to be the blocking of a coordination site, thus preventing binding of the second diene ligand.

Scheme 7.9

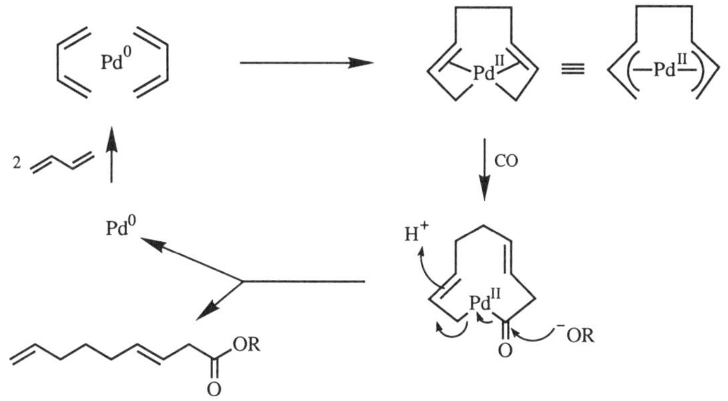

Preparation of Isopropyl 3,8-Nonadienoate[279]

Palladium acetate (0.30 g, 1 mmol), triphenylphosphine (0.70 g, 2.67 mmol), and isopropanol (30 cm^3) are placed in a 200-cm^3 rocking autoclave, and butadiene (20 g, 37 mmol) and carbon monoxide (50 bar) are then introduced. The vessel is shaken at 110°C for 16 h, then cooled and the product (32.5 g, 90% yield) isolated by distillation under reduced pressure (bp 72–76°C at 1.5 mbar).

Cyclic, branched, and linear alkenes may be converted to esters under essentially ambient conditions using the hydrido-zirconium reagent $Zr(C_5H_5)_2(H)Cl$ (Scheme 7.10). This type of chemistry has already been discussed (Sec. 6.5) in the context of carboxylic acid synthesis, and apart

Scheme 7.10

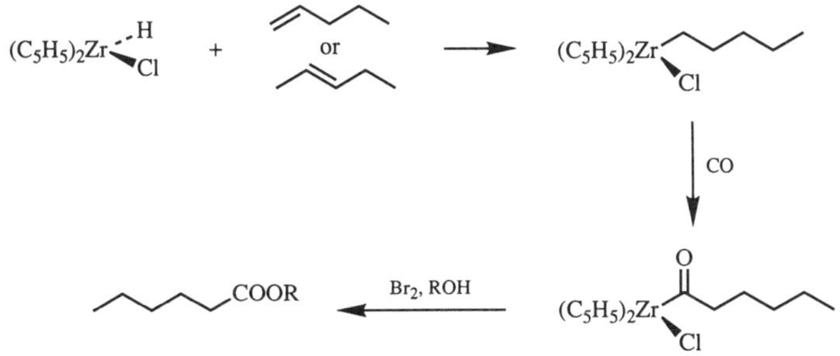

from the use of alcoholic rather than aqueous bromine in the final cleavage reaction, the same procedures are applicable.

Palladium-catalyzed hydroesterification of styrene gives predominantly the "branched" ester, in which a new chiral center has been created at the 2-position (Scheme 7.11). Carrying out this reaction at low pressure (1–2 bar), in the presence of a bulky, chiral ligand (neomenthyldiphenyl-phosphine) and trifluoroacetic acid leads to appreciable asymmetric induction ($>50\%$ e.e.),[280] but this work seems to represent the only significant advance towards stereocontrol of catalytic hydroesterification reactions.

Scheme 7.11

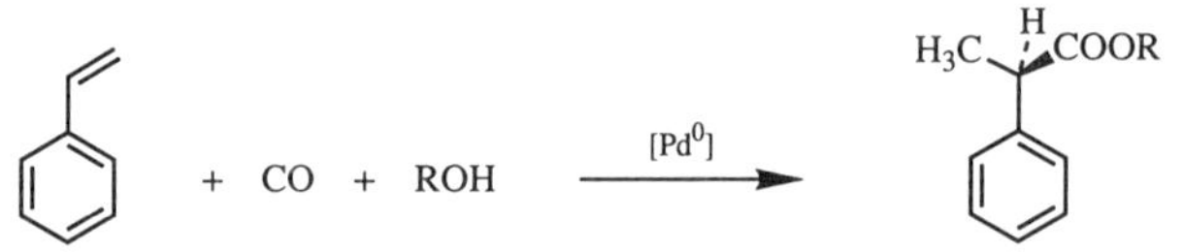

Alkene hydroesterification with palladium catalysts (either homogeneous or heterogeneous) under *oxidative* conditions (e.g., in the presence of Cu(II) salts which can reoxidize Pd(0) to Pd(II), as discussed in Sec. 2.6.5) results in formation of either unsaturated monoesters, saturated diesters, or β-alkoxy esters (Scheme 7.12).[281] Under neutral conditions linear 1-alkenes afford principally β-alkoxy esters, but under more basic conditions 1,2-diesters predominate. Careful optimization of this chemistry (25°C, 3–4 bar, methanol as solvent, $CuCl_2$ as reoxidant, sodium butyrate as buffer) has resulted in a valuable and resonably general synthesis of

Scheme 7.12

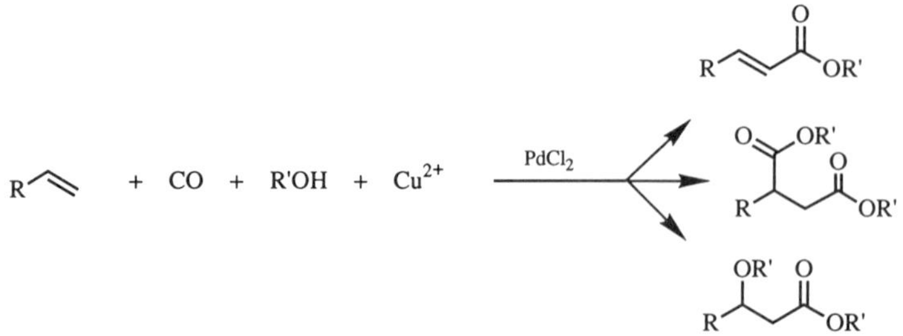

diesters,[282] and in fact the latter are produced in high yield from cyclo-alkenes even in the absence of base or buffer. The synthetic utility of such reactions is demonstrated in Scheme 7.13, which shows the preferred route to a novel bis-diene for multiple Diels–Alder reactions.[283]

Scheme 7.13

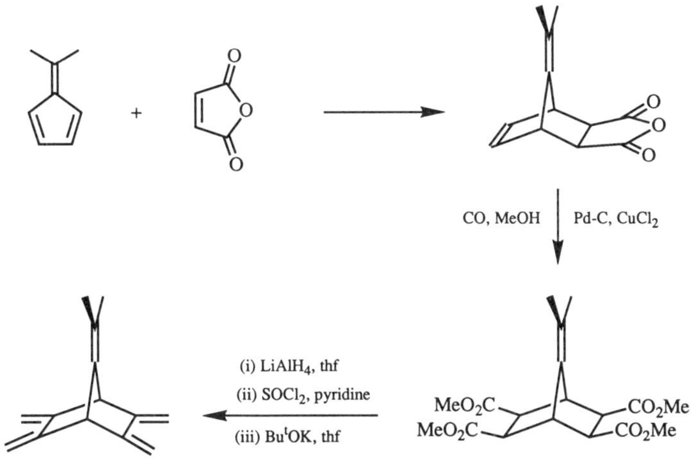

CO, MeOH | Pd-C, CuCl$_2$

(i) LiAlH$_4$, thf
(ii) SOCl$_2$, pyridine
(iii) Bu^{t}OK, thf

Preparation of Tetramethyl 7-Isopropylidenebicyclo[2.2.1]heptane-2-exo, 3-exo,5-exo,6-exo-tetracarboxylate[283]

The 6.6-dimethylfulvene/maleic anhydride Diels–Alder adduct (15 g, 0.073 mol), anhydrous CuCl$_2$ (49.4 g, 0.367 mol), 10% Pd/C catalyst (1.8 g, 1.75 mmol), and anhydrous methanol (250 cm^3) are placed in a 1-liter glass-lined Parr hydrogenator and the mixture is degassed and the vessel pressurized with carbon monoxide (3 bar). After stirring at 20°C for 48 h, with the pressure being maintained between 2 and 4 bar, the vessel is vented and the solvent removed on the rotary evaporator. The residue is triturated with a mixture of CHCl$_3$ (150 cm^3) and saturated NaHCO$_3$ solution (150 cm^3), the extract is filtered, and the organic layer is washed with further saturated NaHCO$_3$ solution (3 × 40 cm^3) and dried over magnesium sulfate. Evaporation of the solvent and recrystallization from ether affords pure tetraester (22.2 g, 82%), mp 108–110°C.

Conjugated dienes generally undergo oxidative carbonylation to give mixtures of 1,4-addition products (Scheme 7.14), but again conditions may be optimized to give high yields of a single product. Carbonylation of 1,3-

butadiene (CuCl$_2$-PdCl$_2$ catalyst system) in the presence of benzyl alcohol, for example, affords dibenzyl *trans*-hex-3-ene-1,6-dioate in 90% yield.[282]

Scheme 7.14

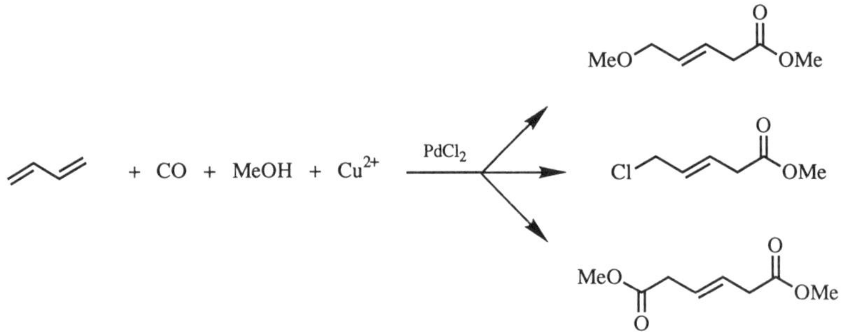

Oxidative carbonylation of alkenes as a specific route to α,β-unsaturated esters has not been widely investigated, but there are indications that this could be developed into a useful procedure. Methyl cinnamate has, for example, been obtained as the major product from carbonylation of styrene in methanol using PdCl$_2$/MgCl$_2$ as catalyst, CuCl$_2$ as reoxidant, and sodium acetate as buffer (Scheme 7.15). The formation of diester is minimized by working at a carbon monoxide partial pressure of only 0.2 bar (CO/N$_2$ mixture).[284]

Scheme 7.15

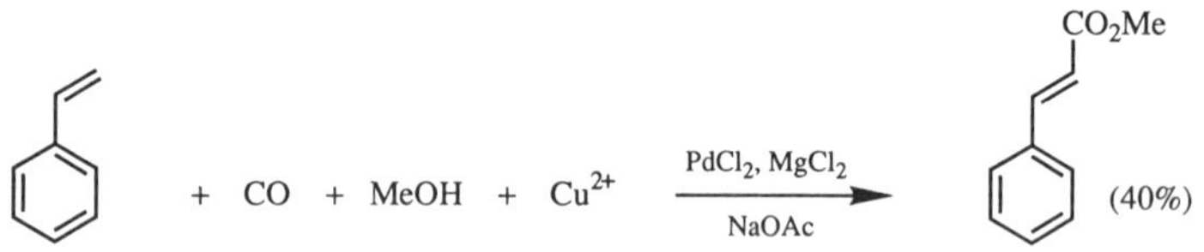

Curiously, conditions that might be expected to lead to *oxidative* carbonylation (PdCl$_2$/CuCl$_2$/O$_2$/MeOH/HCl) seem to allow simple nonoxidative hydroesterification of alkenes under very mild conditions.[285] It seems possible that this type of reaction is more closely related to Koch-type (carbonium ion) chemistry (Sections 6.4 and 6.5) than to conventional palladium-catalyzed hydroesterification since (a) strong-acid conditions (concentrated hydrochloric acid) are necessary, and (b) branched-chain esters are produced *exclusively*. Further work is clearly required to resolve the nature of these reactions.

A noncatalytic, but nevertheless remarkable, alkoxy–ester synthesis involves reaction of carbon monoxide (1 bar) with Pd(norbornadiene)Cl$_2$ in methanol at 25°C.[286] The product, which is formed in high yield, is thought to arise by the sequence shown in Scheme 7.16, and a catalytic version of this reaction seems possible since only the reoxidation of Pd(0) to Pd(II) is required. Carbonylation of π-allyl palladium complexes

Scheme 7.16

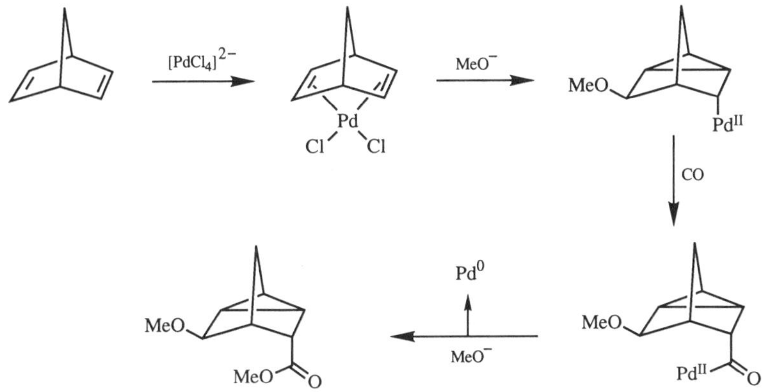

(accessible directly from alkenes, as shown in Scheme 7.17) can also be achieved under relatively mild conditions (3 bar, 25°C).[287] β,γ-Unsaturated esters are obtained in excellent yield. In a development of this

Scheme 7.17

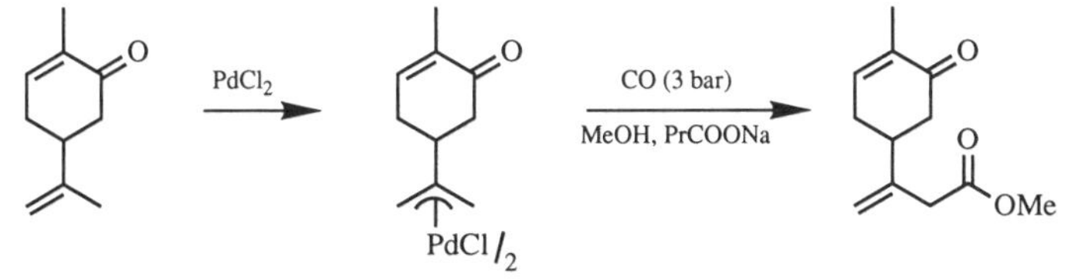

chemistry, substituted 1,4-cyclohexadienes (readily available via Birch reduction of arenes) undergo sequential palladation and carbonylation, as shown in Scheme 7.18, to form methoxycyclohexene carboxylate esters.[288] Very high regio- and stereoselectivity are observed, and yields of up to 90% are obtainable.

Scheme 7.18

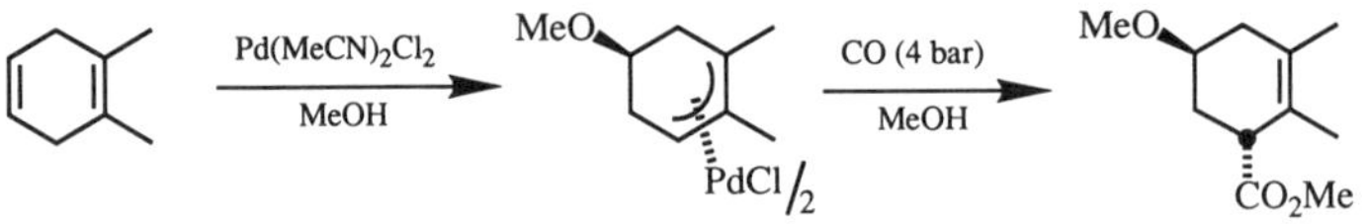

Catalytic *intramolecular* β-alkoxy ester formation can be achieved by oxidative carbonylation, provided the alkene and hydroxyl groups are in an appropriate stereochemical relationship for cyclization to occur. The reaction sequence almost certainly involves pyran ring formation by intramolecular nucleophilic attack on the coordinated double bond, followed by carbon monoxide insertion into the resulting Pd−C σ-bond and alcoholysis of the resulting acyl palladium complex (Scheme 7.19).[289] The effects of substituents on ring-size preference (e.g., where cyclization can lead to either furan or pyran ring systems) have been explored in some

Scheme 7.19

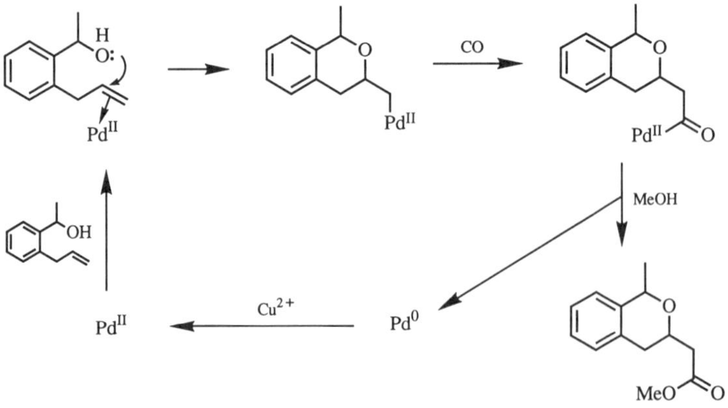

detail.[290] This reaction provides a crucial step in a recently reported synthesis of the natural antibiotic frenolicin (Scheme 7.20).[289]

Scheme 7.20

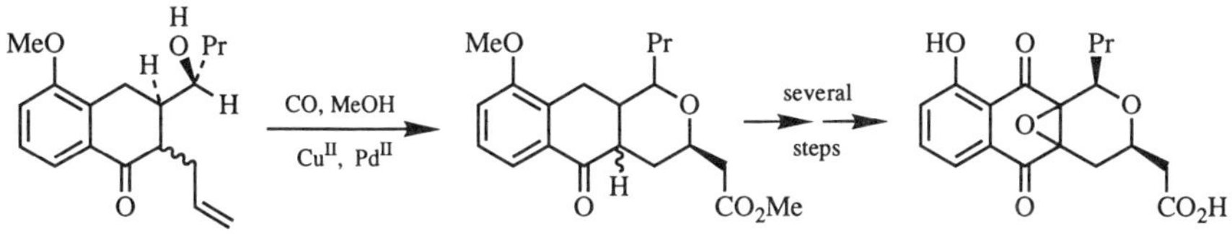

(87%)   frenolicin

A reaction closely related to that shown in Scheme 7.17 results in the transformation of allenic amines or amides to alkyl esters of α-heterocyclic acrylic acids (Scheme 7.21). Reactions are carried out under essentially ambient conditions and (nonoptimized) yields of 60%–70% are generally obtained.[291]

Scheme 7.21

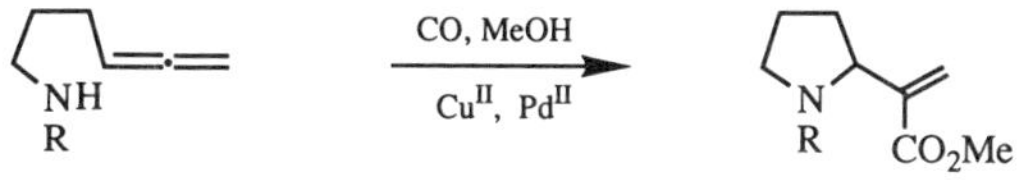

An elegant extension of the chemistry outlined in Scheme 7.19 allows the synthesis of spirocyclic esters by catalytic oxidative carbonylation of hydroxyenones.[292] Here, formation of the alkylpalladium intermediate requires a *double* cyclization, with subsequent carbonyl insertion and alcoholysis leading to the product spiroacetal in around 80% yield (Scheme 7.22).

The carbonylations outlined in Schemes 7.16–7.22 are all dependent on the initial creation of a palladium–carbon σ-bond by nucleophilic attack on a coordinated alkene. When the attacking nucleophile is a *carbanion*, the overall process is generally referred to as "carboacylation,"[293] and this type of reaction has recently been applied to the derivatization of enamides (Scheme 7.23), en route to the β-lactam (±)-thienamycin.[294]

Scheme 7.22

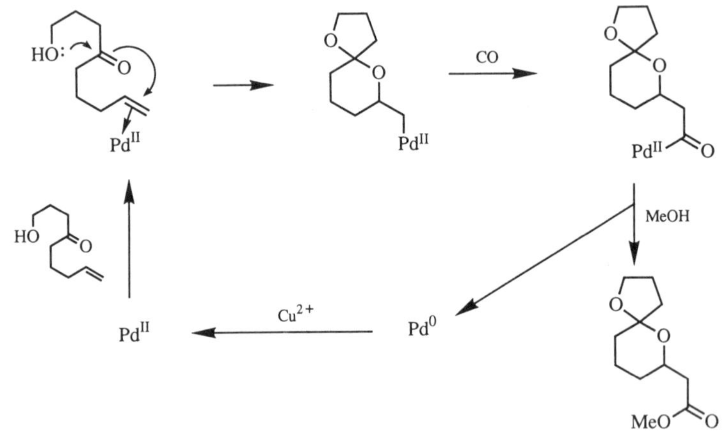

Scheme 7.23

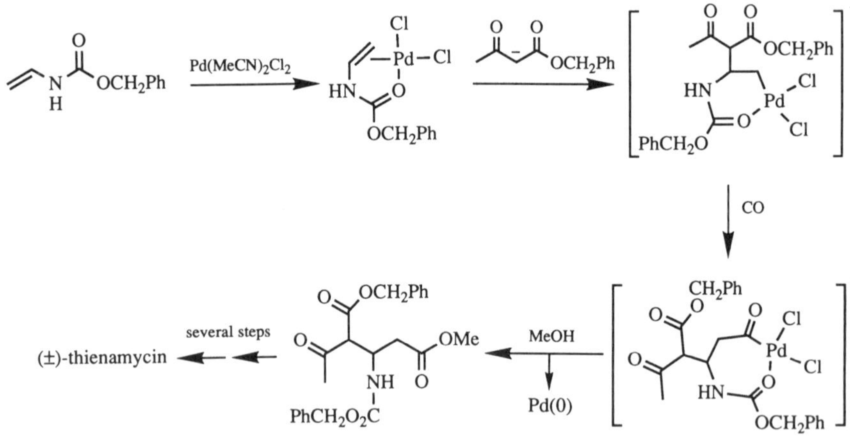

The conversion of 1-iodo-1,4-dienes to unsaturated cyclic keto esters, in the presence of homogeneous palladium catalysts, provides a splendid example of the ability of homogeneous transition metal catalysts to promote a complex sequence of metal-centered reactions yielding a single product with a high degree of specificity.[295] Two carbonylation steps are involved as shown in Scheme 7.24, the first leading to formation of a cyclic

Scheme 7.24

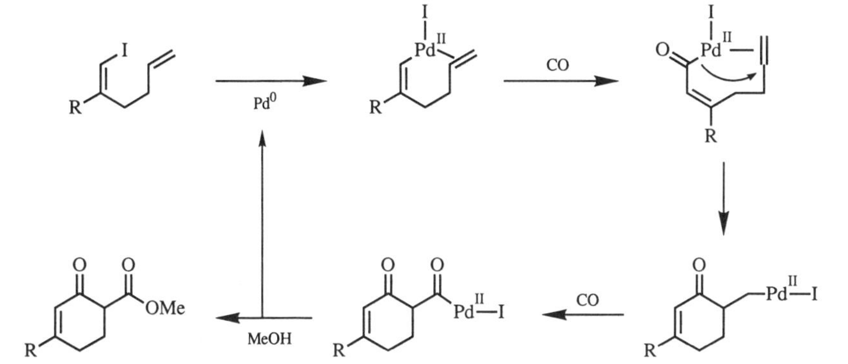

keto-alkyl ligand, and the second to the introduction of an ester function and release of the product from the metal. A wide variety of keto esters have been synthesized in good yield (Table 7.2), reactions normally being

Table 7.2. Palladium-Catalyzed Carbonylative Cyclizations[a]

Substrate	Product	Yield (%)
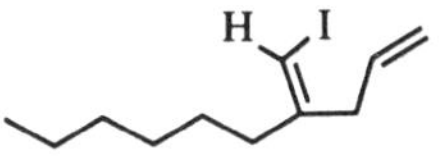	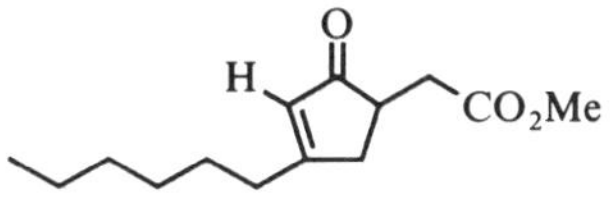	90
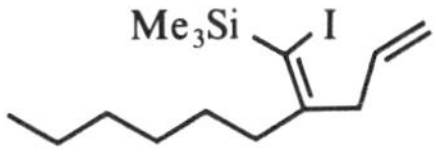	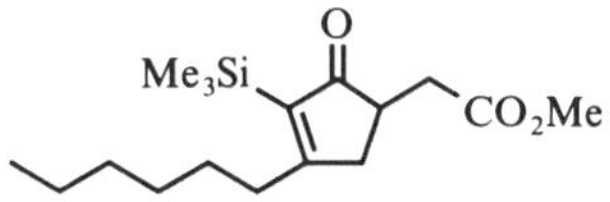	75
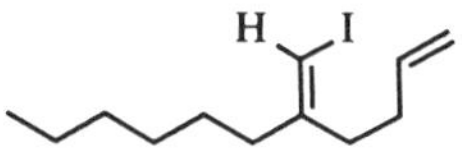	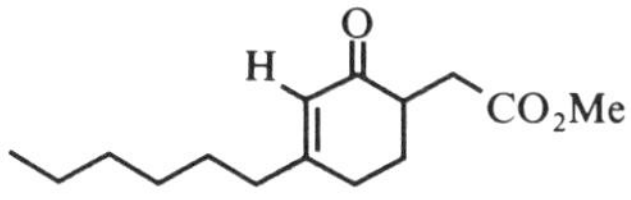	66
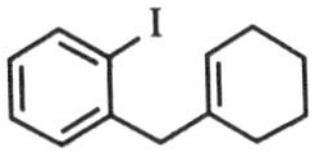	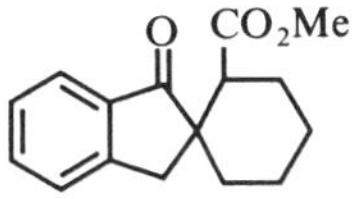	67
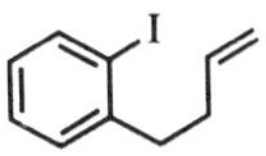	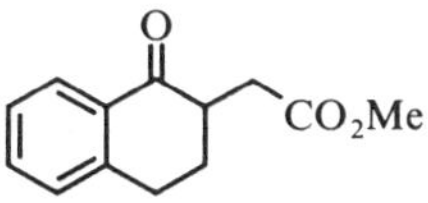	73

[a] Reference 295.

carried out in the presence of 5 mol% of a palladium catalyst (typically $Pd(PPh_3)_2Cl_2$), under 40 bar of carbon monoxide, in benzene/acetonitrile (1:1) at 100°C.

7.4. Esters: From Alkynes

Simple hydroesterification of alkynes to α,β-unsaturated esters occurs in the presence of nickel-, cobalt-, or palladium-based catalyst systems, and the reaction of alcohols with carbon monoxide and acetylene, catalyzed by nickel tetracarbonyl, has in the past been applied to the industrial synthesis of acrylate esters. A substantial body of early experimental data in this field, mostly relating to very high-pressure reactions (100–1000 bar) has been reviewed elsewhere,[296] and in this section we concentrate on more recent, lower-pressure chemistry.

With $Ni(CO)_4$ as catalyst, addition of the carboalkoxy group nor- mally occurs at the 2-position of terminal alkynes (Markownikov addi- tion), but this selectivity is reversed to give predominantly linear products when the phosphine-stabilized palladium–tin catalysts described in Section 7.3 are used.[297] Selectivity falls off as the pressure of carbon monoxide increases, so that although at 70 bar the carbonylation of 1-heptyne in the presence of $[(p\text{-}MeC_6H_4)_3P]_2PdCl_2\text{-}SnCl_2$ [Eq. (7.5)] is three times faster than at atmospheric pressure, the selectivity for methyl 2-octenoate falls from 93% to 78%.

$$\text{~~~~} + \text{ CO } + \text{ MeOH } \xrightarrow[\text{SnCl}_2]{[(p\text{-tol})_3P]_2PdCl_2} \text{~~~~} + \text{~~~~} \qquad (7.5)$$

Platinum–tin catalysts such as $(PPh_3)_2PtCl_2\text{-}SnCl_2$ are reported to be more active for hydroesterification of internal alkynes than the correspond- ing palladium complexes,[298] allowing conversion of diphenylacetylene to methyl α-phenyl cinnamate in 85% yield at 100°C and 80 bar [Eq. (7.6)].

$$\text{Ph}\text{—}\equiv\text{—Ph } + \text{ CO } + \text{ MeOH } \xrightarrow[\text{SnCl}_2]{(PPh_3)_2PtCl_2} \quad \underset{\text{Ph}}{\overset{\text{H}}{>}}=\underset{\text{Ph}}{\overset{\text{CO}_2\text{Me}}{<}} \qquad (7.6)$$

Alper and co-workers have recently shown that certain internal alkynes (2-alkynes) can be carbonylated to α,β-unsaturated esters under essentially *ambient conditions*, using a catalyst system containing $PdCl_2$,

$CuCl_2$, HCl, and O_2. As with alkene substrates (Section 7.3), this apparently oxidizing system again leads to regioselective, *nonoxidative* carbonylation, which is also now specific for the *cis*-product. 2-Heptyne, for example, affords the *n*-propyl ester of *cis*-2-methylheptenoic acid in over 75% isolated yield [Eq. (7.7)].[299] With terminal alkynes under similar

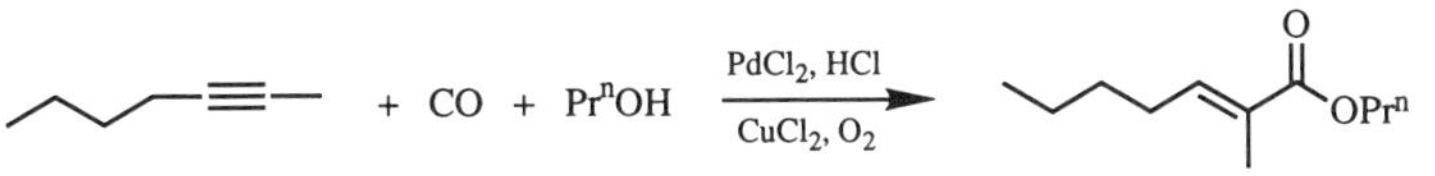

$$\text{CH}\equiv\text{CH} \quad + \quad CO \quad + \quad Pr^nOH \quad \xrightarrow[\text{CuCl}_2,\ O_2]{\text{PdCl}_2,\ \text{HCl}} \quad \overset{O}{\underset{}{\text{OPr}^n}} \qquad (7.7)$$

conditions, however, oxidative carbonylation *is* observed, giving unsaturated diesters in essentially quantitative yield (Scheme 7.25).[299] The ratio of *cis* to *trans* diester depends on the substituent R, varying from 4:1 for linear alkyl to more than 8:1 for cyclohexyl. With acetylene itself this ratio is 6.1:1 (86% total yield of maleate and fumarate esters), although an alternative catalyst comprising palladium chloride and thiourea gives a similar yield with very much greater selectivity for maleate (about 45:1).[300]

Scheme 7.25

$$RC\equiv CH \quad + \quad CO \quad + \quad R'OH \quad \xrightarrow[\text{CuCl}_2,\ O_2]{\text{PdCl}_2,\ \text{HCl}} \quad \underset{R'O_2C}{\overset{R}{>}}\!\!=\!\!\underset{CO_2R'}{\overset{H}{<}} \quad + \quad \underset{R'O_2C}{\overset{R}{>}}\!\!=\!\!\underset{H}{\overset{CO_2R'}{<}}$$

Oxidative carbonylation can, on occasion, occur even in the absence of any apparent source of oxidizing power. Palladium-catalyzed carbonylation of diphenylacetylene in methanol at 100°C/100 bar gives the dimethyl ester of diphenyl maleic acid in 34% yield.[301] This compound, however, seems to behave as its own oxidizing agent since the coproduct diphenylcrotonolactone (60% yield) obviously arises from reduction of one carbomethoxy group with subsequent cyclization (Scheme 7.26).

Scheme 7.26

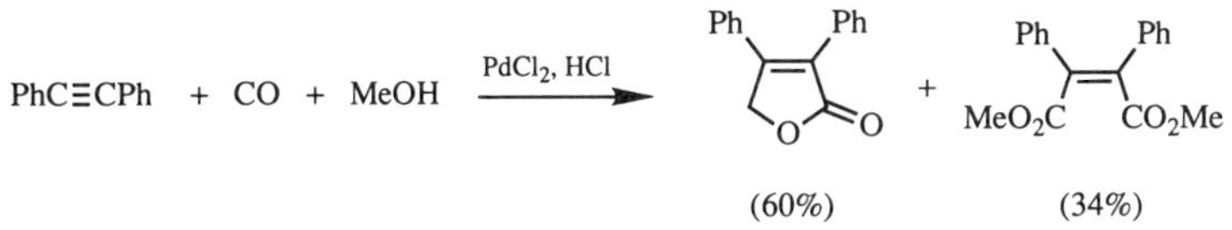

$$PhC\equiv CPh \quad + \quad CO \quad + \quad MeOH \quad \xrightarrow{\text{PdCl}_2,\ \text{HCl}} \quad \overset{Ph\quad Ph}{\underset{O\quad O}{\bigcirc}} \quad + \quad \underset{MeO_2C}{\overset{Ph\quad Ph}{>}}\!\!=\!\!\underset{CO_2Me}{<}$$

$$\qquad\qquad\qquad\qquad\qquad\qquad (60\%) \qquad\qquad (34\%)$$

When the anionic nickel complex $[Ni(CO)_3I]^-$ is used as catalyst precursor, hydroesterification of acetylene at atmospheric pressure leads to sequential dimerization and carbonylation of the substrate,[302] with formation of 2,4-pentadienoate esters in good yield [Eq. (7.8)]. Surprisingly perhaps, acrylate esters are not formed as by-products, suggesting that the reaction follows a very different path from that of conventional hydroesterification, but mechanistic details are still unclear.

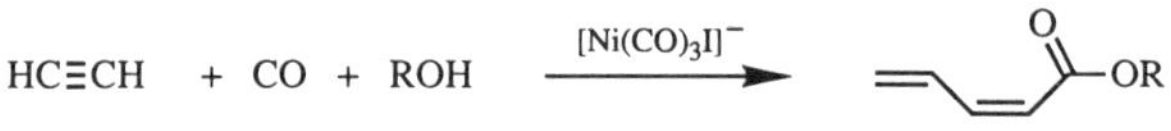

$$ HC\equiv CH \ + \ CO \ + \ ROH \ \xrightarrow{\ [Ni(CO)_3I]^-\ } \ \text{(structure)} \qquad (7.8) $$

Since alkynes "insert" not only into metal–hydrogen but also metal–carbon bonds, it has proved possible to develop multicomponent carbonylations in which oxidative addition of an organic halide to a low-valent transition metal is followed by sequential insertions of an alkyne and carbon monoxide. In particular, allylic halides react readily with acetylene, carbon monoxide, and methanol in the presence of zerovalent nickel, as outlined in Scheme 7.27, to give cis-dienoate esters in 70%–80% yield.[303]

Scheme 7.27

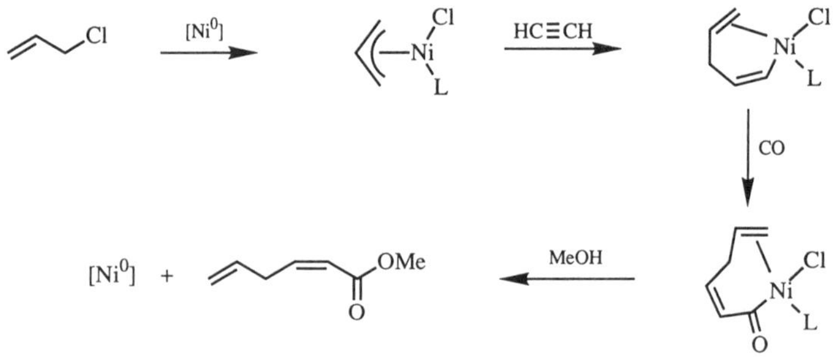

More recently it has been shown that the palladium-catalyzed reaction of perfluoroalkyl iodides with terminal alkynes, carbon monoxide (1 bar), and alcohols, leads to β-perfluoroalkyl alkenoate esters in 50%–60% yield.[304] The perfluoroalkyl group invariably adds to the terminal carbon of the alkyne, and in view of the presence of significant quantities of perfluoroalkylalkenes as by-products, a mechanism involving initial formation of perfluoroalkyl radicals has been proposed (Scheme 7.28).

Scheme 7.28

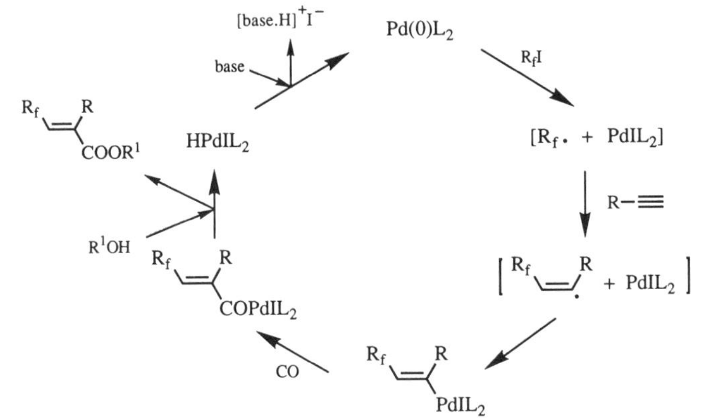

7.5. Esters: From Main-Group Organometallic Compounds

7.5.1. From Organolithium and Grignard Reagents[305]

Alkyl and aryl Grignard reagents add readily to a carbon monoxide ligand of iron pentacarbonyl to give acyl tetracarbonyl anions $[RCOFe(CO)_4]^-$ which on reaction with alcoholic iodine, as shown in Scheme 7.29, afford esters in high yield.[306] Alkynyl lithium compounds (from 1-alkynes and *n*-butyl lithium) behave similarly, so providing a novel route to alkyne carboxylate esters (Scheme 7.30).[307]

Scheme 7.29

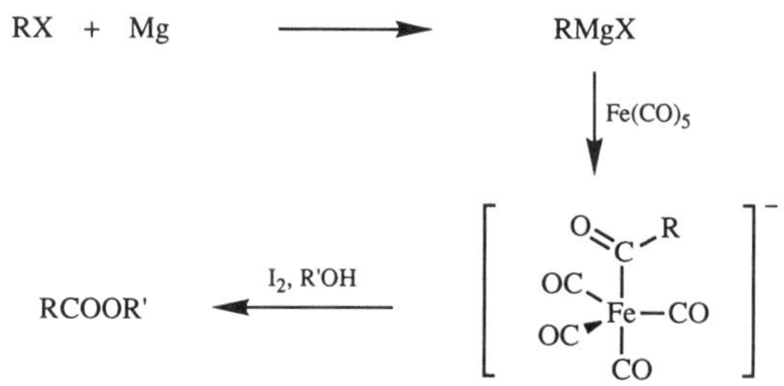

Scheme 7.30

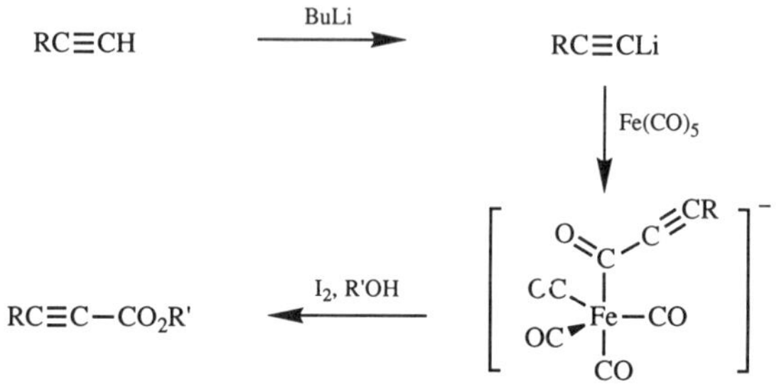

$$RC\equiv CH \xrightarrow{\text{BuLi}} RC\equiv CLi$$

$$\downarrow \text{Fe(CO)}_5$$

$$RC\equiv C-CO_2R' \xleftarrow{\text{I}_2,\ R'OH} \left[\begin{array}{c} O\diagdown \\ C \\ C.C \diagdown \mid \\ OC \diagup Fe-CO \\ \mid \\ CO \end{array} \begin{array}{c} C\diagdown ^{CR} \\ \end{array} \right]^{-}$$

7.5.2. From Organomercury Compounds

Although it has been known for over 20 years that transition metal salts and their complexes promote the carbonylation of organomercurials to carboxylic esters,[308] only in 1985 were procedures described for a simple, catalytic, and reasonably general ester synthesis based on this chemistry [Eq. (7.9)].[309] Most organomercurials (alkyl, aryl, allyl, and

$$RHgX + CO + R'OH \longrightarrow RCOOR' + Hg + HX \qquad (7.9)$$

vinyl) can be carbonylated in good yield at 80–100°C and 3–5 bar of carbon monoxide, using about 1 mol% of $Pd(PPh_3)_2Cl_2$ or $Rh(PPh_3)_3Cl$ as catalyst. Mixtures of ester and carboxylic acid are usually formed in proportions that depend on the solvent and catalyst. Ester formation is apparently maximized when methanol is used as solvent and the palladium complex as catalyst, and since many arylmercury derivatives are available by direct metallation of arenes, this chemistry provides an effective route to substituted benzoate and phthalate esters from the corresponding hydrocarbons. Metallic mercury released in the reaction is easily separated from the product, and direct air oxidation of this in a carboxylic acid medium regenerates a mercury(II) carboxylate suitable for reconversion to an organomercurial. Earlier work in this area tended to emphasize reactions which were stoichiometric in palladium, as in Eq. (7.10), though attempts were made to recycle Pd(0) to Pd(II) *in situ* by addition of reoxidants such as $CuCl_2$. This work led to remarkably facile stoichiometric and catalytic synthesis of α,β-unsaturated esters from vinyl mercuric halides,[310] which are themselves readily accessible from alkynes

$$RHgX + CO + R'OH + Pd^{2+} \longrightarrow RCOOR' + Hg^{2+} + Pd + HX \quad (7.10)$$

via hydroboration and mercuration.[311] Reactions proceed at low temperature and 1 bar pressure, with stereospecific replacement of the chloro-mercury group (Scheme 7.31), and are tolerant of a wide variety of functional groups including halogens, cyano, and carboalkoxy.

Scheme 7.31

$$R\diagup\diagdown\diagup^{HgCl} \ + \ Pd^{2+} \ + \ CO \ + \ R'OH \ \longrightarrow \ R\diagup\diagdown\diagup^{CO_2R'} \ + \ Pd^0$$

Stoichiometric Synthesis of Methyl trans-Cyclohexylacrylate[310]

Anhydrous lithium chloride (20 mmol), $PdCl_2$ (10 mmol), and methanol (100 cm^3), are placed in a well-dried 250-cm^3 flask fitted with a septum inlet and carbon monoxide inlet. The flask is cooled to $-78°C$ and *trans*-cyclohexylethenylmercury(II) chloride (10 mmol) is added. The flask is flushed thoroughly with carbon monoxide, and the reaction mixture stirred as the flask is warmed to room temperature over about 4 h. After stirring overnight under a slight positive pressure of carbon monoxide, ether, and activated charcoal are added, and the mixture is filtered, washed with saturated aqueous ammonium chloride, and dried over sodium sulfate. Evaporation to remove the solvent leaves essentially pure ester (1.61 g, 96 % yield).

7.5.3. From Organothallium Compounds

Organothallium(III) reagents can be carbonylated to esters [Eq. (7.11)] under very mild (essentially ambient) conditions in the presence of $PdCl_2$ as catalyst (0.1 moles/mole of starting material) and stoichiometric quantities of magnesium oxide (1 mole) and lithium chloride (2 moles).[312]

$$ArTl(O_2CCF_3)_2 \ + \ CO \ + \ ROH \ \xrightarrow[\text{MgO, LiCl}]{PdCl_2} \ ArCOOR \ + \ TlO_2CCF_3 \ + \ CF_3COOH \quad (7.11)$$

This process takes advantage of the inherently high regioselectivity of electrophilic arene thallation, which affords essentially pure starting materials for carbonylation. Thallation of *t*-butyl benzene with $Tl(OOCCF_3)_3$, for example, followed by carbonylation of the crude product, gives an 80 % yield of $p\text{-}Me_3CC_6H_4COOMe$ containing less than 2 % of the ortho-isomer. Conversely, many *coordinating* functional groups (amide and ketone for example) direct the incoming Tl(III) electrophile exclusively to the *ortho*-position, leading to carbonylation at the same site (Scheme 7.32).

This approach has been extended to *ortho*-thallation/carbonylation of benzylic alcohols, where intramolecular reaction at the final stage results in lactone formation (see also Section 9.3).[312]

Scheme 7.32

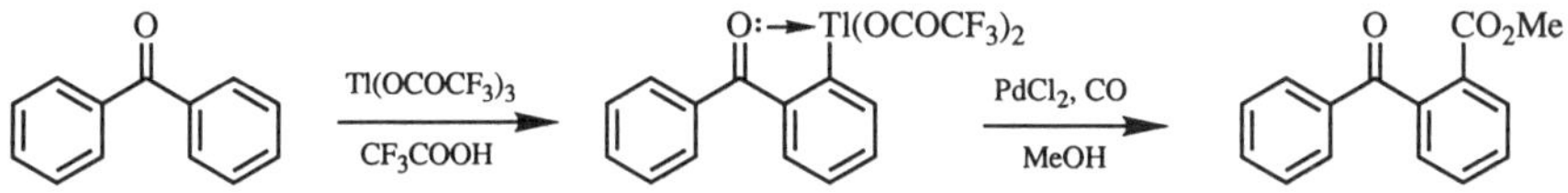

7.5.4. From Organoboron and Organosilicon Compounds

1-Alkenyl boranes, which are readily prepared by hydroboration of alkynes, react readily with carbon monoxide (1 bar) and methanol, in the presence of palladium chloride and sodium acetate, to give α,β-unsaturated esters with retention of geometry at the double bond.[313] Very high yields (greater than 90%) are achievable, and the reaction can be carried out either stoichiometrically in palladium, or catalytically by using 1,4-benzoquinone as reoxidant (Scheme 7.33). Alkenyl pentafluorosilicates (accessible via hydrosilylation of alkynes) behave similarly as substrates in stoichiometric carbonylation,[314] but a catalytic version of this reaction does not appear to have been investigated.

Scheme 7.33

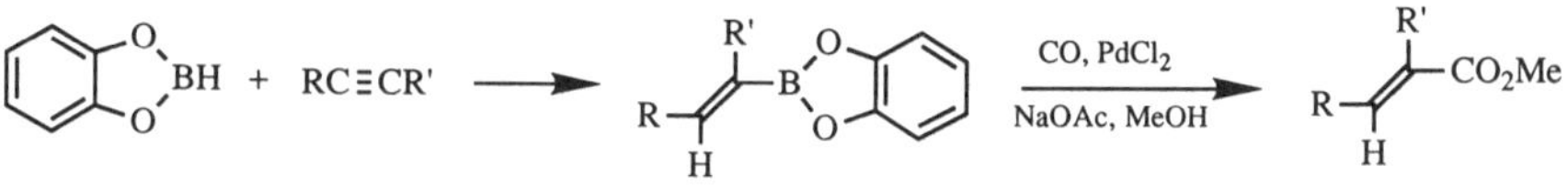

7.5.5. From Organosulfides, -selenides, and -tellurides

Diaryl chalcogenides react with carbon monoxide at high temperatures and pressures (typically 20–100 bar and 100–200°C), in the presence of dicobalt octacarbonyl as catalyst, to give thio-, seleno-, and telluro-esters in moderate yield [Eq. (7.12)].[315] Catalytic activity is highest for sulfides, the cobalt complex being susceptible to poisoning by

$$2\text{ArMMAr} + 3\text{CO} \xrightarrow{\text{Co}_2(\text{CO})_8} \text{ArCOMAr} + \text{ArMAr} + 2\text{MCO} \qquad (7.12)$$

(M = S, Se, or Te)

small amounts of elemental selenium and tellurium formed during the reaction. Addition of triphenylphosphine as a chalogen scavenger prevents this type of catalyst deactivation with selenium, but activity remains low with diaryl tellurides, for which essentially stoichiometric amounts of $\text{Co}_2(\text{CO})_8$ are required.

7.6. Esters: From Aryl and Vinyl Triflates

The recently discovered ability of the trifluoromethanesulfonate ("triflate") group $(\text{CF}_3\text{SO}_2\text{O}^-)$ to behave as a pseudohalogen in many palladium-catalyzed organic reactions marks a significant step forward in homogeneous catalysis.[316] In particular, it allows direct, high-yield carbonylation of the readily accessible triflate derivatives of phenols[317, 318] (Scheme 7.34a) and enols[319] (Scheme 7.34b) to give esters of aromatic and α,β-unsaturated carboxylic acids. Ketone-derived enyl

Scheme 7.34

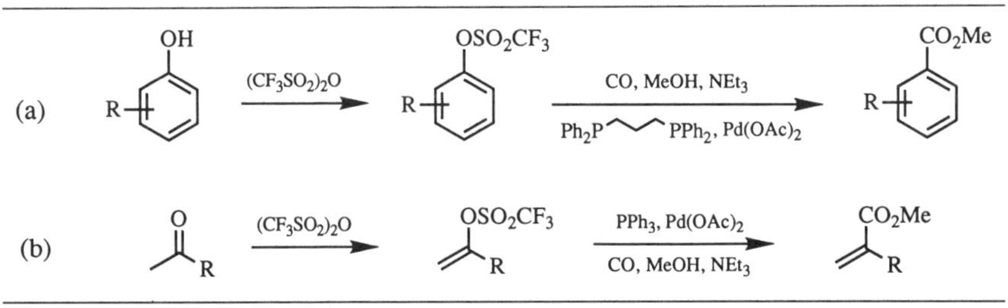

triflates react readily (25°C, 1 bar) using the conventional combination of triphenylphosphine and palladium acetate as catalyst (Scheme 7.34b), but this phosphine does not seem to be the optimum ligand for carbonylation of phenyl triflates. Instead, it appears that a chelating phosphine is necessary. The original report of this chemistry[317] specified the relatively exotic ligand 1,1'-bis(diphenylphosphino)ferrocene, except for one or two substrates such as 2-naphthyl triflate, where triphenylphosphine proved adequate. More recently however it has been shown that 1,3-bis(diphenyl

Table 7.3. Carbomethoxylation of Aryl Triflates[a]

Substrate	Product	Yield (%)
F, OSO$_2$CF$_3$, F	F, CO$_2$Me, F	93
OSO$_2$CF$_3$ (naphthalene)	CO$_2$Me (naphthalene)	96
OSO$_2$CF$_3$ (pyridine)	CO$_2$Me (pyridine)	72
CF$_3$SO$_2$O—biphenyl—OSO$_2$CF$_3$	MeO$_2$C—biphenyl—CO$_2$Me	85
CF$_3$SO$_2$O—phenyl—NHCO$_2$CH$_2$Ph, CO$_2$Me	MeO$_2$C—phenyl—NHCO$_2$CH$_2$Ph, CO$_2$Me	80
CF$_3$SO$_2$O—steroid—OSiMe$_2$Bu^t	MeO$_2$C—steroid—OSiMe$_2$Bu^t	91

[a] Reference 318.

phosphino)propane is also extremely effective (60–70°C, 1 bar, 1–2 h), and that moreover this ligand produces significant rate enhancements even for palladium-catalyzed carbonylations of aryl bromides and iodides.[318] Results for carbonylation of a range of aryl triflates, using the latter catalyst system, are given in Table 7.3.

7.7. Esters: From Ethers

A new approach to the naturally occurring 2-alkyl cyclopentenone skeleton depends on the cobalt-catalyzed carbonylation of an epoxide to a β-hydroxy ester (Scheme 7.35).[320] Mild conditions (40°C, 1 bar), high selectivity and applicability to a wide range of epoxides are features of the reaction, despite early reports relating to this type of chemistry which suggested that high temperatures (130°C) and very high pressures (240 bar) would be necessary.[321, 322]

Scheme 7.35

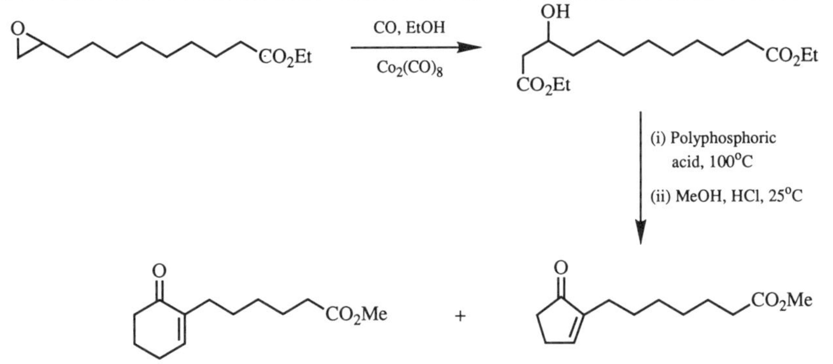

Direct carbonylation of an acyclic ether to an ester [Eq. (7.13)] is one of the simplest carbonylation reactions, and has been achieved under vigorous conditions (typically 200°C, 300 bar), using cobalt, nickel, or rhodium catalysts.[323] Cyclic ethers are rather more easily converted to lactones (see Chapter 9), but with both cyclic and acyclic substrates the product ester is readily carbonylated further to anhydride (see Chapter 8) so that a mixture of products generally results.

$$ROR + CO \longrightarrow RCOOR \qquad (7.13)$$

Allylic ethers have long been known to be convertible to β,γ-unsaturated esters by palladium-catalyzed carbonylation.[324] Recent work has shown that the long-chain allylic ethers resulting from telomerization of butadiene with methanol can be carbonylated to the linear esters.[325] A detailed study of the carbonylation of methoxyoctadienes to methyl nona-3,8-dienoate (130 bar, 100°C) indicates that the reaction is strongly promoted by hydrogen chloride, and that selectivity is improved by addition of ionic species such as tetrabutylammonium tetrafluoroborate.[326]

7.8. Esters: From Alcohols or Phenols

The high-pressure carbonylation of alcohols in the presence of oxygen and a palladium catalyst affords dialkyl esters of oxalic acid, although a dehydrating agent such as an orthoformate ester is necessary to remove the water formed in the reaction, and a copper cocatalyst is required to

promote reoxidation of Pd(0) to Pd(II).[327] The key intermediate appears
to be a bis(carboalkoxy)palladium(II) complex,[328] which releases oxalate
by reductive elimination as shown in Scheme 7.36. A closely related, but

Scheme 7.36

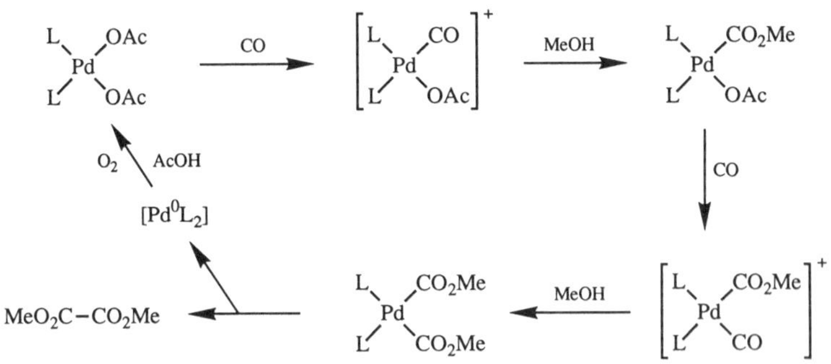

higher-yielding, two-stage process involves palladium-catalyzed carbonylation of alkyl nitrites to oxalate esters [Eq. (7.14)]. The nitric oxide
coproduct is used with oxygen in a separate step to convert further alcohol
to alkyl nitrite.[329] Carbonate esters are usually formed as significant by-

$$2RONO + 2CO \xrightarrow{\text{Pd(II)}} ROOCCOOR + 2NO \qquad (7.14)$$

products of oxidative alcohol carbonylation, and it has recently been
shown than carbonylation of *phenol* in the presence of oxygen, using a
palladium–manganese catalyst system, affords diphenyl carbonate (an
intermediate in the production of polycarbonate resins) with good selectivity.[330] Although yields and turnover numbers for palladium can be high
(up to 75% and 500, respectively), the reaction is very slow and requires
significant improvement in this respect if it is ever to be an industrially
viable process.

7.9. Esters: From Allyl or Propargyl Carbonates

Alkyl allyl carbonates have recently become established as valuable
starting materials for a number of palladium-catalyzed syntheses. Their
principal virtues are ease of oxidative addition to Pd(0) (forming a π-allyl

complex) and liberation of an alkyl carbonate anion which spontaneously decarboxylates to give the corresponding strongly basic and nucleophilic alkoxide ion (Scheme 7.37).[331] This alkoxide ion can be used *in situ* to

Scheme 7.37

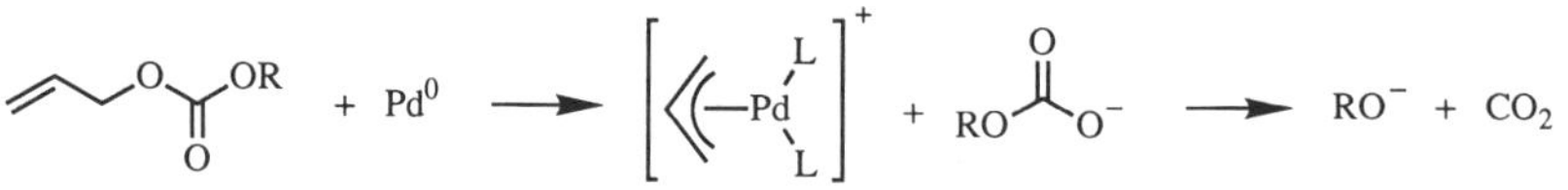

generate stabilized carbanions by deprotonation of suitable precursors, hence promoting catalytic allylic alkylation.[331] In the present context, however, the role of the alkoxide is as a nucleophile,[332] attacking a carbonylated π-allyl complex in a catalytic synthesis of β,γ-unsaturated esters [Eq. (7.15)]. Reactions are best carried out at 50°C under 5–10 bar of carbon monoxide, though acceptable yields can be obtained at atmospheric

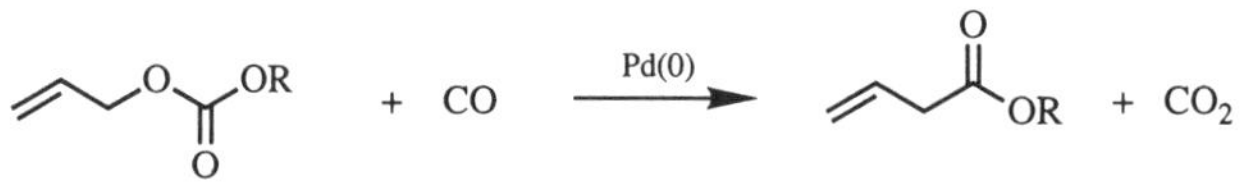

pressure. Whether ester formation occurs via carbon monoxide insertion into the allyl–metal bond followed by attack of alkoxide (Scheme 7.38a) or by alkoxide attack on a carbonyl ligand followed by reductive elimination (Scheme 7.38b) remains to be established.

Scheme 7.38

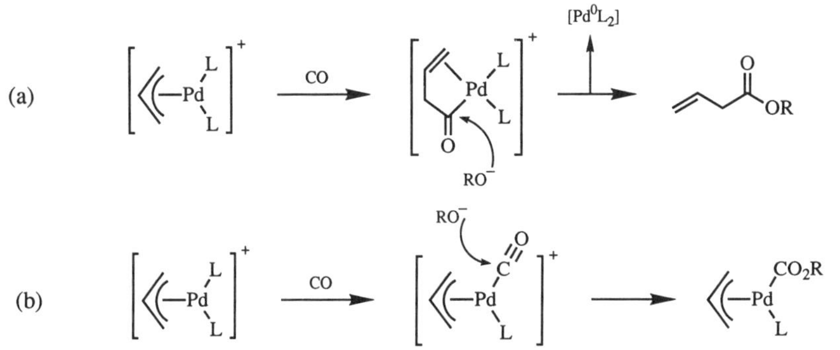

Very recently, alkyl *propargyl* carbonates have been shown to undergo a complex series of carbonylation reactions, yielding allenic or 2,4-dienoic esters depending on reaction conditions (Scheme 7.39).[333] The allenic ester is always the primary product, but when diethyl ether rather than

Scheme 7.39

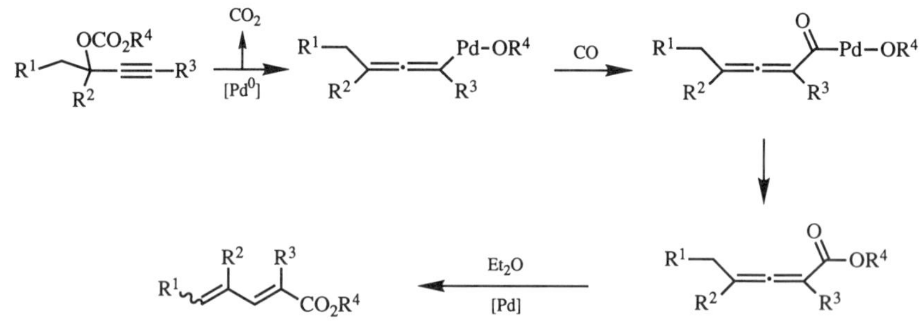

Table 7.4. Palladium-Catalyzed Carbomethoxylation of Propargylic Carbonates[a]

Substrate	Product	Yield (%)
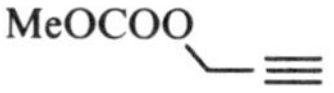	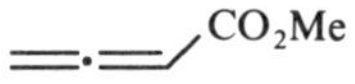	50
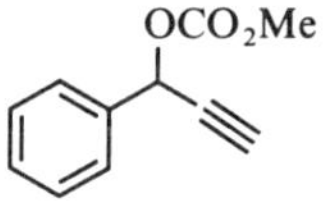	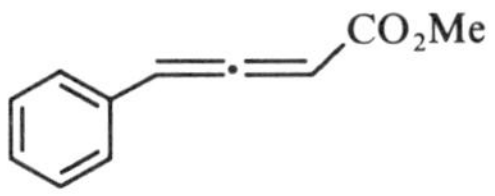	71
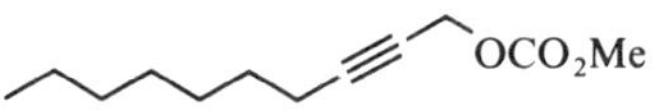	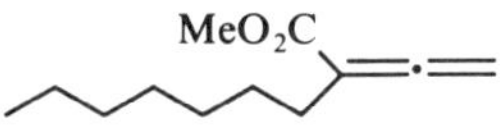	82
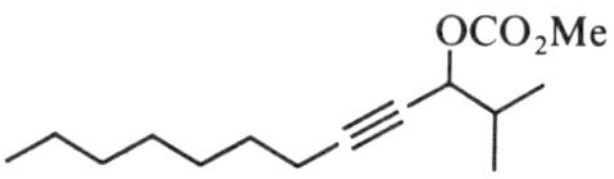	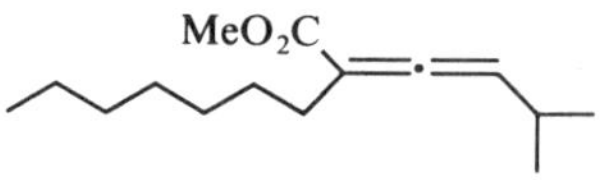	96
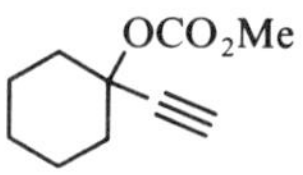	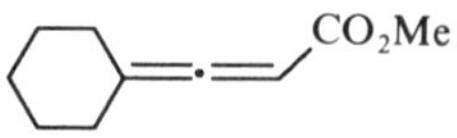	92
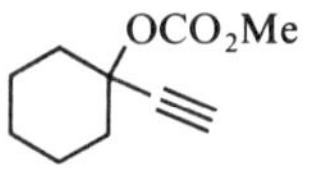 [b]	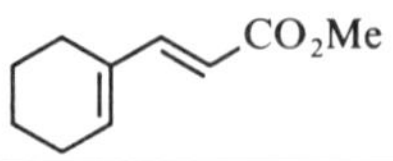	85

[a] Reference 333.
[b] Reaction carried out in diethyl ether.

methanol is used as solvent, isomerization to the corresponding 2,4-dienoate follows rapidly. There is an obvious relationship to the stoichiometric, nickel carbonyl-promoted synthesis of allenic acids from propargyl halides, described in the previous chapter (Sec. 6.2), but the present synthesis of allenic esters is far less hazardous and affords very much better yields. The results of a number of syntheses are summarized in Table 7.4.

Synthesis of Allenic Esters from Methyl Propargyl Carbonates[333]

In a 50-cm^3 stainless steel autoclave are placed the propargylic carbonate (3 mmol), Pd$_2$(dibenzylideneacetone)$_3$ (0.03 mmol), and a solution of PPh$_3$ (0.24 mmol) in methanol (6 cm^3). The autoclave is pressurized with carbon monoxide (15 bar), heated to 50°C, and the reaction mixture stirred for several hours until gas absorption ceases. The product is isolated by column chromatography on silica gel.

Chapter 8
Synthesis of Amides and Further Carboxylic Acid Derivatives

8.1. Introduction

Chapters 6 and 7 are devoted, respectively, to syntheses of carboxylic acids and esters, and the present chapter covers carbonylation-based routes to other carboxylic acid derivatives. Included here are amides, anhydrides, acyl halides, urethanes, ureas, and isocyanates. The last three types of compound can be viewed, respectively, as the products of esterification, amidation, and dehydration of carbamic acids ($RNHCOOH$), and carbonylative syntheses of such materials are currently of growing scientific and technological interest.

8.2. Amides

The synthesis of acyclic amides via carbonylation chemistry is not especially well developed, although, as described in Chapter 10, there are many catalytic reactions leading to *cyclic* amides (lactams) which have no counterpart in acid or ester synthesis. Early catalytic work on the conversion of alkenes to open-chain amides (hydroamidation, Scheme 8.1) depended on relatively unsophisticated catalyst systems such as unliganded transition metal salts [RuCl$_3$, Ni(CN)$_2$] or simple metal carbonyls [Fe(CO)$_5$, Co$_2$(CO)$_8$], and involved extremely high temperatures (up to 350°C) and pressures (up to 1000 bar).[334] Yields and selectivities were often good, but the extreme conditions required seem to have inhibited further development of this chemistry. In view of recent successes in achieving amide synthesis by *substitutive* carbonylation under very mild conditions, using halocarbons or triflate esters as starting materials with liganded palladium- or rhodium-based catalysts, reinvestigation of hydro-amidation reactions may well be timely.

Scheme 8.1

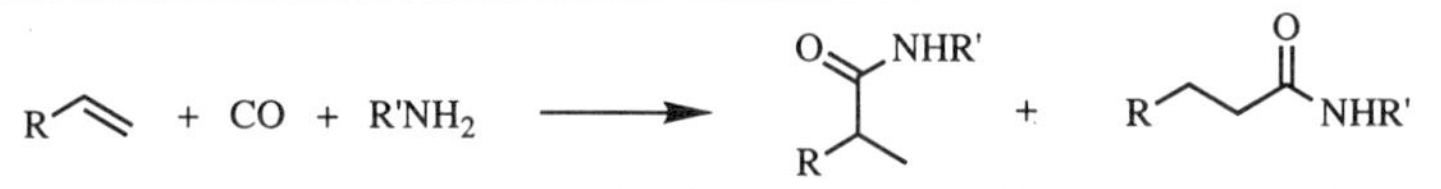

8.2.1. Amides: From Halocarbons

Several of the carbonylative routes from halocarbons to esters, described in Section 7.2, can instead give high yields of amide if the alcohol present in the reaction is replaced by an amine. Thus aryl, vinyl, and heteroaryl bromides or iodides react cleanly with carbon monoxide and primary or secondary amines in the presence of a dihalobis(triphenyl-phosphine)palladium(II) catalyst (1 bar, 60–100°C) to give secondary or tertiary amides.[335] Unless the amine used is only very weakly basic (e.g., a primary aromatic amine) addition of the tertiary amine required as an acid acceptor in the corresponding ester synthesis is unnecessary, since an excess of the reacting amine can normally be used as the base [Eq. (8.1)].

$$\text{ArX} + \text{CO} + 2\text{RNH}_2 \xrightarrow{\text{(PPh}_3\text{)}_2\text{PdCl}_2} \text{ArCONHR} + \text{RNH}_3\text{X} \qquad (8.1)$$

Amidation proceeds much more rapidly than esterification under similar conditions, so that, for example, the reaction of bromobenzene with carbon monoxide and benzylamine (giving N-benzyl benzamide) is some 17 times faster than the corresponding reaction with carbon monoxide and *n*-butanol, which yields *n*-butyl benzoate. Stereospecificity in carbonylation of *cis* or *trans* vinylic halides is also significantly greater for the amidation reaction, where essentially 100% retention of geometry is obtained compared with only 90%–95% for comparable esterifications. Both observations suggest that the rate-determining step in this type of reaction is nucleophilic attack on an acylpalladium(II) intermediate, since (a) amines are generally much more nucleophilic than alcohols, and (b) the consequent rapid removal of acyl complex limits the equilibrium concentration of its precursor *alkenyl* complex, which is believed[335] to be the isomerizable intermediate in the catalytic cycle (Scheme 8.2).

Scheme 8.2

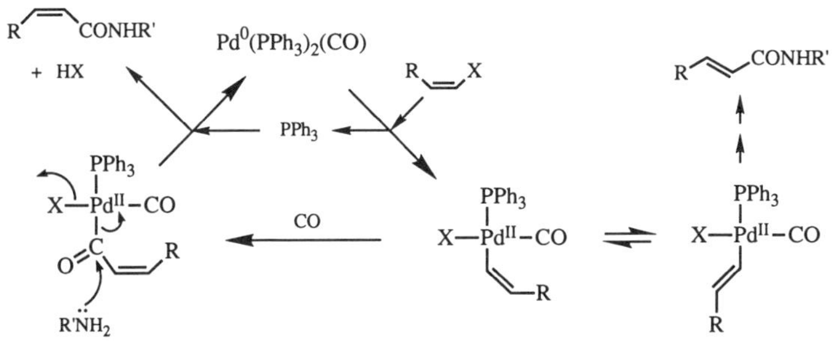

Preparation of trans-N-Cinnamoyl Pyrrolidine[335]

Pyrrolidine (50 mmol), *trans-β*-bromostyrene (17.2 mmol), and $Pd(PPh_3)_2Br_2$ (0.23 mmol, 1.5 mol% on halide) are placed under an atmosphere of carbon monoxide in a flask fitted with a magnetic stirrer, thermometer, and condenser connected to a carbon monoxide gas burette. The mixture is stirred at 60°C under a constant 1 bar pressure until a total of 423 cm^3 of carbon monoxide have been absorbed (2-3 h). The reaction mixture is extracted with ether, and the extracts are washed several times with water. The organic layer is evaporated to give a solid which on recrystallization from heptane (with charcoal treatment and hot filtration) affords 3.15 g (91% yield) of *trans*-N-cinnamoyl pyrrolidine, mp 100.5°C.

As in the corresponding acid and ester syntheses (Secs. 6.2 and 7.2), it is sometimes possible to achieve high yields of α-keto amides from aryl halides [Eq. (8.2)] by working at higher pressures (10-80 bar) with a slightly modified palladium-based catalyst [PPh_3 is replaced by PPh_2Me or $Ph_2P(CH_2)_4PPh_2$].[336, 337] Strongly nucleophilic amines are required for successful α-keto amide synthesis (aniline, for example, gives only monocarbonylation), and a comprehensive study of this reaction has shown

$$ArX + 2CO + 2R_2NH \xrightarrow{(PPh_2Me)_2PdCl_2} ArCOCONR_2 + R_2NH_2X \qquad (8.2)$$

that selectivity for keto amide over simple amide formation decreases in the order $Pr_2NH > Et_2NH >$ piperidine $> Me_2NH >$ pyrrolidine, apparently reflecting an order of decreasing steric bulk.[338] Primary amines *can* be used, but they tend to condense with the keto amide to give the corresponding imine as the final product (Scheme 8.3). An exception is *t*-butylamine, which (again presumably for steric reasons) does not form the imine, but gives keto *t*-butylamides in excellent yield. Electron-donating substituents (Me, OR, NR_2) in the aryl halide do not significantly reduce selectivity for the keto amide, though reactivity is somewhat diminished. Electron-withdrawing groups (Cl, CF_3, CN), on the other hand, enhance the overall carbonylation rate at the expense of a substantial drop in selectivity.

Scheme 8.3

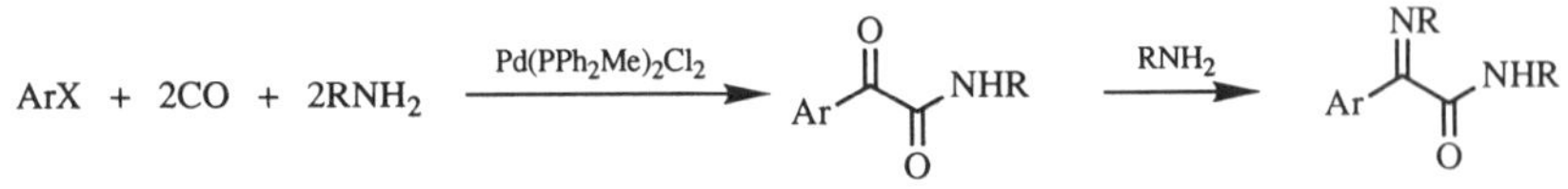

Careful mechanistic investigations[338, 339] indicate that keto amide formation proceeds as shown in Scheme 8.4, where key steps are nucleophilic addition of amine to the CO ligand of a cationic aroyl carbonyl complex, followed by reductive elimination of the product. The requirement for high

Scheme 8.4

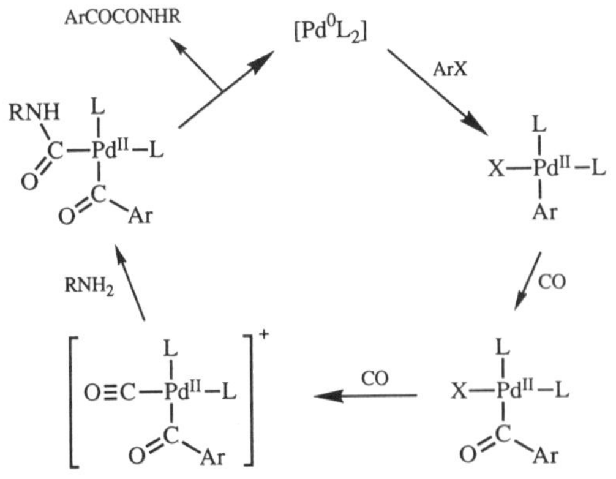

carbon monoxide pressure presumably indicates that the equilibrium concentration of cationic aroyl carbonyl complex is strongly pressure dependent, and the need for the amine to be a reasonably strong nucleophile stands out as a consequence of its required ability to add to coordinated carbon monoxide.

Preparation of N,N-Diethyl-α-(4-methoxyphenyl)glyoxylamide [Eq. (8.2); Ar = 4-MeOC$_6$H$_4$, R = Et].[338]
     4-Bromoanisole (9.6 mmol), diethylamine (29 mmol), and [Pd(PMePh$_2$)$_2$Cl$_2$] (0.096 mmol), are placed under nitrogen in a 100 cm^3 magnetically stirred stainless steel autoclave. After evacuating the system, 10 bar of carbon monoxide is introduced at room temperature, and the temperature is raised to 100°C. After stirring at this temperature for 50 h the system is cooled, purged with nitrogen, and the reaction mixture extracted with ether. The products are separated by column chromatography on silica-gel (hexane-ether as eluent), affording the keto amide (ν[CO] 1680, 1640 cm^{-1}) in 81% yield.

Double carbonylation of 2-bromoacetanilides has recently been used to generate intermediates for the synthesis of a series of substituted quinolines via the Pfitzinger condensation,[340] as shown in Scheme 8.5.

Scheme 8.5

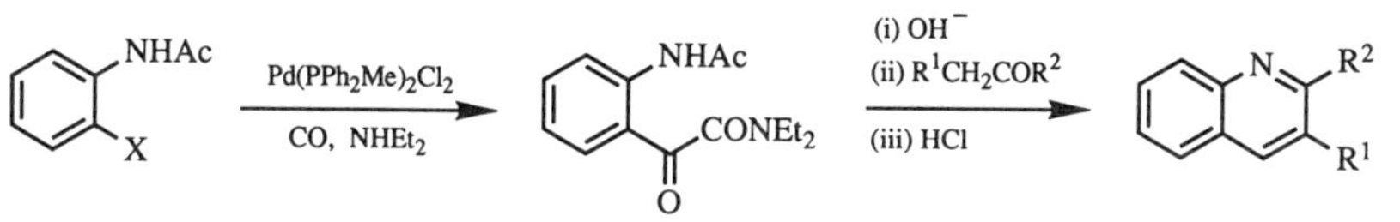

Carbonylations of aryl halides to amides have until recently been restricted to bromo- and iodo-arene substrates, but chlorobenzene has been shown to undergo this type of reaction in the presence of Pd(dippp)$_2$ (dippp = 1,3-bis(di-isopropyl)propane; see also Sec. 7.2).[341] An alternative approach to carbonylation of chlorobenzene involves conversion to its more reactive tricarbonyl chromium complex.[342] With this complex as substrate, use of the double-carbonylation catalyst Pd(PPh$_2$Me)$_2$Cl$_2$ leads to formation of both amide and keto amide. Decomplexation of the Cr(CO)$_3$ fragment occurs spontaneously so that metal-free products are isolated (Scheme 8.6). Selectivity for keto amide increases with pressure, reaching over 50% at 30 bar.

Scheme 8.6

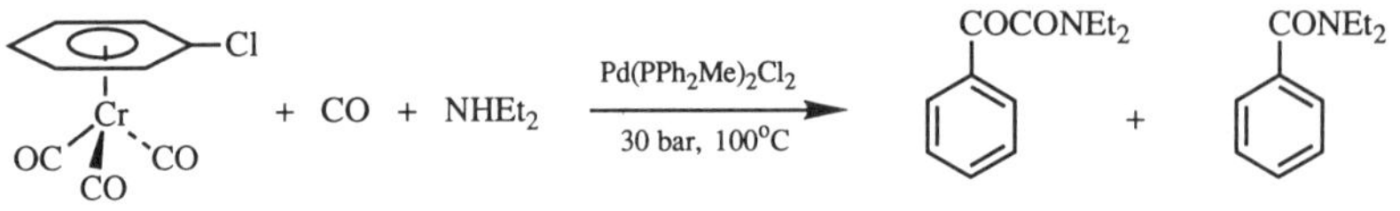

Conversions of alkyl or benzyl halides to amides are possible using NaCo(CO)$_4$ or Na$_2$Fe(CO)$_4$ as stoichiometric reagents,[343, 344] but there is as yet no general methodology for catalytic amidation of these substrates. This may be related to the relatively high nucleophilicity of amines, which could lead to direct alkylation, rather than carbonylative acylation, at nitrogen. However, *tertiary benzylic* amines can themselves undergo direct carbonylation,[345] as shown in Eq. (8.3), so that alkylation need not always preclude amide formation.

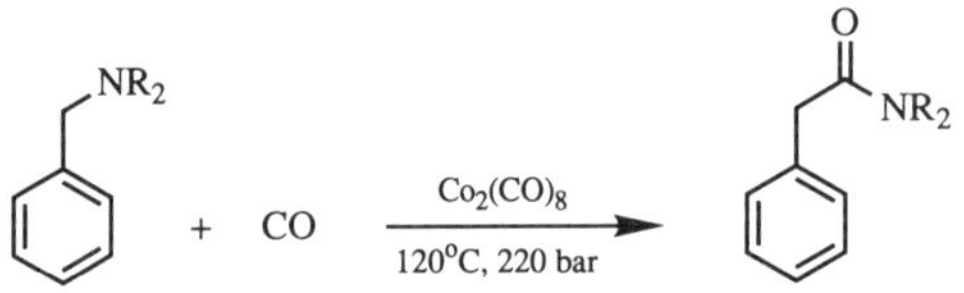

$$(8.3)$$

Nickel tetracarbonyl reacts stoichiometrically with 1,1-dibromo-cyclopropanes in the presence of amines to give cyclopropanecarboxamides in moderate yield (Scheme 8.7). The mechanism of this reductive carbonylation reaction is not yet well understood, but a nickel–enolate complex has been postulated as a key intermediate.[346] Related reactions, using 1,1-

Scheme 8.7

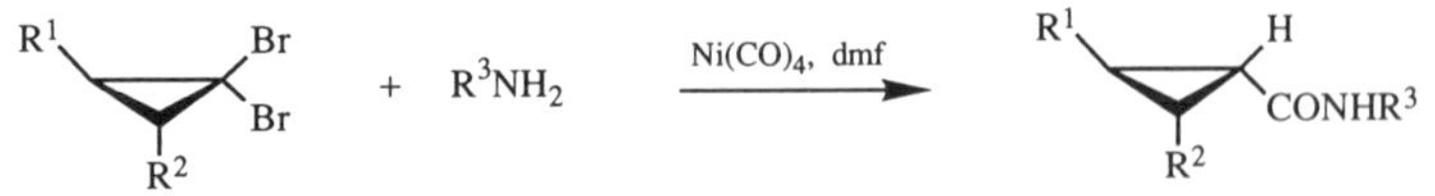

dibromo-2-chloromethyl-cyclopropanes as substrates, afford ring-opened amides (Scheme 8.8) with reasonable yield and selectivity.[347]

Scheme 8.8

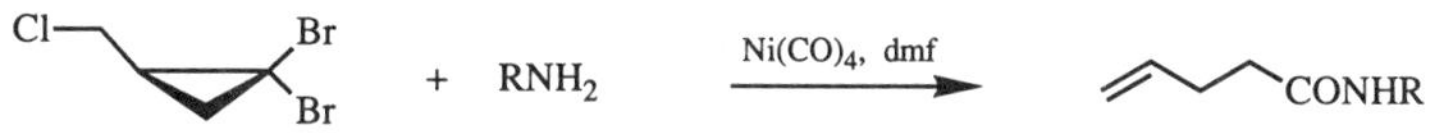

8.2.2. Amides: From Triflate Esters

Vinyl and aryl esters of trifluoromethanesulfonic acid are readily carbonylated to amides using the same catalyst systems (palladium acetate and either triphenylphosphine or 1,1'-bis(diphenylphosphino)-ferrocene) as in the corresponding ester syntheses (Sec. 7.6). The starting materials are accessible in a single step from phenols and ketones, respectively, and amides are formed in excellent yield (Scheme 8.9) at 60°C and 1 bar of carbon monoxide.[348, 349] The chelating phosphine 1,4-

Scheme 8.9

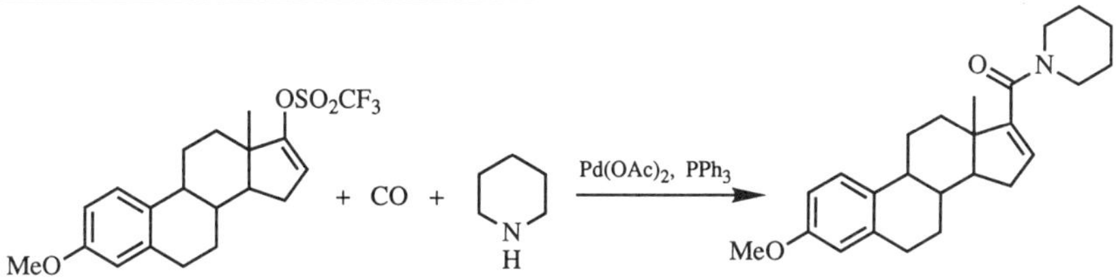

bis(diphenylphosphino)butane, which was so successfully used in the synthesis of esters from aryl triflates (Sec. 7.6),[350] has apparently not yet been explored as a ligand for the corresponding preparation of amides.

8.2.3. Amides: From Allylic Derivatives

During the course of work aimed at synthesizing esters from allylic carbonates via palladium-catalyzed carbonylation (Sec. 7.9),[351] it was found that high yields of a β,γ-unsaturated amine were obtained under relatively mild conditions (50°C, 8 bar) if a secondary amide (diethyl-

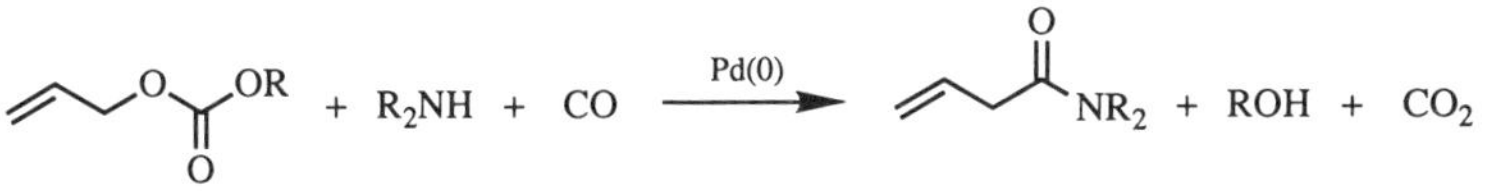

$$\text{allyl-O-CO-OR} + R_2NH + CO \xrightarrow{Pd(0)} \text{allyl-CO-NR}_2 + ROH + CO_2 \qquad (8.4)$$

Table 8.1.   Rhodium-Catalyzed Amidocarbonylation of Allylic Phosphates[a]

Reactants	Product	Yield (%)
~~OPO(OEt)$_2$ + NHEt$_2$	~~~C(O)NEt$_2$	81
~~OPO(OEt)$_2$ + morpholine NH	~~C(O)–N(morpholine)	74
~~OPO(OEt)$_2$ + HN–tetrahydroisoquinoline	~~~C(O)–N(tetrahydroisoquinoline)	84
Ph~~~OPO(OEt)$_2$ + NHEt$_2$	Ph~~~C(O)NEt$_2$	77
~~OPO(OEt)$_2$ + H$_2$N~~	~~C(O)N(H)~~	62

[a] Reference 352.

amine) was included in the system [Eq. (8.4)]. However, only a single example of amide formation was reported.

Allylic phosphates are cleanly carbonylated to secondary or tertiary amides in the presence of Rh$_6$(CO)$_{16}$. Conditions are somewhat more forcing (50°C, 20 bar) than in the palladium-catalyzed conversion of allylic carbonate discussed above, but the rhodium-catalyzed reaction appears to be quite general and gives good yields (75%–85%) with high selectivity for β,γ-unsaturated amide (Table 8.1).[352]

At still higher temperatures and pressures (110–150°C, 50 bar) tertiary allylamines may be carbonylated *directly* to β,γ-unsaturated amides,[353] as shown in Eq. (8.5). The preferred catalyst is a combination of palladium acetate and 1,3-bis(diphenylphosphino)propane (dppp) in a 1:2 mole ratio, and yields of amide are typically 80%–90%.

$$R^3\!\!\diagup\!\!\diagdown\!\!\underset{R^4\ R^5}{\diagdown}\!\!N\!\!\underset{R^2}{\overset{R^1}{\diagdown}} \;+\; CO \;\xrightarrow{Pd(OAc)_2 \text{ - dppp}}\; R^3\!\!\diagup\!\!\diagdown\!\!\underset{R^4\ R^5}{\diagdown}\!\!\overset{O}{\overset{\|}{C}}\!\!N\!\!\underset{R^2}{\overset{R^1}{\diagdown}} \qquad (8.5)$$

8.2.4. Amides: From Alkenes and Alkynes

As previously noted, early work on alkene hydroamidation gave reasonably promising yields and selectivities,[334] but the extremes of temperature and pressure involved seem to have discouraged further investigation. One of the few systematic early studies[354] showed that hydroamidation of propene in the presence of aniline, with dicobalt octacarbonyl as catalyst, gave both *n*- and *iso*-butyranilide in a ratio of about 5:1 (Scheme 8.10). At 180°C and 336 bar, a combined yield of 90% was achieved, and results were reported over the temperature range 150–290°C, at alkene/amine ratios of 0.1 to 9.1, and in various solvents including methanol, dioxan, and benzene.

Scheme 8.10

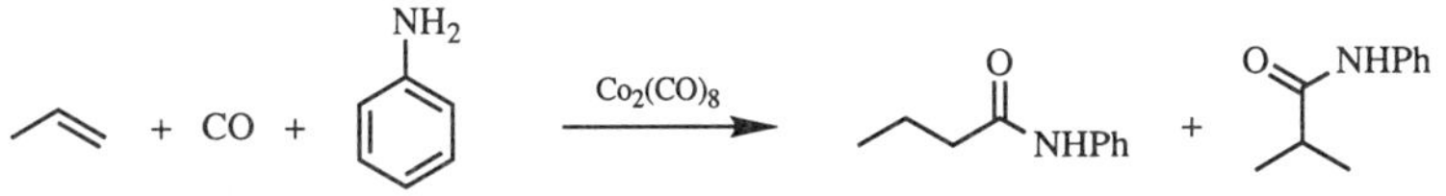

Under similar conditions (210°C, 400 bar), butadiene gave a mixture of adipic and ethylsuccinic dianilides,[355] but a more recently discovered, stoichiometric reaction of 1,3-dienes with secondary amines and nickel(II) salts proceeds under very mild conditions (less than 1 bar partial pressure of carbon monoxide at 70°C) and affords mono-carbonylated products in reasonable yield (Scheme 8.11).[356]

Scheme 8.11

A related nickel-promoted, oxidative carbonylation of 1,2-dienes (allenes) gives unsaturated amino-amides (Scheme 8.12) in around 50% yield.[357] Very mild conditions (20°C, 1 bar) are again sufficient, but extreme caution is required during workup since stoichiometric quantities of nickel tetracarbonyl are produced in such reactions (see also Chapter 3).

Scheme 8.12

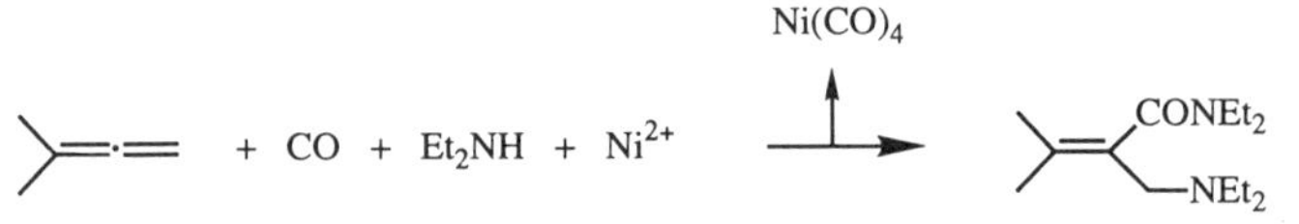

Aminopalladation of 1-alkenes (Scheme 8.13) affords two isomeric β-aminoacyl complexes which undergo further carbonylation on reaction with piperidine and carbon monoxide (50 bar at 25°C). Keto amides are formed in good yield, though the reaction is stoichiometric in palladium. Small quantities of the corresponding oxamide are sometimes produced, but simple amide formation is not observed.[358]

Scheme 8.13

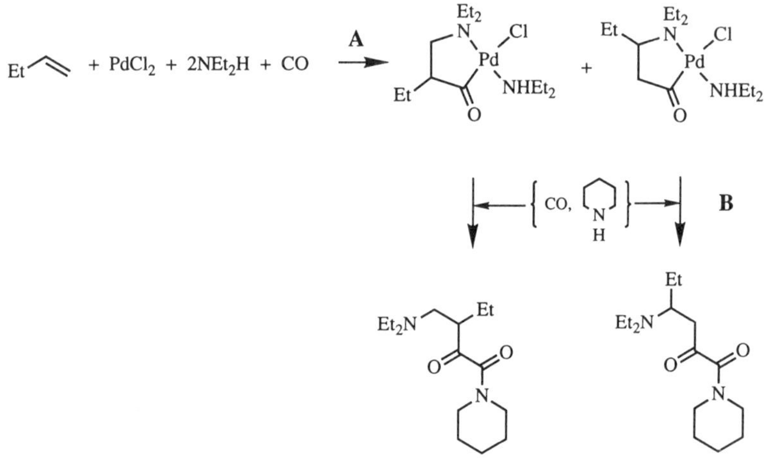

Although hydroamidation of acetylene to N-alkyl acrylamides and succinamides was first observed many years ago by Reppe and co-workers,[359] this type of reaction has not yet achieved much synthetic importance. There have, however, been one or two more recent investigations that seem to offer potentially useful transformations. Carbonylation of 2-butyne in the presence of a secondary amine and a nickel(II) salt can be either stoichiometric or catalytic in transition metal (Scheme 8.14). At

70°C and 15 bar of carbon monoxide, stoichiometric double carbonylation occurs to give an unsaturated diamide in high yield. At 100°C and atmospheric pressure, however, a catalytic reaction produces an aminobutenolide in about 30% yield, representing a turnover number for nickel of around 15.[360] Since both reactions involve double carbonylation of the alkyne, it seems likely that the temperature difference is the more important factor in achieving a catalytic reaction.

Scheme 8.14

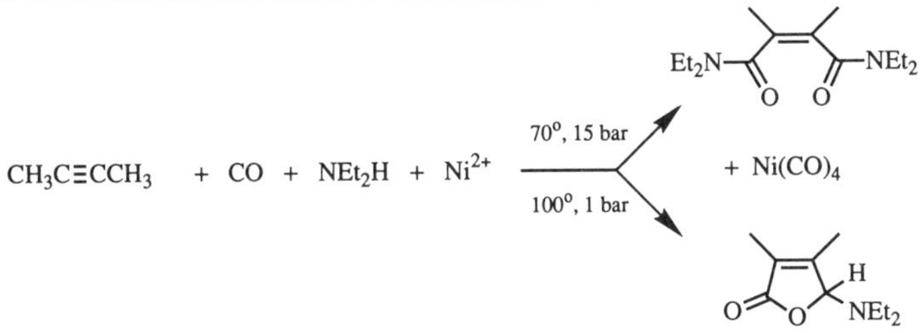

8.2.5. *Amides: From Aminolithium Reagents*

Lithium dialkylamides absorb carbon monoxide at low temperatures (0 to −75°C) and pressures (less than 1 bar) to give thermally unstable carbamoyl–lithium derivatives.[361] These compounds can be trapped in reactions with electrophiles to give amides and keto-amides in reasonable yields,[362] as shown in Scheme 8.15. Use of carbon monoxide labeled with

Scheme 8.15

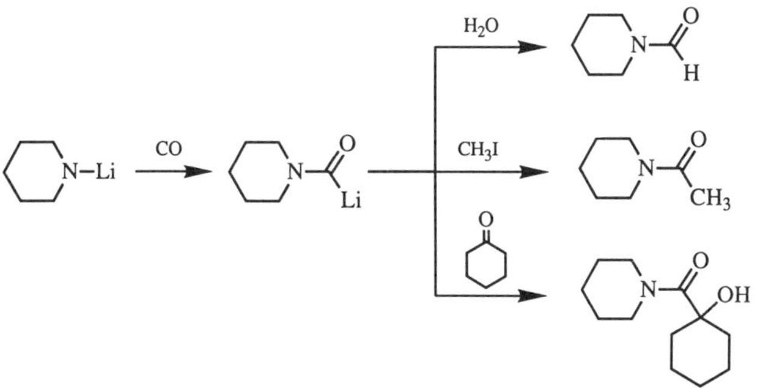

the short-lived isotope ^{11}C has allowed rapid synthesis of radioactive amides for medical applications such as positron emission tomography.[363]

A more stable aminocarbonylation reagent is obtained by treatment of the amino–lithium compound with copper(I) chloride (0.5 equivalents), followed by carbon monoxide at room temperature and atmospheric pressure. The resulting bis(carbamoyl) cuprate complex reacts under carbon monoxide (50 bar) with alkyl, allyl, aryl, and acyl halides, at temperatures up to 80°C, to give good yields of amide (Scheme 8.16). Even with acyl halides as electrophiles, amides rather than keto-amides are formed, probably via decarbonylation of an intermediate acyl–copper complex.[364]

Scheme 8.16

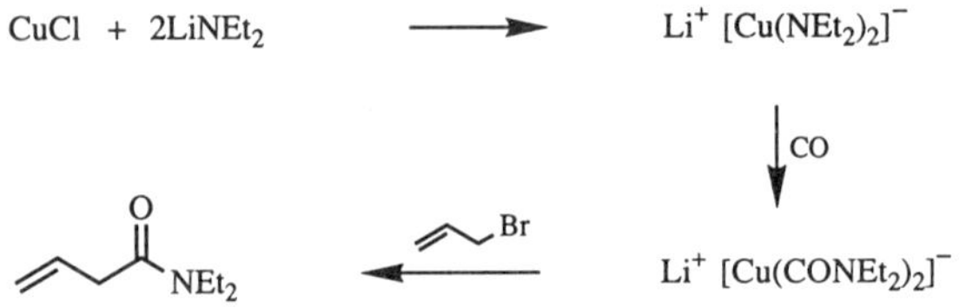

$$CuCl \; + \; 2LiNEt_2 \longrightarrow Li^+ \; [Cu(NEt_2)_2]^-$$

An alternative approach to obtaining stable aminocarbonylation reagents involves carbonylation of a lithium amide followed by reaction of the resulting carbamoyl–lithium with trimethyltin chloride. The resulting carbamoylstannane undergoes palladium-catalyzed cross-coupling with alkenyl-, aryl-, and heteroaryl halides, as shown in Scheme 8.17, to give amides in fair to excellent yield.[365]

Scheme 8.17

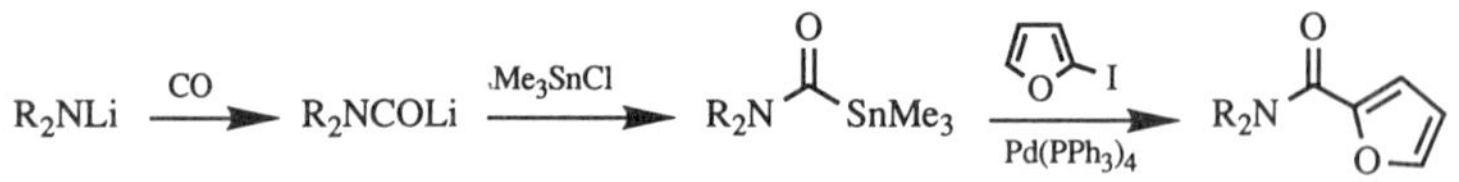

8.2.6. Amides: From Aldehydes

Aromatic aldehydes containing an appropriately sited N-donor group undergo metallation on reaction with salts of palladium(II), to give chelated aroylpalladium complexes as shown in Scheme 8.18. Such complexes have recently been shown to react with carbon monoxide (80 bar) and secondary amines to give high yields of keto amides.[366] Amide formation is negligible, but traces of oxygen in the system can lead to production of significant levels of ditertiary oxamides.

Scheme 8.18

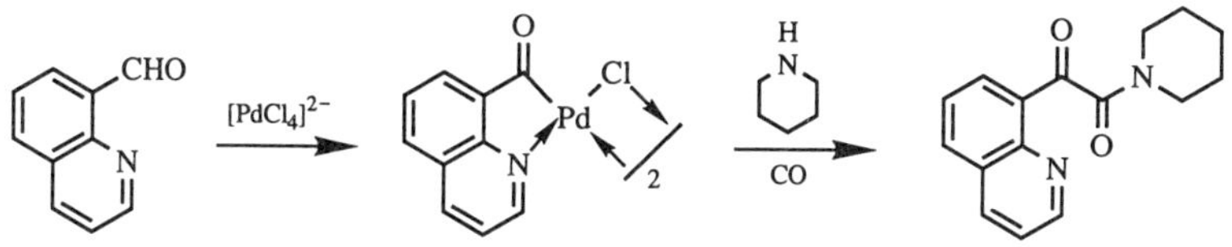

8.2.7. Amides: From Cyclic Ethers

Cyclic ethers including epoxides, oxetane, and tetrahydrofuran, react with carbon monoxide and N-(trimethylsilyl)amines in the presence of $Co_2(CO)_8$ as catalyst, to form β-, γ-, or δ-siloxyamides, respectively, in good yields (Scheme 8.19).[367] The mechanism of this reaction has not yet been investigated, although the presence of the N-silyl group is known to be essential for carbonylation to occur.

Scheme 8.19

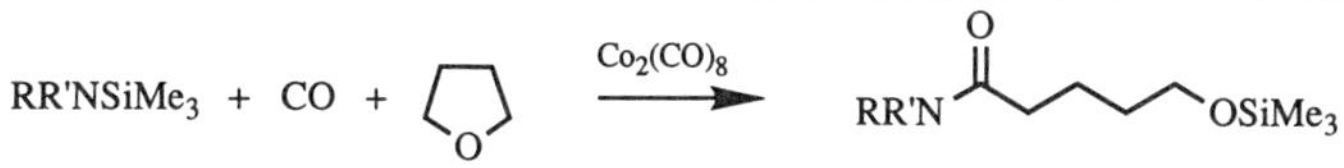

8.3. Anhydrides

8.3.1. Anhydrides: From Esters

Carbonylation of esters to anhydrides was mentioned earlier (Sec. 7.7) only as a side-reaction in the preparation of esters from ethers, but the production of *acetic* anhydride from methyl acetate by this type of reaction is now in fact a major commercial process.[368] A plant based on coal-gasification was commisioned by Tennessee Eastman in 1983 to convert some 300,000 tons of coal per annum to approximately the same quantity of acetic anhydride for cellulose acetate production. The overall process is outlined in Scheme 8.20.

Scheme 8.20

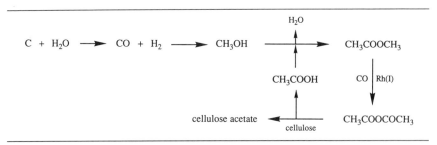

The homogeneous carbonylation catalyst is almost certainly an iodide-promoted rhodium complex, as in the synthesis of acetic acid from methanol (see Sec. 6.4). The reaction is strongly promoted by triphenylphosphine, and although chromium hexacarbonyl has also been claimed to act as a catalyst promoter,[369] careful investigations have shown that this compound has no effect on the reaction rate, but may act as a reducing agent to shorten the induction period if rhodium(III) chloride is used as the catalyst precursor.[370] The reaction mechanism is probably very similar to that for the carbonylation of methanol, so that initial ester cleavage by HI to give acetic acid and iodomethane is followed by catalytic carbonylation of the latter to acetyl iodide, which in turn reacts with acetic acid to form the product and re-form hydrogen iodide (Scheme 8.21). Reported conditions for this type of reaction are generally in the range

Scheme 8.21

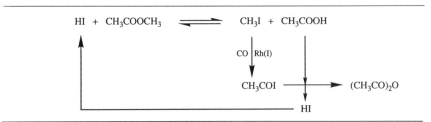

150–200°C and 20–80 bar of carbon monoxide,[369] but kinetic studies[370] indicate that the reaction rate is in fact independent of carbon monoxide pressure above about 15 bar.

Early work on carbonylative anhydride syntheses concentrated on the carbonyls of cobalt and nickel as catalysts, and although requiring more forcing conditions than the rhodium-based system, a number of interesting conversions were discovered. Strained cyclic esters such as β-propiol-actone[371] or γ-butyrolactone[372] were, for example, carbonylated at 150°C and several hundred bar of carbon monoxide to give the corresponding cyclic anhydrides in moderate yield.

Esters of allyl alcohol can be converted to 3-butenoic anhydrides in 60% yield by carbonylation at 100°C and 100 bar, in the presence of palladium chloride as catalyst [Eq. (8.6)].[373]

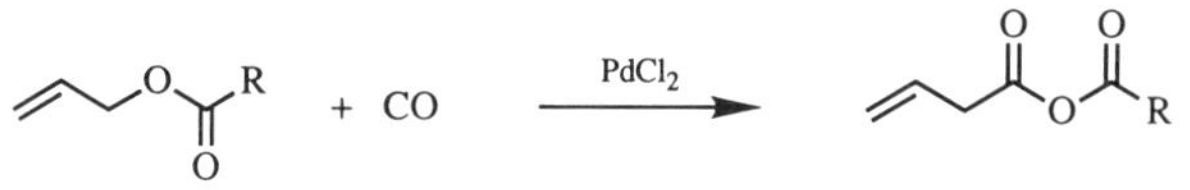

$$\text{CH}_2=\text{CHCH}_2-O-\underset{O}{\overset{}{\text{C}}}-R \; + \; CO \; \xrightarrow{\text{PdCl}_2} \; \text{(product)} \tag{8.6}$$

8.3.2. Anhydrides: From Arenediazonium Ions and Aryl Halides

The palladium-catalyzed carbonylation of arenediazonium salts was originally reported as a route to aromatic carboxylic acids[374] (see Sec. 6.7), but the authors suggested that the primary products were mixed benzoic–acetic anhydrides which underwent hydrolysis on workup (Scheme 8.22). This has since been confirmed[375] and the reaction extended to give a wide range of mixed aromatic–aliphatic anhydrides in good yields. Such compounds are often thermally unstable to disproportionation, and it is only the high reactivity of the diazonium ion, permitting carbonylation to occur rapidly (20 min) under mild conditions (25°C, 9 bar), that allows isolation of a pure product. Iodoarenes have recently been shown to undergo a similar carbonylation reaction, though because of the higher temperatures required (50–90°C) disproportionation of mixed anhydride products *does* occur in this system.[376]

Scheme 8.22

$$ArN_2^+ \; + \; CO \; + \; CH_3CO_2^- \; \xrightarrow{\text{Pd(OAc)}_2} \; ArCOOCOCH_3 \; \xrightarrow{\text{H}_2\text{O}} \; ARCOOH \; + \; CH_3COOH$$

8.3.3. Anhydrides: From Alkynes

Substituted propargylic alcohols, though not the parent compound, undergo palladium-catalyzed carbonylation under vigorous conditions to give modest yields of alkylidenesuccinic anhydrides. As shown in Scheme 8.23, the difunctional 2,5-dimethyl-3-hexyne-2,5-diol, for example, gives diisopropylidenesuccinic anhydride in 49% yield.[377] Since carbonylation

Scheme 8.23

of propargylic compounds under milder conditions is known to give derivatives of allenic acids (Sec. 7.8),[378, 379] these may well be intermediates in the present anhydride synthesis (Scheme 8.24).

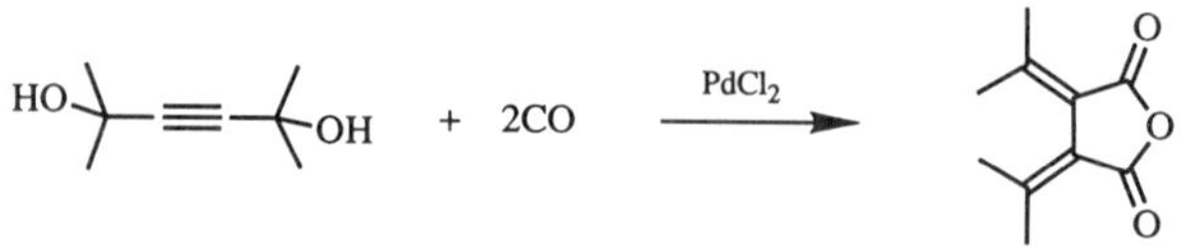

Scheme 8.24

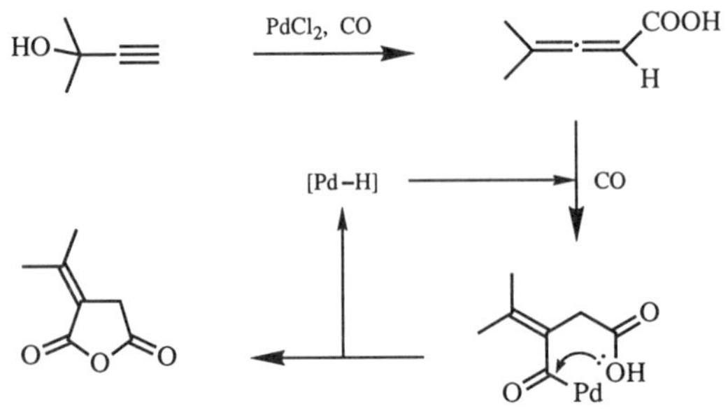

8.4. Acyl Halides

The literature provides relatively few reports of carbonylation reactions leading to acyl halide formation, perhaps because such products react readily with any nucleophile present in the system. Nevertheless, there are indications that this could be a fruitful area for research since, when nucleophilic species are absent, acyl halides have sometimes been isolated in good yield.

8.4.1. Acyl Halides: From Halocarbons

An excellent example of a carbonylation reaction that, in the absence of nucleophiles, yields acyl halide is the rhodium-catalyzed conversion of iodomethane to acetyl iodide.[380] This is a crucial step in the catalytic cycle of the "Monsanto" process for acetic acid production from methanol (Sec. 6.4), where the acyl halide is normally hydrolyzed *in situ* as it is formed.

Allylic chlorides are catalytically converted to 3-alkenoyl chlorides in high yield by reaction with carbon monoxide under vigorous conditions (110°C, 500 bar) in the presence of palladium(II) chloride. The reaction almost certainly proceeds via a π-, rather than a σ-allyl palladium complex since, as shown in Scheme 8.25, both 1-chloro-2-butene and 3-chloro-1-butene (which would give the same π-complex but different σ-complexes) afford the same carbonylation product.[381] A closely related stoichiometric

Scheme 8.25

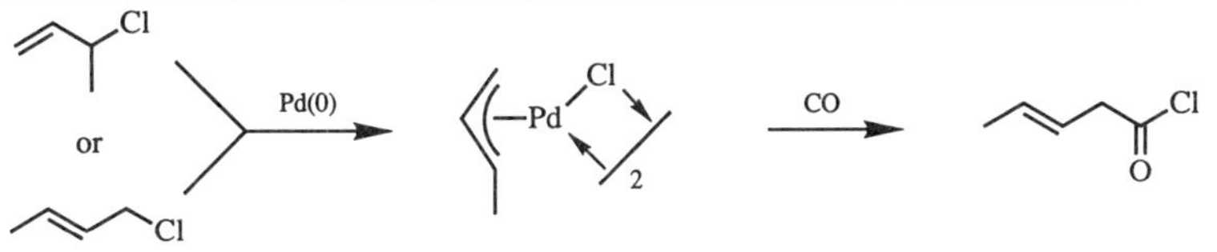

reaction (Scheme 8.26) converts π-allyl nickel halides to 3-alkenoyl halides.[382] The very mild conditions under which this reaction proceeds (0°C, 1 bar of carbon monoxide) suggest that the palladium-catalyzed synthesis (Scheme 8.25) may not in fact require quite such extreme conditions as were reported.

Scheme 8.26

The palladium-catalyzed carbonylation of aryl halides to esters (Sec. 7.3) and amides (Sec. 8.1) has recently been extended to the synthesis of aroyl fluorides [Eq. (8.7)]. An essentially quantitative conversion of iodobenzene to benzoyl fluoride is obtained at 80°C and 1 bar pressure of carbon monoxide, using caesium fluoride as the source of F$^-$ and propionitrile as solvent. Although triphenylphosphine complexes of rhodium, platinum, and cobalt are also active in this reaction, the palladium-based catalyst Pd(PPh$_3$)$_2$Cl$_2$ is superior in terms both of activity and selectivity.[383]

$$\text{ArI} + \text{CO} + \text{CsF} \xrightarrow{\text{Pd(PPh}_3)_2\text{Cl}_2} \text{ArCOF} + \text{CsI} \qquad (8.7)$$

Arenesulfonyl halides have also been converted to aroyl fluorides by palladium-catalyzed carbonylation [Eq. (8.8)].[384, 385] The reaction is, however, complicated by the requirement that the liberated sulfur dioxide must be removed by continuous oxidation and acidification to FSO$_3$H.

$$\text{ArSO}_2\text{Cl} + \text{CO} + \text{F}^- \xrightarrow{\text{PdCl}_2} \text{ArCOF} + \text{Cl}^- + \text{SO}_2 \qquad (8.8)$$

A remarkably useful technique for isotopic labeling of both aromatic and aliphatic acyl chlorides depends on the rhodium-catalyzed exchange, under mild conditions (100°C, 1 bar), of labeled carbon monoxide (^{13}CO or ^{14}CO) for the acyl carbonyl group [Eq. (8.9)].[386] The reaction is

$$\text{RCOX} + \overset{*}{\text{C}}\text{O} \xrightarrow{\text{RhCl(CO)(PPh}_3)_2} \text{R}\overset{*}{\text{C}}\text{OX} + \text{CO} \qquad (8.9)$$

$$(\text{*} = 13 \text{ or } 14)$$

closely related to the decarbonylation of acyl halides described in Chapter 11. The labeling process involves, as outlined in Scheme 8.27, oxidative addition of the acyl halide, migration of the hydrocarbon group to the metal, exchange of the resulting CO ligand with free, labeled carbon monoxide, migration of the organic ligand to carbonyl, and finally reductive elimination of the labeled acyl halide. Overall, decarbonylation of the substrate is not observed, and the reaction appears general for aroyl halides and both saturated and unsaturated aliphatic acyl halides.

8.4.2. Acyl Halides: From Alkenes and Alkynes

Alkenes react with carbon monoxide and carbon tetrachloride, in the presence of binuclear metal carbonyl catalysts such as [(C$_5$H$_5$)Fe(CO)$_2$]$_2$,

Scheme 8.27

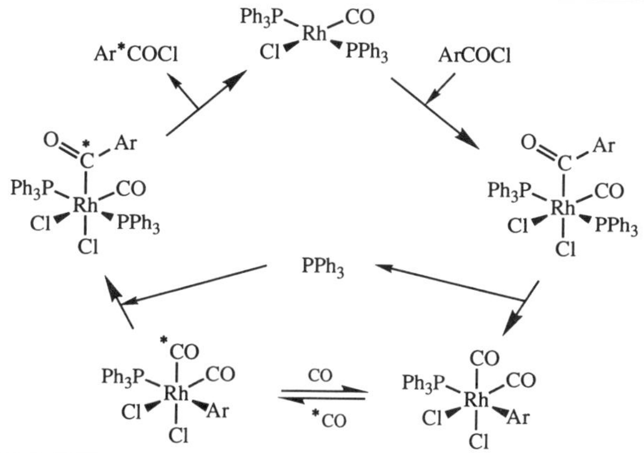

to give 2-alkyl-4,4,4-trichlorobutanoyl chlorides in fair yield [Eq. (8.10)].[387] Simple addition of carbon tetrachloride is also observed, espe-

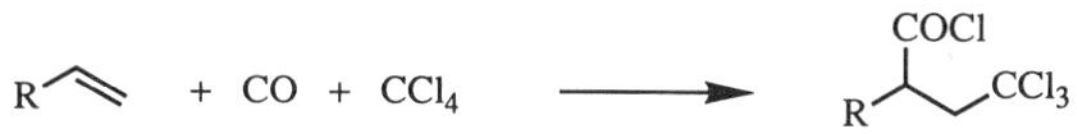

$$R \diagdown \diagdown \quad + \quad CO \quad + \quad CCl_4 \quad \longrightarrow \quad R \diagdown \diagup \diagdown CCl_3 \quad \overset{COCl}{} \tag{8.10}$$

cially for the higher alkenes, and to achieve selectivity for acyl chloride the reaction must be run at very high pressures of carbon monoxide (150–200 bar). By analogy with the known metal-mediated radical-chain mechanism for addition of CCl_4 to alkenes, the carbonylation reaction is thought to proceed as shown in Scheme 8.28.

Scheme 8.28

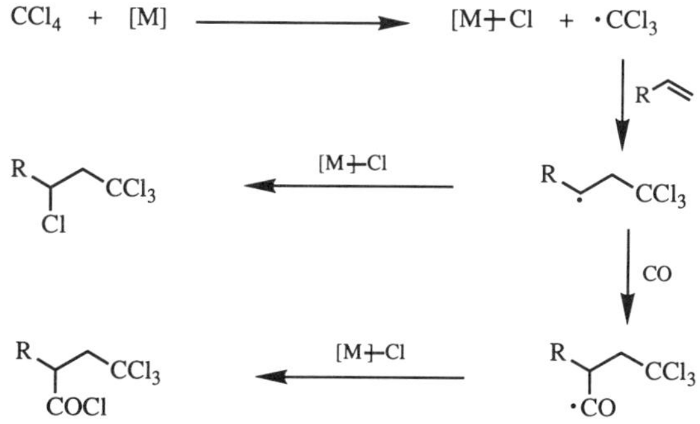

Stoichiometric carbonylations of alkenes with palladium(II) chloride yield β-chloroacyl chlorides, although yields are not high.[388, 389] The corresponding reaction with acetylene (100°C, 100 bar) gives principally muconyl chloride, together with maleyl and fumaryl chlorides, in about 70% total yield (Scheme 8.29).

Scheme 8.29

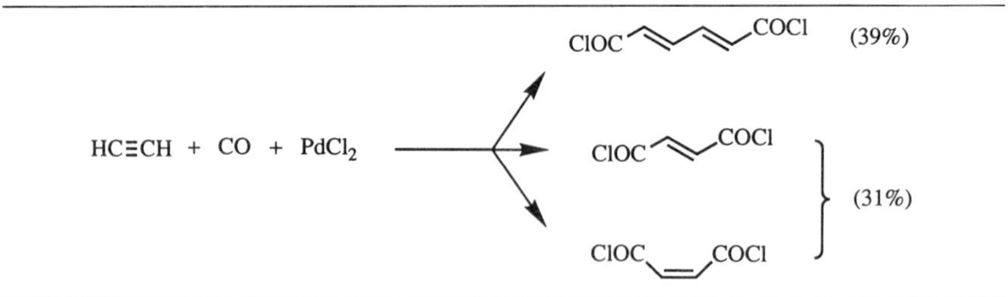

Finally, acetylene reacts stoichiometrically with carbon monoxide and π-allyl nickel complexes at 0°C and 1 bar pressure,[382] to form 2,5-hexadienyl halides in good yield [Eq. (8.11)].

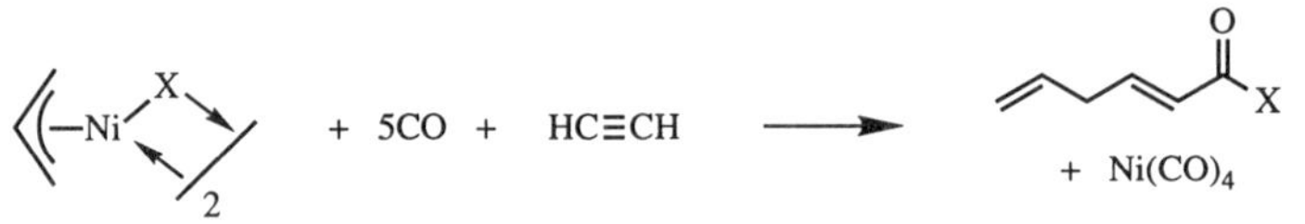

8.5. *Carbamic Acid Derivatives (Isocyanates, Urethanes, and Ureas)*

These compounds are conveniently treated under a single heading, not only because their structures can all be formally derived from carbamic acids, RNHCOOH (by dehydration, esterification, and amidation, respectively), but also because in practice all three types of compound can often be obtained from the same type of N-carbonylation reaction, e.g., the reductive carbonylation of a nitroarene. Thus, in the absence of nucleophilic species isocyanates are isolated; the presence of an alcohol leads to urethane formation, and addition of amines to the system generates ureas, as shown in Scheme 8.30.

Scheme 8.30

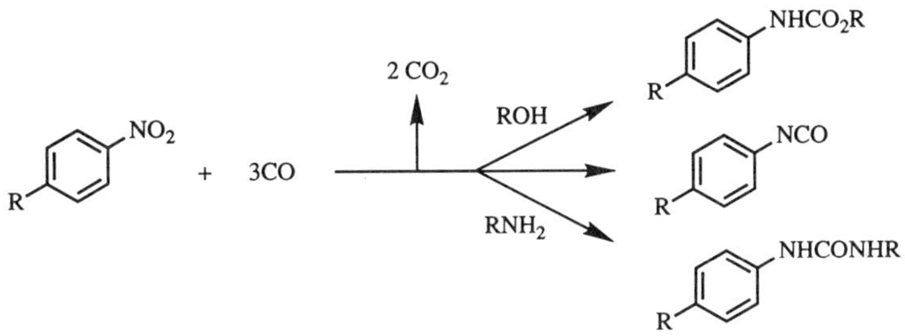

The chemistry described in this section is unusual in that it is rich in empirical observation, but very poorly understood. This state of affairs has undoubtedly arisen because of the considerable industrial interest in replacing current phosgene-based processes for isocyanate manufacture, coupled with the relative difficulty of obtaining mechanistic information about reactions that generally only proceed under conditions of elevated temperature and pressure. Nevertheless, since more than 20 years have elapsed since this type of chemistry was discovered,[390] it seems remarkable that in a major review of the area,[391] published in 1988, it could be noted that "... in practice no studies of the mechanisms of the N-carbonylation of aromatic nitro compounds with alcohols, leading to carbamates (urethanes), have been carried out." [Since that review was written, a small amount of mechanistic information for one particular type of catalyst has in fact been published (see Sec. 8.5.3).]

Much of the early literature and patent data in the N-carbonylation area have been systematically reviewed elsewhere,[391, 392] and in this section we concentrate on more recent work aimed especially at broadening the applicability of such chemistry to organic synthesis.

8.5.1. Carbamic Acid Derivatives: From Azides

The conversion of aryl azides to isocyanates by reaction with carbon monoxide is conceptually much the simplest type of N-carbonylation reaction, involving only the isoelectronic exchange of N_2 for CO as shown in Eq. (8.12). The uncatalyzed reaction nevertheless requires drastic conditions (160–180°C and 150–300 bar of carbon monoxide)[393] and the selectivity for isocyanate is only 60%.

It has recently been shown, however, that such reactions are efficiently catalyzed under very mild conditions (1 bar, 25–50°C) by a variety of rhodium(I) complexes including a reusable, heterogenized catalyst derived from $RhCl(CO)(PPh_3)_2$ and phosphinated polystyrene.[394] Isocyanates are obtained in quantitative yield, and the catalysts maintain their activity even in the presence of alcohols and amines, conditions that lead to formation of urethanes and ureas, respectively. The reaction has been extended to allow controlled, stepwise carbonylation of the *ortho* di-azide 2,3-diazidonaphthalene,[395] thus providing a novel route to naphthimidazolinones (Scheme 8.31).

Scheme 8.31

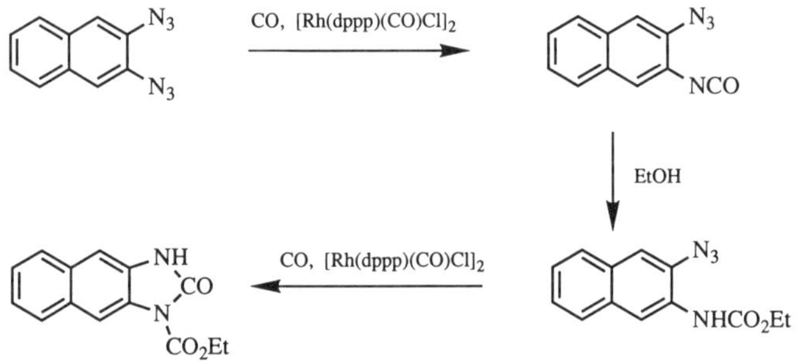

Preparation of N-(4-Methoxyphenyl)-N'-(4-nitrophenyl)urea[394]

To a stirred flask containing toluene (25 cm³), through which carbon monoxide is bubbled, are added sequentially 4-nitrophenyl azide (267 mg, 1.63 mmol) and [Rh{1,3-bis(diphenylphosphino)propane}(CO)Cl]₂ (37 mg, 0.032 mmol). After 30 min the infrared spectrum shows complete conversion of the azide to isocyanate (ν[NCO] at 2260 cm⁻¹). Addition of 4-methoxyaniline (205 mg, 1.67 mmol) gives a quantitative precipitate of N-(4-methoxyphenyl)-N'-(4-nitrophenyl) urea, which is filtered off and recrystallized from ethanol.

8.5.2. Carbamic Acid Derivatives: From Azirines

Ring-opening carbonylation of aryl azirines to give isocyanates [Eq. (8.13)] occurs under mild conditions in the presence of both rhodium and palladium complexes. Stoichiometric quantities (0.5 mole equivalents) of $[Rh(CO)_2Cl]_2$ are required,[396] but $Pd(dba)_2$ (dba = dibenzylidene-acetone) is a genuine catalyst,[397] giving essentially quantitative yields of isocyanates (isolated and characterized by conversion to the methyl urethane) at substrate:catalyst ratios of 10:1. The course of the palladium-catalyzed reaction is remarkably ligand dependent, giving only isocyanates when dba is used, but affording moderate yields of β-lactams (see Chapter 10) when the diene ligand is replaced by triphenylphosphine.

$$\text{(structure)} + \text{CO} \xrightarrow{\text{[Pd(dba)}_2\text{], 40°C}} \text{(structure)} \quad (8.13)$$

8.5.3. Carbamic Acid Derivatives: From Nitro- and Nitroso-arenes

The field of N-carbonylation first attracted extensive investigation in the mid-1960s, when ICI's patents[390, 398] disclosed the formation of urethanes from nitroarenes by high-pressure carbonylation in alcoholic solution in the presence of rhodium complexes such as $[RhCl(CO)_2]_2$ [Eq. (8.14)]. Subsequent work has shown that complexes of palladium,

$$\text{ArNO}_2 + 3\text{CO} + \text{ROH} \xrightarrow[150°C, 100 \text{ bar}]{[RhCl(CO)_2]_2} \text{ArNHCO}_2\text{R} + 2\text{CO}_2 \quad (8.14)$$

particularly those with heterocyclic ligands (pyridine, isoquinoline, etc.), are also effective catalysts for this type of transformation.[399] In the presence of such ligands even palladium on carbon is effective,[400] and at least one company (Mitsui Toatsu) has developed a viable production process based on carbonylation of 2,4-dinitrotoluene in the presence of a heterogeneous palladium catalyst, pyridine, and methanol.[401] The product bis-urethane is subsequently thermolyzed to generate toluene-2,4-diiso-cyanate (Scheme 8.32), and although this material can be produced *directly* by carbonylation of 2,4-dinitrotoluene,[402] the two-stage process apparently operates under milder conditions and gives higher overall selectivity.

Substrates more complex than dinitrotoluene have only rarely been investigated for this type of N-carbonylation reaction, but a platinum-based catalyst system $[PtCl_2(PhCN)_2 + PPh_3 + SnCl_4 + NEt_3]$ *has* been used to carbonylate chloro- and alkoxy-substituted nitroarenes to urethanes without apparent interference from these functional groups.[403]

Scheme 8.32

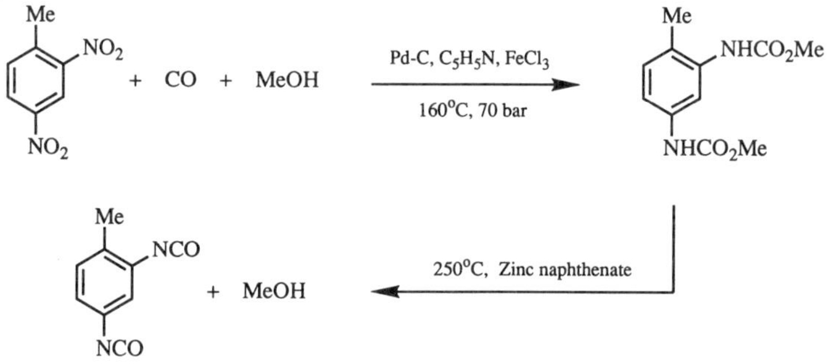

Selenium catalysis for carbonylation of nitroarenes to isocyanates and urethanes has been the subject of extensive patent coverage,[404] but it is not yet clear whether this technology will be commercialized. Although catalyst costs are clearly much less than for a rhodium- or palladium-based process, the toxicological and environmental problems posed by hydrogen selenide (a key intermediate in the process; see Scheme 8.33 for a possible catalytic sequence) may outweigh the obvious cost advantage.

Scheme 8.33

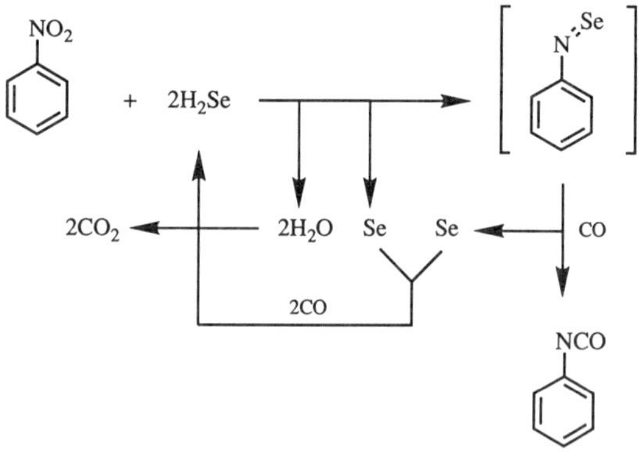

Recent work has shown that the ruthenium carbonyl $Ru_3(CO)_{12}$, in the presence of $[NEt_4]^+Cl^-$, catalyzes the reductive carbonylation of nitrobenzene and substituted mononitrobenzenes with high selectivity.[405]

Conditions are similar to those required for other catalyst systems (160°C, 60 bar), but the importance of this system stems from the isolation, for the first time,[406] of imido complexes such as $PhNRu_3(CO)_{10}$ and $(PhN)_2Ru_3(CO)_9$, which are plausible intermediates in a catalytic cycle leading to isocyanate or urethane formation.[406] Indeed, it has now been shown that the closely related iron complex $(PhN)_2Fe_3(CO)_9$ forms an anionic carbomethoxy derivative on treatment with methoxide ion,[407] and that this eliminates methyl phenylurethane upon thermal or oxidative degradation, as shown in Scheme 8.34. This work seems to represent the first success in bringing at least a little mechanistic understanding to the field of N-carbonylation.

Scheme 8.34

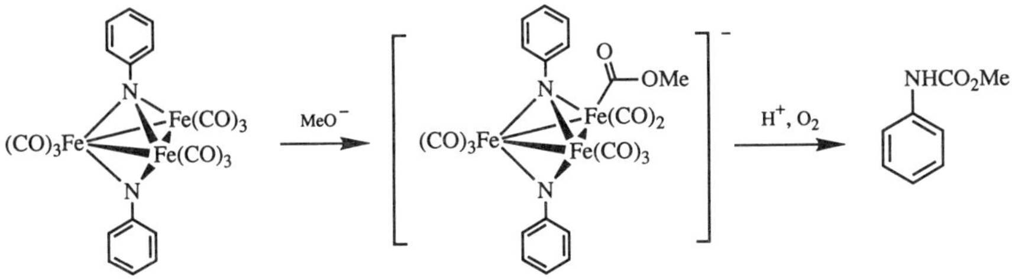

In view of the high temperatures and pressures required for carbonylation of nitro-arenes, it is of considerable interest that *nitroso*-arenes can be converted to urethanes in very high yield under essentially ambient conditions [Eq. (8.15)]. The preferred catalyst is derived from palladium acetate, copper(II) acetate, and hydrochloric acid, and the presence of oxygen appears necessary for successful reaction.[408]

$$ArNO + ROH + 2CO \xrightarrow[O_2, HCl]{Pd(OAc)_2, Cu(OAc)_2} ArNHCOOR + CO_2 \quad (8.15)$$

8.5.4. Carbamic Acid Derivatives: From Amines

Formation of urethanes by the oxidative carbonylation of aromatic and aliphatic amines is efficiently catalyzed by metallic palladium in the presence of iodide ion [Eq. (8.16)].[409] This chemistry has been successfully developed by Asahi Chemical into a potential industrial process for

$$RNH_2 + CO + 0.5O_2 + R'OH \xrightarrow{Pd, I^-} RNHCO_2R' + H_2O \quad (8.16)$$

the synthesis of ethyl phenylurethane and thence methylenediphenyldi-isocyanate (MDI), an important intermediate for polyurethane manufacture.[410] It is claimed that the new process, outlined in Scheme 8.35, not only avoids the use of toxic and corrosive phosgene, but also enables production of higher purity MDI than is possible with conventional technology.

Scheme 8.35

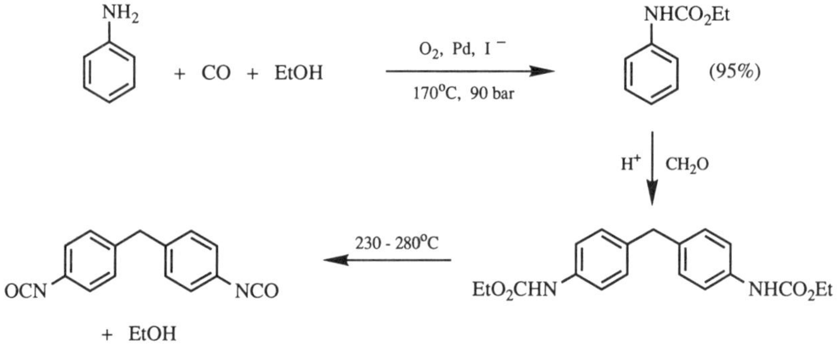

Although an apparently viable industrial process, the very high temperatures and pressures required for this type of reaction (170°C, 90 bar total pressure of oxygen and carbon monoxide)[409] obviously limit its applicability as a general synthetic method. It is therefore of interest that recent work indicates the same chemistry to be possible under ambient

Scheme 8.36

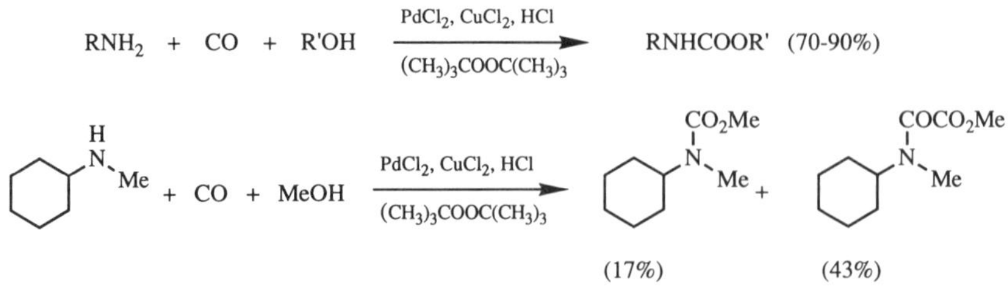

conditions,[411] using an organic peroxide as reoxidant in place of oxygen, and a *homogeneous* catalyst system containing palladium(II) chloride, copper(II) chloride, and hydrochloric acid (Scheme 8.36). A wide range of primary aromatic and aliphatic amines can be successfully carbonylated to urethanes using this system, but secondary amines give mixtures of urethanes and oxamates, the latter arising from the novel double carbonylation reaction shown in Scheme 8.36.

Chapter 9
Synthesis of Lactones

9.1. Introduction

Lactones are often obtained by carbonylation reactions closely related to the acyclic ester syntheses described in the previous chapter. Although only limited data are available for carbonylative routes to four- and six-membered lactone rings, the preparation of five-membered (γ) lactones has been one of the most intensively studied areas of carbonylation chemistry in recent years. Palladium catalysis in particular has provided a number of extremely clean and facile syntheses, including routes to a wide range of α-methylene-γ-lactones.

9.2. Four-Membered Lactones

Oxidative carbonylation of alkenes (see Chapter 2) in the presence of water can sometimes lead to formation of β-lactones. Ethylene, for example, is converted to β-propiolactone by carbonylation in aqueous acetonitrile at $-20°C$, in the presence of palladium(II) chloride as catalyst and a stoichiometric amount of copper(II) chloride as reoxidant for palladium [Eq. (9.1)].[412] Reaction of a suspension of the cyclo-octadiene palladium complex (1) with carbon monoxide in aqueous sodium acetate

$$CH_2{=}CH_2 \;+\; CO \;+\; H_2O \quad \xrightarrow[\text{CuCl}_2]{\text{PdCl}_2} \quad \text{（β-lactone）} \qquad (9.1)$$

similarly produces the fused-ring lactone (2) in about 80% yield [Eq. (9.2)].[413]

$$
\text{1} \quad + \quad CO \quad + \quad H_2O \quad \longrightarrow \quad \text{2} \qquad (9.2)
$$

Palladium-catalyzed carbonylation of halogeno-alcohols can also be used to synthesize β-lactones under very mild conditions,[414] as illustrated in Eq. (9.3) and (9.4).

$$
\underset{Br}{\overset{Ph}{\bigwedge}}\text{OH} \quad + \quad CO \quad \xrightarrow{Pd(PPh_3)_2Cl_2} \quad \text{(65\%)} \qquad (9.3)
$$

$$
\underset{OH}{\overset{Cl}{\bigcirc}} \quad + \quad CO \quad \xrightarrow{Pd(PPh_3)_2Cl_2} \quad \text{(52\%)} \qquad (9.4)
$$

Preparation of 2-Phenyl-propanolide[414]

A mixture of Pd(PPh₃)₂Cl₂ (0.029 g, 0.041 mmol), 2-bromo-2-phenyl-ethanol (2.51 g, 12.5 mmol), and K₂CO₃ (1.73 g, 12.5 mmol) is placed in a flask fitted with a serum cap, and the flask evacuated and refilled with carbon monoxide. The solvent (DMF, 10 cm³) is added by syringe through the serum cap, followed by addition of one drop of hydrazine hydrate, and the mixture is stirred at room temperature under carbon monoxide for 24 h. After extraction with ether and distillation, the product lactone is isolated in 63% yield.

9.3. Five-Membered Lactones

One of the simplest conceivable routes to a five-membered lactone is the direct insertion of carbon monoxide into four-membered cyclic ether (oxetane). Early work in the patent literature indicated that this type of reaction could be achieved for oxetane itself using a cobalt catalyst under forcing conditions (250 bar, 200°C).[415] Recent investigations have shown that a wide range of oxetanes and thietanes may similarly be carbonylated under less vigorous (though still scarcely mild) conditions.[416] The reactions are regiospecific, carbonylation occurring at the less-substituted carbon–heteroatom bond [Eq. (9.5)], and the preferred catalyst appears to be a mixture of cobalt and ruthenium carbonyls.

$$(9.5)$$

Carbonylation of ethynyl alcohols is an attractive route to α-methylene-γ-lactones, and although the original methodology involved stoichiometric quantities of nickel tetracarbonyl,[417] a more recent technique[418] uses palladium(II) chloride (liganded with thiourea) in only catalytic amounts [Eq. (9.6)].

$$(9.6)$$

Table 9.1. *Cyclocarbonylation of Ethynyl Alcohols[a]*

Entry	Ethynyl alcohol	α-Methylene lactone	Yield (%)
1			31
2			85
3			83
4			77
5			93

[a] References 418, 419.

An improved version of the catalyst system consists of $PdCl_2$ and a tertiary phosphine (two equivalents) in the presence of one equivalent of tin(II) chloride, with anhydrous acetonitrile as the solvent.[419] Both catalytic reactions give good yields of α-methylene-γ-lactones (Table 9.1), and the approach has been extended to carbonylation of *cis*-ethynylcyclo-alkanols as a route to *cis*-fused α-methylenelactones, an important group of natural products (Table 9.1).

Cyclocarbonylation of trans-2-Ethynylcyclohexanol (Table 9.1, Entry 2)[419]

To a solution of palladium(II) chloride (81 mg, 0.46 mmol) and tri-*n*-butylphosphine (186 mg, 0.92 mmol) in acetonitrile (5 cm³) at 75°C is added *trans*-2-ethynylcyclohexanol (149 mg, 1.2 mmol), and the mixture is stirred for 6 h under carbon monoxide pressure (ca. 8 bar). After cooling and venting the autoclave, analysis of the reaction mixture by GLC indicates an 85% yield of α-methylene lactone, which is isolated by preparative GLC.

Such syntheses are thought to proceed as shown in Scheme 9.1, where formation of an alkoxycarbonylpalladium species is followed by intra-molecular insertion of the acetylenic triple bond. Cleavage of the resulting vinyl–palladium bond by the proton generated in the first step completes the catalytic sequence.[420]

Scheme 9.1

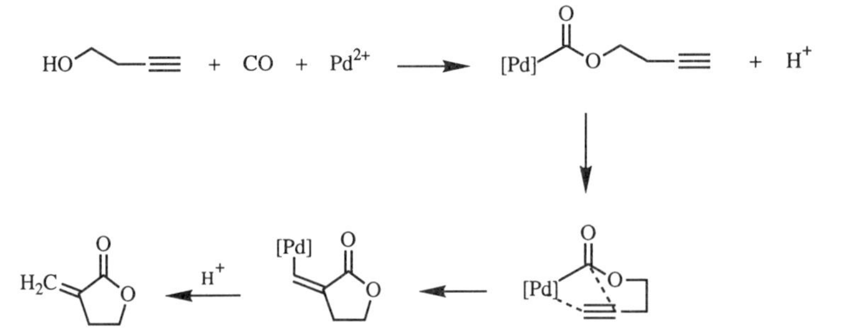

Halogenated homoallylic alcohols can be converted to α-methylene-γ-lactones using stoichiometric quantities of $Ni(CO)_4$ in the presence of base [Eq. (9.7)].[421] The choice of base is particularly important here, since with potasium acetate yields of lactone can be as high as 60%, whereas use of methoxide causes yields to fall to only about 4%. An improved, though still stoichiometric, reagent for this type of synthesis is the relatively involatile and therefore less hazardous $Ni(PPh_3)_2(CO)_2$, in conjunction

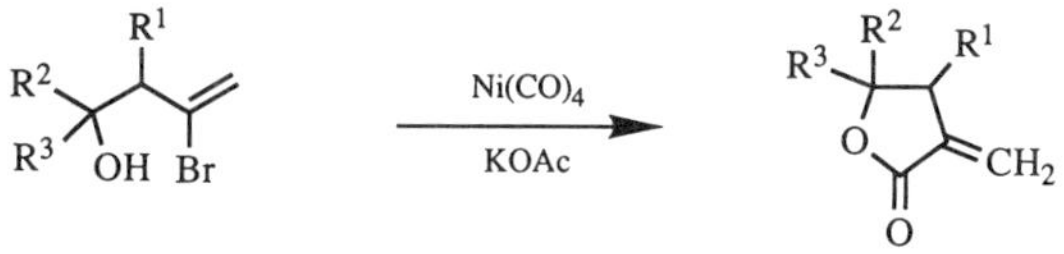

$$R^2 \underset{R^3 \ OH \ Br}{\overset{R^1}{\diagup}} \quad \xrightarrow[\text{KOAc}]{\text{Ni(CO)}_4} \quad R^3 \underset{O}{\overset{R^2 \ R^1}{\diagup}} CH_2 \qquad (9.7)$$

with triethylamine as base.[422] This type of reaction has been used in a synthesis of the fused-ring sesquiterpene-lactone frullanolide (**3**) [Eq. (9.8)].[423]

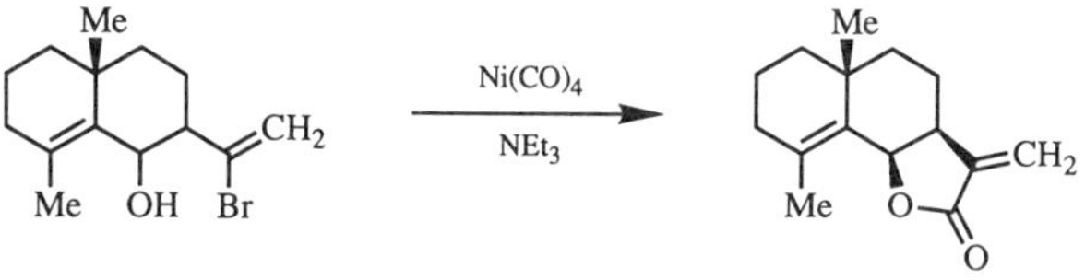

$$\xrightarrow[\text{NEt}_3]{\text{Ni(CO)}_4} \qquad (9.8)$$

3

As in the acetylene-based synthesis [Eq. (9.6)], palladium *catalysts* can replace stoichiometric nickel reagents for cyclocarbonylation of halogenated homoallylic alcohols [Eq. (9.9)].[424] The starting materials

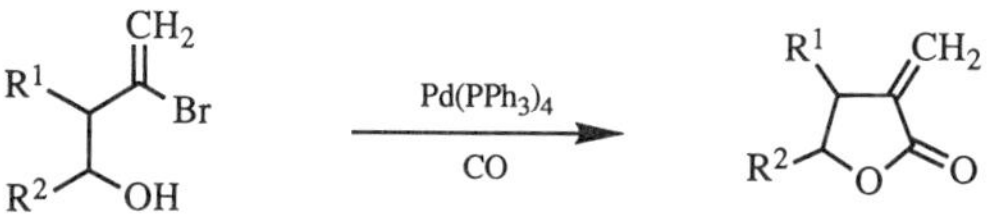

$$R^1 \underset{R^2 \ OH}{\overset{CH_2}{\diagup}} Br \quad \xrightarrow[\text{CO}]{\text{Pd(PPh}_3)_4} \quad R^2 \underset{O}{\overset{R^1 \ CH_2}{\diagup}} O \qquad (9.9)$$

can be obtained from epoxides and α-silylated vinyl Grignard reagents, followed by bromo-desilylation. Because of the regio- and stereospecific nature of the epoxide ring opening, and the noninvolvement of asymmetric carbons in the cyclocarbonylation reaction, optically active lactones can be obtained from the corresponding optically active epoxides, as shown in Scheme 9.2.

A further variation on this theme is the generalized cyclocarbonylation of hydroxy-substituted vinylic halides in the presence of a triphenylphosphine/palladium(II) acetate catalyst system [Eq. (9.10)].

Scheme 9.2

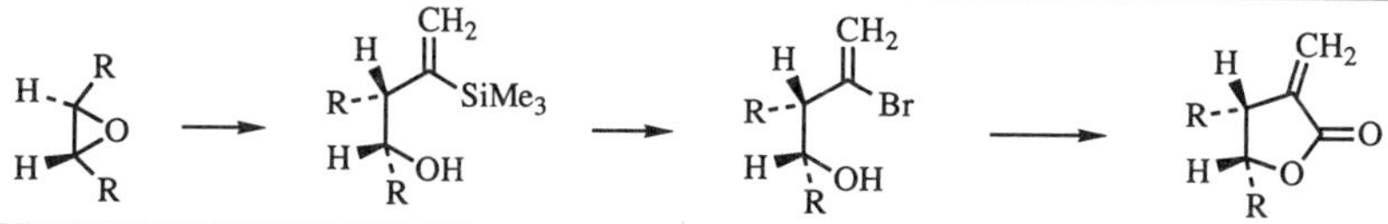

This procedure allows the synthesis of five-, six-, and seven-membered α-methylene lactones.[425]

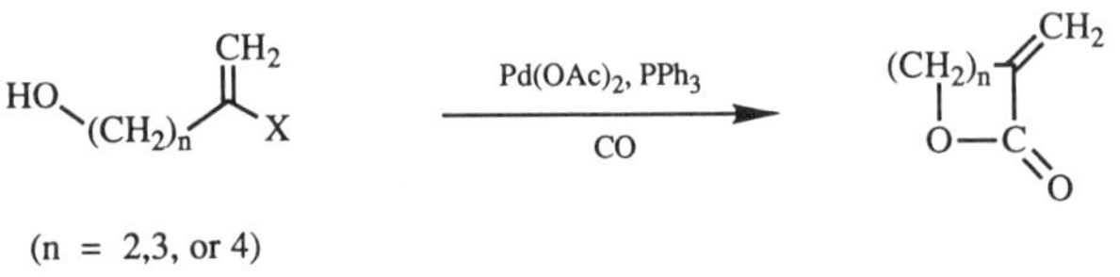

$$
\text{(9.10)}
$$

(n = 2,3, or 4)

Homoallylic chloroformates (readily obtained from phosgene and the corresponding alcohol) undergo a closely related cyclization, with loss of hydrogen chloride, to give α-methylene-γ-lactones in good yield (Scheme 9.3).[426]

Scheme 9.3

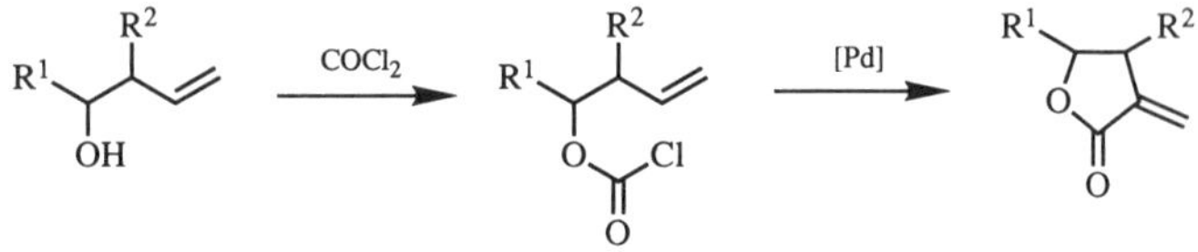

The tetracarbonylcobalt anion, $[Co(CO)_4]^-$, generated by reaction of $Co_2(CO)_8$ with sodium hydroxide and a phase-transfer catalyst in a vigorously stirred two-phase (benzene–water) system, catalyzes the synthesis of α-hydroxybutenolides from alkynes, iodomethane, and carbon monoxide [Eq. (9.11)].[427] Although a potentially very useful reaction, only a limited number of acetylenes have been investigated, and yields seem rather variable. Replacing iodomethane in this type of reaction by an alkyl

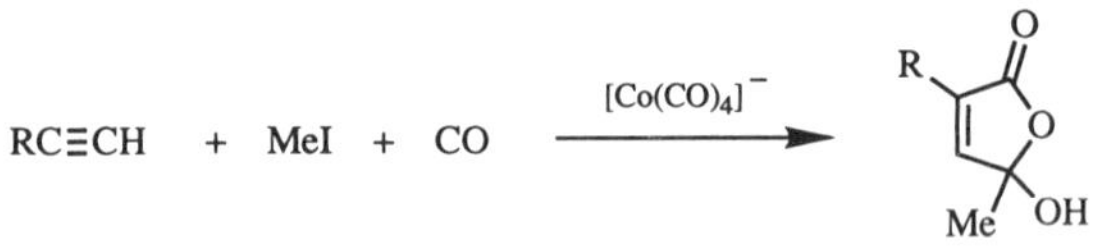

$$
RC\equiv CH \;+\; MeI \;+\; CO \xrightarrow{[Co(CO)_4]^-} \qquad\qquad \text{(9.11)}
$$

halide such as methyl bromoacetate, which bears an *electron-withdrawing* group on the α-carbon, and using a nonnucleophilic base such as a tertiary amine, leads to formation of a 2,4-pentadieno-4-lactone (Scheme 9.4). The reaction remains catalytic in cobalt, and yields are around 60% for a variety of substituted alkynes and alkyl halides.[428]

Scheme 9.4

$$R^1CH_2Br \; + \; NaCo(CO)_4 \xrightarrow{\; CO \;} R^1CH_2COCo(CO)_4 \xrightarrow{\; R^2C \equiv CR^3 \;}$$

Reaction of acetylene with carbon monoxide at 100°C and 100 bar pressure, in the presence of $Co_2(CO)_8$ as catalyst, produces the *trans*-bifurandione (**4**) in 70% yield, and substituted acetylenes behave similarly.[429] At lower temperatures the major products are the *cis*- and *trans*-isomers of 2,4,6,8-decatetraene-1,4,7,10-diolide (**5**), with only small amounts of bifurandione.[430]

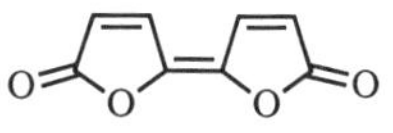

4    **5**

Direct mercuration of propargylic alcohols followed by carbonylation in the presence of a palladium(II) salt and copper(II) chloride leads to formation of chlorobutenolides (Scheme 9.5).[431] The yields of mercurated intermediates are variable, but the carbonylation step is essentially quantitative.

Scheme 9.5

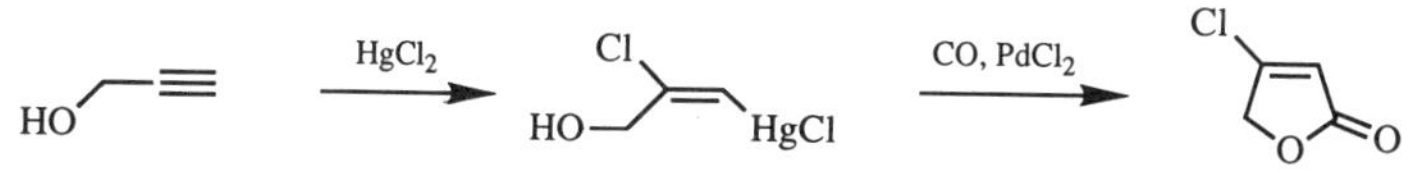

$$HO \equiv \xrightarrow{\; HgCl_2 \;} \; \xrightarrow{\; CO,\; PdCl_2 \;}$$

Preparation of β-Chloro-Δ$^{\alpha,\beta}$-butenolide[431]

To a suspension of palladium(II) chloride (0.84 g, 5 mmol), lithium chloride (4.25 g, 100 mmol), and copper(II) chloride (13.45 g, 100 mmol) in 250 cm³ of diethyl ether at -78°C, is added (*E*)-2-chloro-3-chloromercuri-2-propen-1-ol (16.4 g, 50 mmol). The flask is flushed with carbon monoxide and an atmospheric pressure supply of carbon monoxide is connected to the top of the flask. The cold bath is removed and the reaction mixture stirred at 5°C for 50 h. Saturated ammonium chloride solution (10 cm³) is added and the mixture stirred for a further hour. The resulting suspension is filtered, and the filtrate washed with saturated ammonium chloride and dried over sodium sulfate before evaporation of the solvent. Recrystallization of the residue (6.18 g) from carbon tetrachloride affords β-chloro-$\Delta^{\alpha,\beta}$-butenolide in 78% yield.

Under phase transfer conditions, and in the presence of stoichiometric amounts of $Mn(CO)_5Br$ or $Mn_2(CO)_{10}$, *saturated* γ-butyrolactones are formed by reaction of alkynes with iodomethane and carbon monoxide [Eq. (9.12)].[432] Reaction conditions are mild (35°C, 1 bar) and the

$$RC\equiv CH \; + \; MeI \; + \; CO \; \xrightarrow{\;\;Mn(CO)_5Br\;\;} \; \underset{Me}{\overset{R}{\diagdown}}\!\!\diagdown\!O \qquad (9.12)$$

process is regiospecific, none of the 3,4-disubstituted lactone (6) being detectable.

$$\underset{Me}{\overset{R}{\diagdown}}\!\!\diagdown\!O$$

6

Primary, secondary, and tertiary allylic alcohols react with carbon monoxide and oxygen in the presence of catalytic quantities of palladium(II)- and copper(II) chlorides, to give γ-butyrolactones [Eq. (9.13)].[433] The reaction proceeds at room temperature and atmospheric pressure, and yields are generally in the range 40%–60% (Table 9.2).

$$\underset{R^2 \quad R^3}{\overset{R^1}{\diagdown}}\!\!\diagdown\!OH \; + \; CO \; \xrightarrow[CuCl_2, O_2]{\;\;PdCl_2\;\;} \; \underset{O}{\overset{R^2\diagup R^1}{\diagdown}}\!\!\diagdown\!R^3 \qquad (9.13)$$

Good yields of tetrahydro-2-furanones can be obtained by cobalt-catalyzed hydroformylation of α,β-unsaturated esters,[434] as shown in Eq. (9.14). Conditions for this reaction are rather forcing, however, typically 250°C and 300 bar of $CO-H_2$, using $Co_2(CO)_8$ as catalyst and benzene as solvent.

$$H_2C=C(R)CO_2R^1 \; \xrightarrow[CO, H_2]{\;\;Co_2(CO)_8\;\;} \; \underset{O}{\overset{R}{\diagdown}}\!\!\diagdown\!O \qquad (9.14)$$

Table 9.2. Palladium-Catalyzed Carbonylation of Allylic Alcohols to Lactones[a]

Substrate	Product	Yield (%)
(structure)	(structure)	35
(structure)	(structure)	50
(structure)	(structure)	70
(structure)	(structure)	15
(structure)	(structure)	45
(structure)	(structure)	65
(structure)	(structure)	35
(structure)	(structure)	60
(structure)	(structure)	40

[a] Reference 433.

Carbonylative double-cyclization of 3-hydroxy-4-pentenoic acids (Scheme 9.6) leads to formation of fused-ring bis-lactones.[435] The specific cis stereochemistry of this palladium-catalyzed reaction can be rationalized by assuming that attack of palladium(II) on the alkene is directed by the allylic hydroxyl group, giving rise to the "chelated" alkylpalladium intermediate (7).

Lactones based on cis-3-hydroxy-tetrahydrofuranacetic acid have been prepared in high yield by intramolecular, palladium-catalyzed, oxycarbonylation of 4-penten-1,3-diols, with copper(II) as the source of oxidizing power [Eq. (9.15)].[436] Only mild conditions are required (25°C, 1 bar),

Scheme 9.6

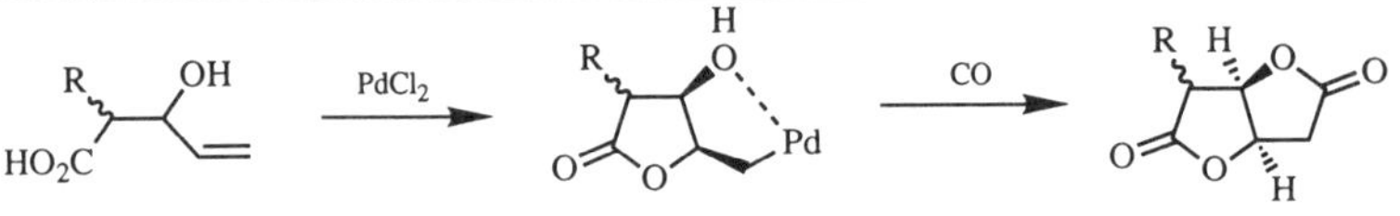

7

and a wide range of the starting diols can be obtained in essentially quantitative yield by crossed-aldol condensations of ketones, esters, or lactones with unsaturated aldehydes, followed by reduction with lithium aluminum hydride.

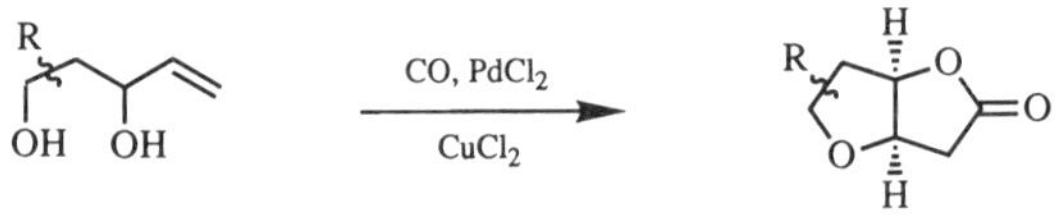

$$(9.15)$$

Oxycarbonylation of 4-Penten-1,3-diols[436]

A flask containing PdCl$_2$ (0.1 mmol), CuCl$_2$ (3 mmol), and sodium acetate (3 mmol) is flushed with carbon monoxide and connected to an atmospheric pressure supply of carbon monoxide. A solution of the diol (1 mmol) in acetic acid (5 cm^3) is introduced via a syringe and the reaction mixture is stirred vigorously overnight at room temperature, during which time the color changes from deep green to pale yellow. Dilution with diethyl ether followed by filtration and evaporation gives a crude product which can be purified by column chromatography on silica gel. Yields are generally good.

In a reaction very closely related to that shown in Eq. (9.15), oxidative carbonylation of N-protected 3-hydroxy-4-pentenylamines at ambient temperature and pressure produces *cis*-3-hydroxypyrrolidine-2-acetic acid lactones with high stereoselectivity (Scheme 9.7).[437] The success of this reaction is strongly dependent on the nature of the protecting group R, the

Scheme 9.7

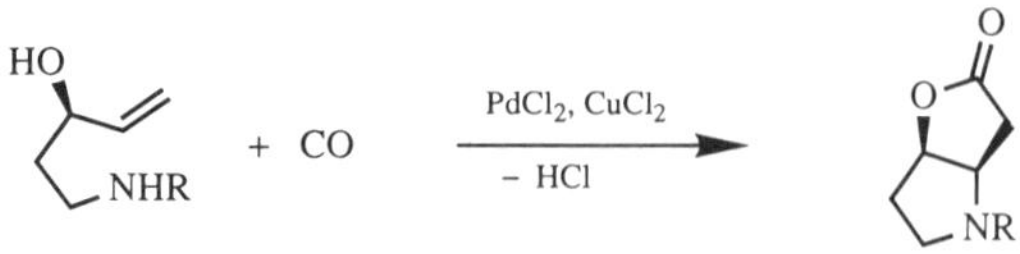

most reactive and versatile being carbamoyl (R = −CONHMe), followed by methoxycarbonyl (R = −COOMe), and arenesulfonyl (R = −SO$_2$Ar). Analogous reactions using N-protected 4-hydroxy-5-hexenylamines also lead to lactone formation (Scheme 9.8), but rates and selectivities are much reduced, to the extent that only carbamoyl-protected substrates undergo reaction, and mixtures of products are obtained.

Scheme 9.8

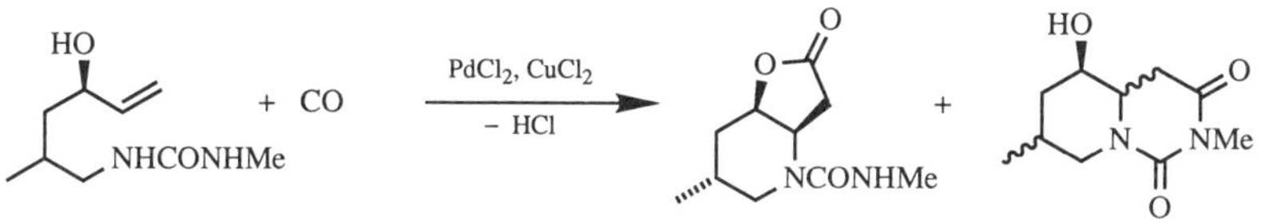

Under two-phase conditions (benzene/NaOH−H$_2$O), styrene oxides may be carbonylated in the presence of iodomethane (NaCo(CO)$_4$ as catalyst) to give the enol tautomers of 4,5-dihydro-4-phenylfuran-2,3-

Scheme 9.9

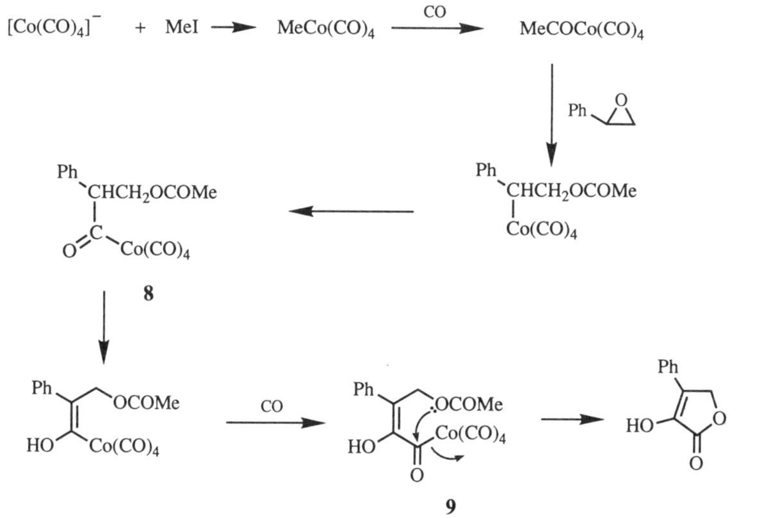

diones [Eq. (9.16)].[438] The anionic cobalt catalyst is generated *in situ* from $Co_2(CO)_8$ and NaOH, and addition of a phase transfer agent is

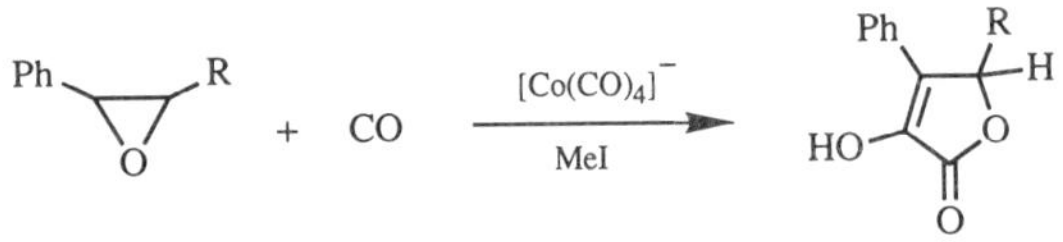

$$\text{(9.16)}$$

necessary. The suggested mechanism, shown in Scheme 9.9, involves ring-opening addition of acetyltetracarbonylcobalt to the epoxide to give a benzylcobalt complex, which, on ligand migration to carbonyl and addition of carbon monoxide to the vacant coordination site, yields the acyl intermediate (8). Enolization of the acyl ligand followed by insertion of a second CO molecule affords (9), which then cyclizes and regenerates acetyltetracarbonylcobalt.

In the presence of stoichiometric quantities of palladium(II) acetate, the diol (10) undergoes intramolecular alkoxycarbonylation to give the *cis*-lactone (11) in 68% yield,[439] as shown in Eq. (9.17).

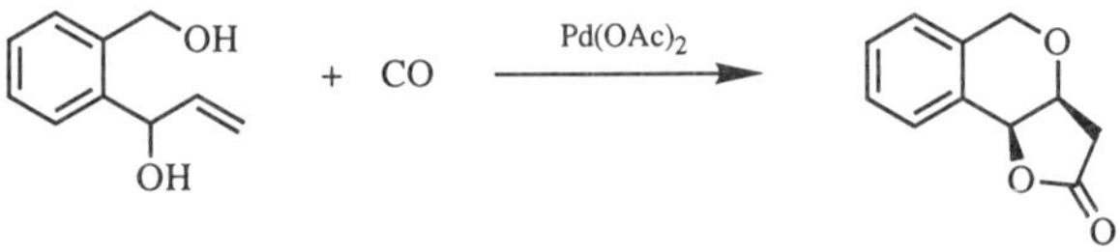

$$\text{(9.17)}$$

Benzylic alcohols with a halogeno- or halogenomethyl-substituent in the 2-position, may be carbonylated to lactones using palladium catalysts such as $Pd(PPh_3)_2Cl_2$ in the presence of a stoichiometric quantity of a tertiary amine [Eq. (9.18)].[414] A proposed catalytic cycle is shown in

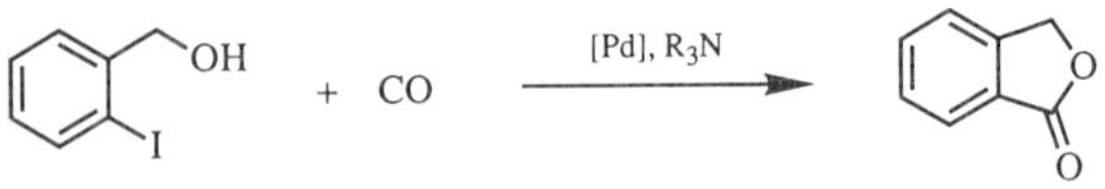

$$\text{(9.18)}$$

Scheme 9.10. Recent work indicates that similar reactions may also be brought about using cobalt-based catalysts.[440]

The direct carbonylation of arylthallium compounds normally requires high temperatures and pressures, but palladium catalysis allows the reaction to proceed in excellent yield under essentially ambient conditions.[441] Thus thallation of 3-methoxybenzyl alcohol, followed by catalytic

Scheme 9.10

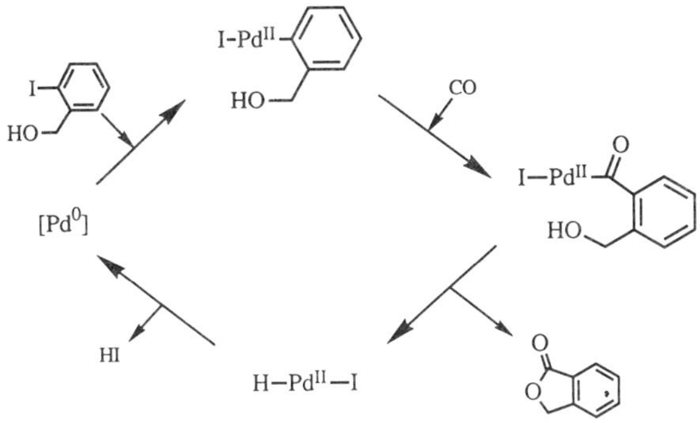

carbonylation, affords 5-methoxyphthalide in 89% yield with only traces of the 7-methoxy isomer (Scheme 9.11).

Scheme 9.11

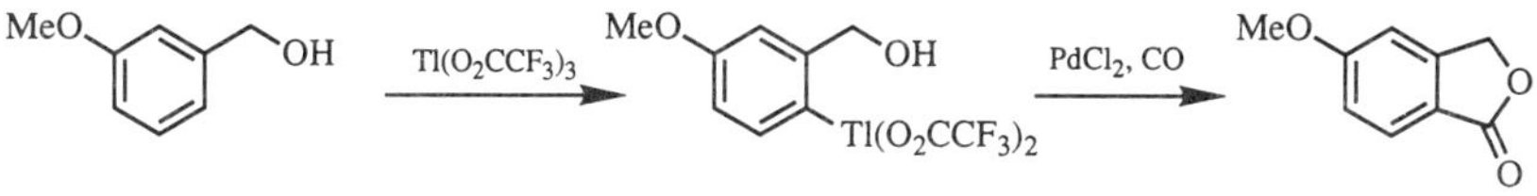

A remarkable keto-lactone synthesis involves the nickel-catalyzed double carbonylation of a bromodiene.[442] The initial product (60% yield, Scheme 9.12) undergoes double-bond migration to give a keto butenolide during column chromatography on silica gel.

Scheme 9.12

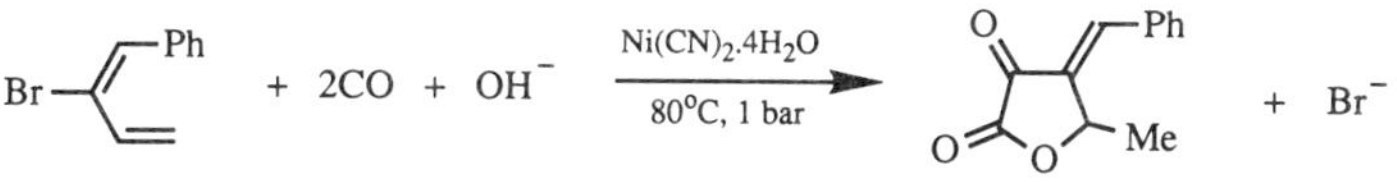

9.4. Six-Membered Lactones

Homoallylic alcohols are efficiently converted into δ-lactones by rhodium-catalyzed hydroformylation followed by oxidation [Eq. (9.19)].[443]

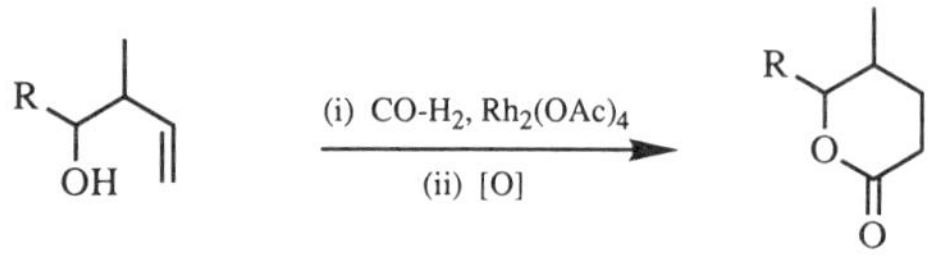

$$
\begin{array}{c}
R \overset{}{\underset{OH}{\diagup}} \diagdown \xrightarrow[\text{(ii) [O]}]{\text{(i) CO-H}_2,\ \text{Rh}_2\text{(OAc)}_4} \quad R \overset{}{\diagdown} \text{(lactone)}
\end{array}
\qquad (9.19)
$$

General Synthesis of δ-Lactones[443]

The homoallylic alcohol (12.8 mmol) triphenylphosphine (7.5 mmol), Rh$_2$(OAc)$_4$ (40 mg), and ethyl acetate are heated for 6 h at 100°C in a steel bomb under 20 bar of CO−H$_2$ (1:1). After cooling and removing the solvent, the crude hemiacetal is dissolved in dichloromethane and treated with pyridinium chlorochromate (25 mmol). The mixture is stirred for 3 h and the product is isolated by column chromatography on silica, with diethyl ether as eluent, in over 80% yield.

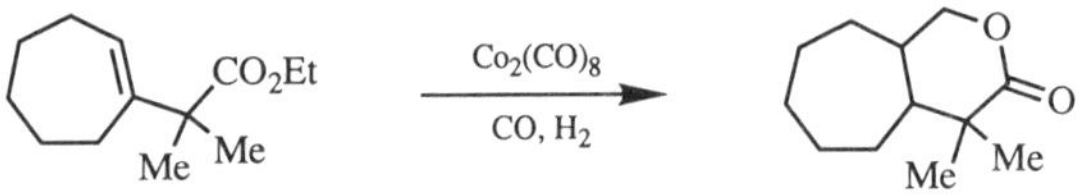

$$
\begin{array}{ccc}
\text{12} & \xrightarrow[\text{CO, H}_2]{\text{Co}_2\text{(CO)}_8} & \text{13}
\end{array}
\qquad (9.20)
$$

Hydroformylation of the unsaturated ester (**12**) in the presence of Co$_2$(CO)$_8$ produces the bicyclic lactone (**13**) in 96% yield [Eq. (9.20)].[434]

Scheme 9.13

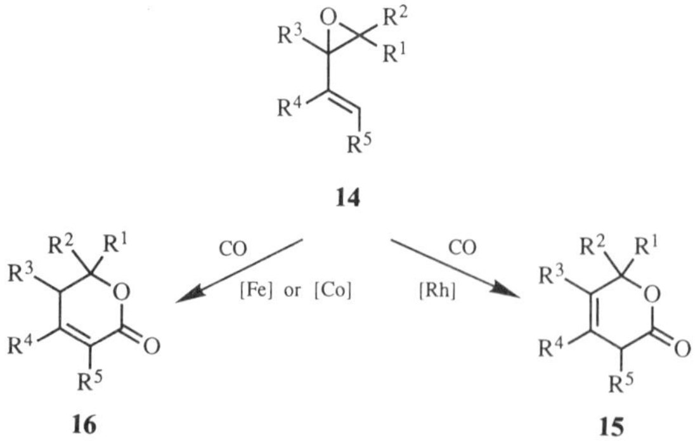

Reaction of the epoxy-alkene (14) with carbon monoxide in the presence of a rhodium catalyst produces the β,γ-unsaturated lactone (15), whereas carbonylation using iron or cobalt catalysts gives an α,β-unsaturated lactone (16) (Scheme 9.13).[444] Yields for these reactions are quoted in the range 10%–75%, but as yet full experimental details are lacking.

Carbonylation of the acetylenic alcohol (17) in the presence of aqueous acidic nickel tetracarbonyl leads to formation of the tetrahydro-pyranone (18) in 20% yield [Eq. (9.21)].[417]

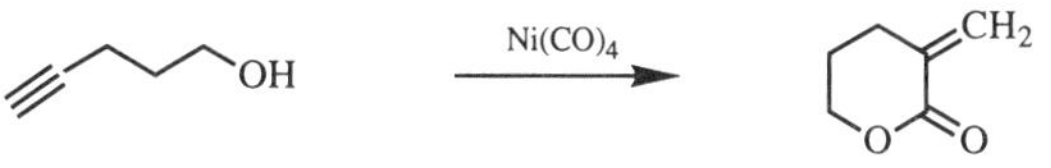

$$\text{17} \quad \xrightarrow{\text{Ni(CO)}_4} \quad \text{18} \qquad (9.21)$$

Palladium-catalyzed carbonylation of the fused-ring system (19), where the ethynyl and hydroxy groups are fixed in the appropriate geometry, produces δ-lactone in good yield [Eq. (9.22)].[445]

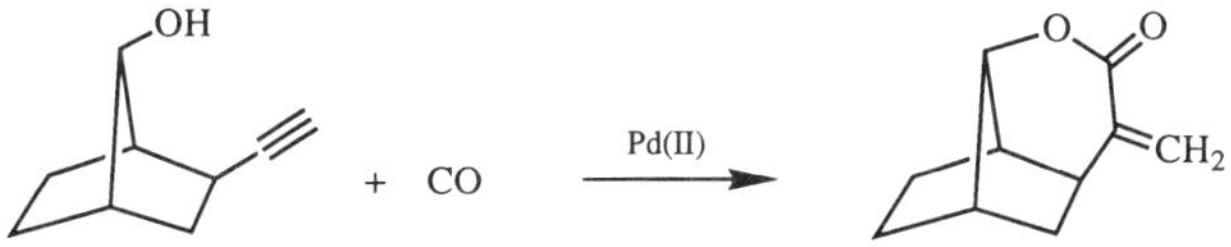

$$\text{19} \quad + \quad \text{CO} \quad \xrightarrow{\text{Pd(II)}} \qquad (9.22)$$

A novel synthesis of coumarins involves the palladium-catalyzed coupling of a 2-iodophenol with norbornadiene and carbon monoxide.[446] The first-formed product undergoes a spontaneous retro-Diels–Alder reaction, as shown in Scheme 9.14, so that the isolated coumarin is effectively derived from acetylene rather than norbornadiene. Interestingly, use of an alkyne rather than norbornadiene does *not* lead to coumarin formation, but instead gives high yields of aurone (Scheme 9.15).[446] Both reactions are clearly initiated by oxidative addition of the C−I bond to Pd(0), but

Scheme 9.14

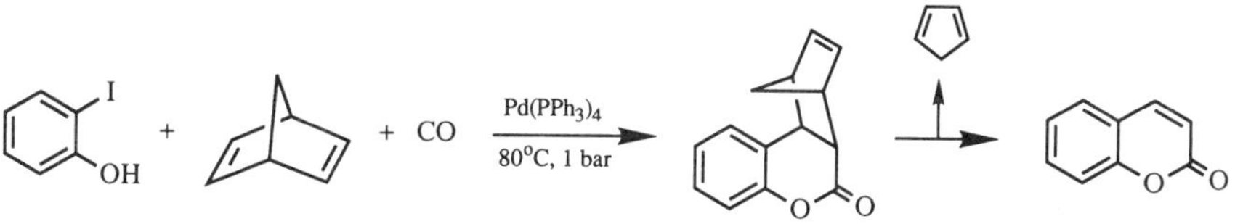

it appears that insertion of norbornadiene into the resulting palladium–carbon bond occurs much more readily than insertion of carbon monoxide, whereas alkyne insertion *follows* that of carbon monoxide.

Scheme 9.15

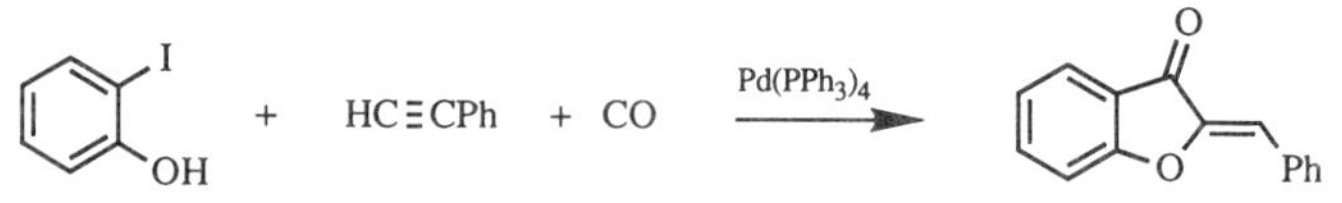

The aromatic thallation and subsequent palladium-catalyzed carbonylation reaction described in Sec. 9.3 can also be applied to the synthesis of six-membered lactones,[441] as shown in Eq. (9.23).

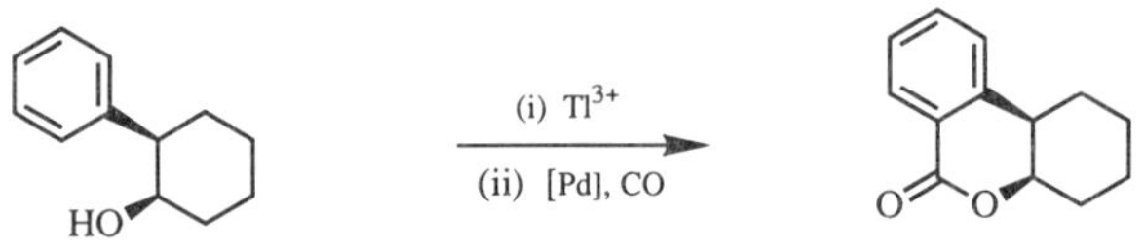

$$(9.23)$$

All the lactone syntheses described so far in this chapter have involved carbonylation as the actual ring-closing step, but a number of lactone-based natural products have also been obtained using carbonylation at an earlier stage of the synthesis. For example, routes to curvularin (Scheme 9.16)[447] and zearalenone (Scheme 9.17)[448] both involve carbonylation steps, with yields of around 70% in both cases.

Scheme 9.16

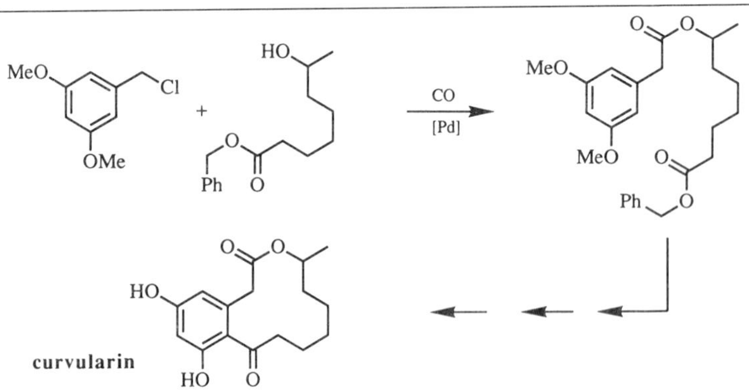

Scheme 9.17

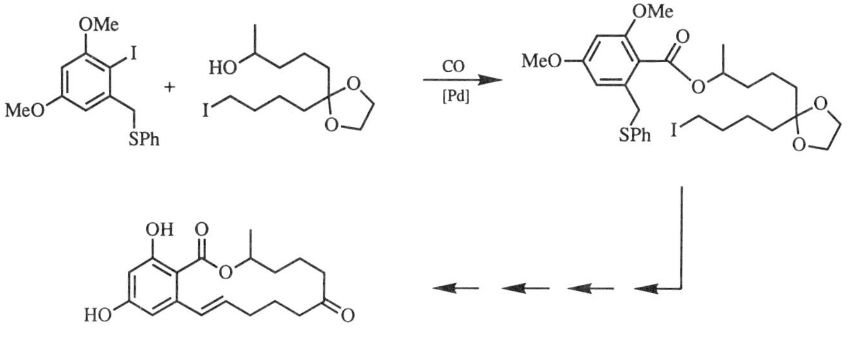

zearalenone

Chapter 10

Synthesis of Lactams and Related N-Heterocycles

10.1. Introduction

In view of the widespread occurrence of lactam rings, particularly the β-lactam ring, in pharmacologically active compounds such as the penicillins and cephalosporins, intensive efforts have been made in recent years to develop carbonylation-based methodologies for constructing this type of subunit.[449] This work has led to the discovery of a number of routes to β-, γ-, and δ-lactams, the more useful of which are outlined in the present chapter.

10.2. Four-Membered Rings

Carbonylation chemistry affords a number of potentially useful β-lactam syntheses, including, as shown in Scheme 10.1, a route to the antibiotic thienamycin. This route[450, 451] is based on the formation of an allyltricarbonyliron complex from a vinyl epoxide (1) and $Fe_2(CO)_9$. The resulting complex (2) reacts with benzylamine to give the aminoacyl derivative (3), and oxidation with Ce(IV) affords a β-lactam which can be further elaborated to thienamycin (4).

Scheme 10.1

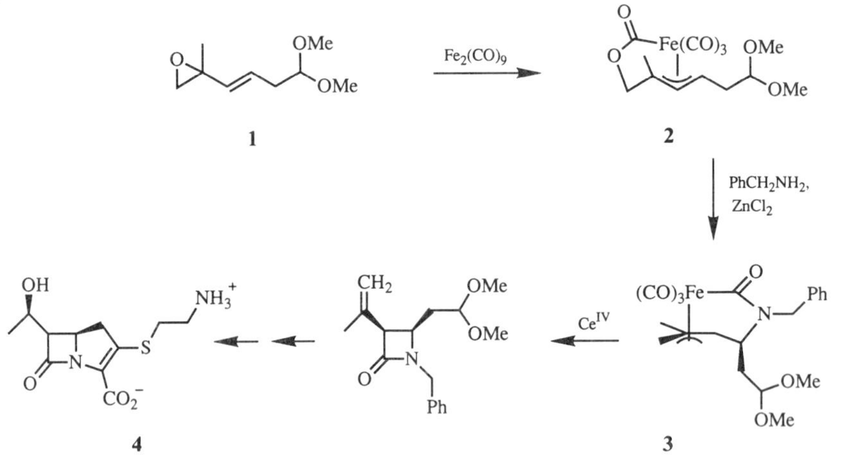

Alkene iron dicarbonyl complexes have also been used as intermediates in carbonylative syntheses of β-lactams,[452] as shown in Scheme 10.2. The alkene complex (5) reacts with ammonia to give a pyrrolidine complex (6) which can be reduced with sodium borohydride to a mixture of stereoisomeric pyrrolidine complexes. Heating leads to insertion of the carbonyl ligand with formation of the diastereomeric chelate complex (7), and finally oxidation in air or with silver oxide yields the β-lactam (8).

Scheme 10.2

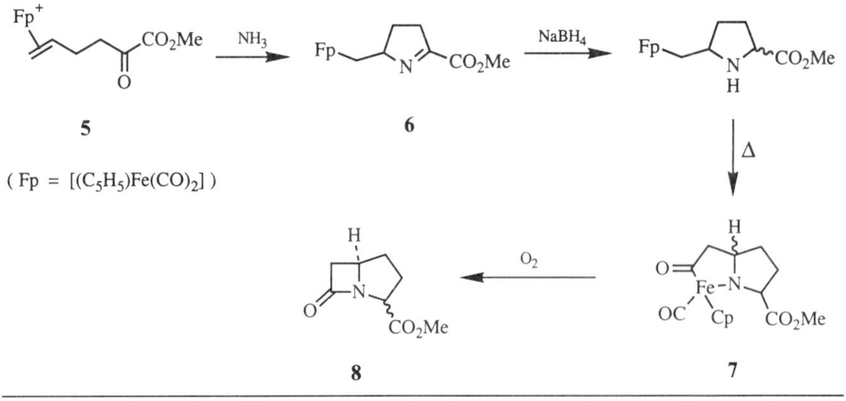

Catalytic carbonylation of 2-bromo-3-propene derivatives (**9**), in the presence of palladium acetate and triphenylphosphine (100°C, 1 bar), affords α-methylene-β-lactams (**10**) in good yields,[453] as shown in Eq. (10.1).

$$
\underset{\mathbf{9}}{\overset{R^2}{\underset{NHR^1}{\diagup}}\hspace{-0.5em}Br} \quad\xrightarrow[\text{Bu}_3\text{N, CO}]{\text{Pd(OAc)}_2, \text{PPh}_3}\quad \underset{\mathbf{10}}{\overset{R^2}{\underset{R^1}{N}}\hspace{-0.5em}O} \tag{10.1}
$$

Synthesis of N-Benzyl-α-methylene-β-lactam ([10]; $R^1 = CH_2Ph$, $R^2 = H$)[453]
A mixture of 2-bromo-3(N-benzyl)aminopropene (9; $R_1 = CH_2Ph$, $R^2 = H$; 1.17 g, 5.16 mmol) and tri-*n*-butylamine (1.19 g, 6.45 mmol) is stirred at 100°C under carbon monoxide (1 bar), in the presence of palladium acetate (0.022 g, 0.1 mmol) and triphenylphosphine (0.10 g, 0.4 mmol) for 5 h. Workup affords the product β-lactam in 67% yield.

Exposure of an azirine to carbon monoxide in the presence of $Pd(PPh_3)_4$ as catalyst affords, under mild conditions (40°C, 1 bar), bicyclic β-lactams (**11**) in reasonable yield [Eq. (10.2)].[454]

$$
\underset{R^2}{\overset{N}{\diagup\!\!\!\diagdown}}\!\!\!\underset{H}{\diagup}R^1 \;+\; CO \quad\xrightarrow{\text{Pd(PPh}_3)_4}\quad \underset{\mathbf{11}}{\overset{H\;\;R^2}{\underset{O}{R^1\diagdown}}} \tag{10.2}
$$

Attempts to obtain *monocyclic* β-lactams from azirines have not been successful, but 2-arylaziridines may be carbonylated in the presence of rhodium-based catalysts to give this type of lactam (Scheme 10.3).[455] The reaction is invariably regiospecific, with no detectable amount of the isomeric lactam (**12**) being formed. Moreover, this type of carbonylation proceeds with complete retention of stereochemistry at both aziridine ring carbons and in the presence of a large excess of a chiral auxiliary (*d*- or

$$
\underset{\mathbf{12}}{\overset{R^2}{\underset{O}{\diagup\!\!\!\!N}}\hspace{-0.5em}R^1}
$$

Scheme 10.3

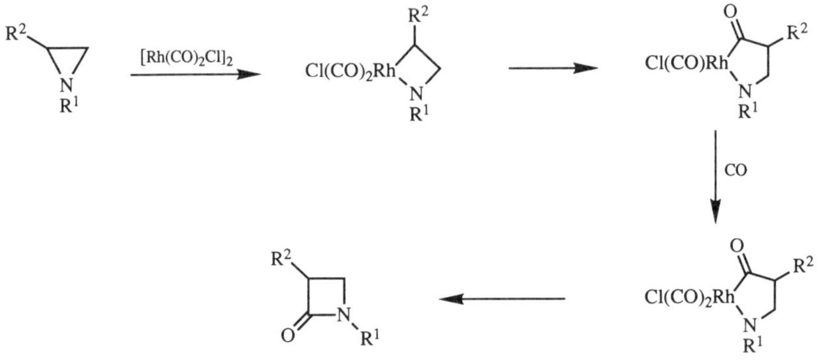

14   **13**

l-menthol), the reaction can even discriminate between enantiomeric aziridines. Chiral β-lactams can thus be obtained with very high enantiomeric excess (over 98%) from racemic starting materials, although chemical yields are obviously limited to a 50% maximum (Scheme 10.4).

Scheme 10.4

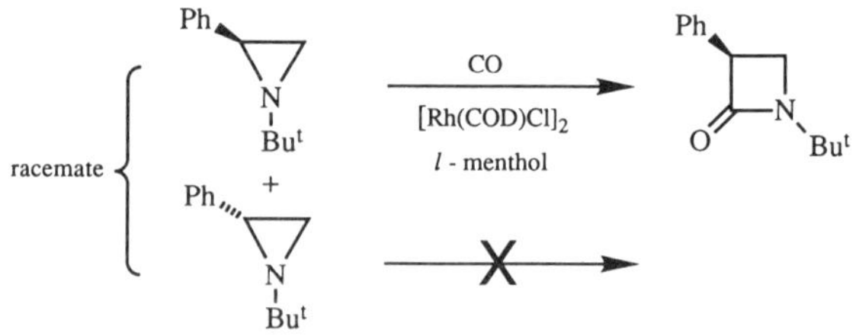

The proposed mechanism, shown in Scheme 10.3, involves oxidative addition of the more substituted $C-N$ bond of the aziridine to rhodium(I), followed by ligand migration to give the acyl rhodium(III) complex (**13**). Reductive elimination affords the β-lactam (**14**) and completes the catalytic cycle.

Azetidine-2,4-diones can be prepared in high yield from aziridinones by carbonylation,[456] using either a rhodium catalyst at 40°C and 30 bar [Eq. (10.3)] or a stoichiometric quantity of $Co_2(CO)_8$ in the absence of carbon monoxide. The mechanism of the catalytic carbonylation is probably similar to that shown in Scheme 10.3 for conversion of aziridines to β-lactams.

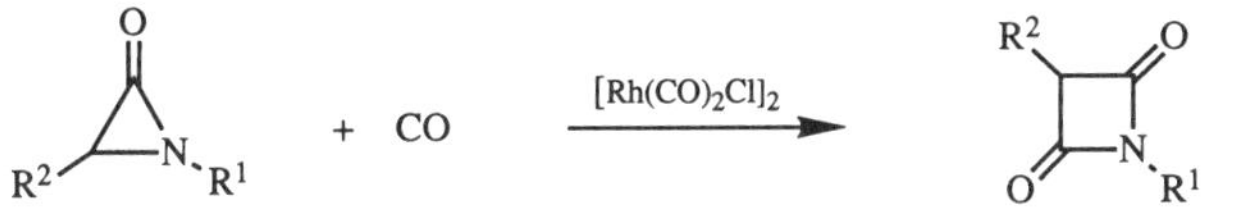

$$(10.3)$$

An interesting, though noncatalytic, approach to the synthesis of mono- and bicyclic β-lactams involves the photochemical reaction of methoxy-carbene–chromium complexes with amines in the presence of sunlight (Scheme 10.5).[457] Yields of β-lactam are generally in the range 40%–80%, and the starting chromium complexes are readily obtained by reaction of chromium hexacarbonyl with alkyl- or aryl-lithium reagents, followed by alkylation with trimethyloxonium tetrafluoroborate.

Scheme 10.5

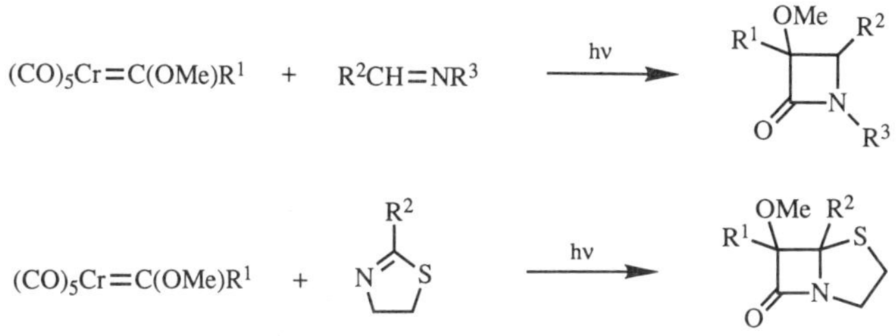

10.3. Five-Membered Rings

Direct carbonylation of four-membered cyclic amines (azetidines), in the presence of cobalt carbonyl as catalyst, has recently been shown to give high yields of pyrrolidinones under relatively mild conditions (90°C, 3.4 bar).[458] As in the corresponding ring expansion of aziridines (Sec. 10.1) the reaction proceeds with high regio- and stereospecificity (Scheme 10.6). Carbonylation of 2-*vinyl* azetidines under these conditions, however, leads to further ring expansion, with formation of seven-membered lactams in good yield (see Sec. 10.5).

Scheme 10.6

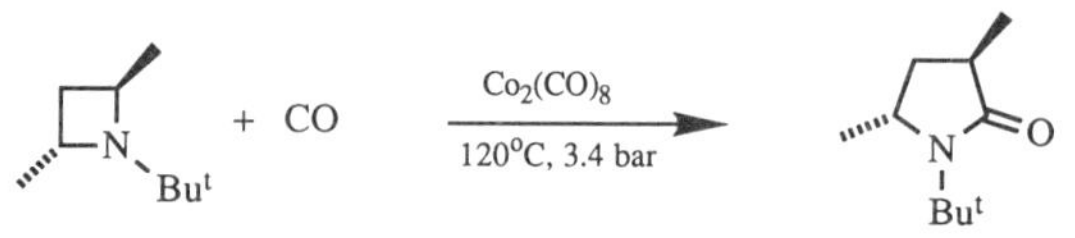

Carbonylation of allylamines has been shown to give pyrrolidinones,[459, 460] as shown in Eq. (10.4). Early work concentrated on the use of soluble cobalt catalysts and generally required vigorous conditions (280°C, 300 bar), although the products were obtained in reasonable yield.[460] More recently, homogeneous rhodium-based catalysts have been found that allow the same type of reaction to proceed under less vigorous conditions (150°C, 140 bar), giving pyrrolidinones in 60%–70% yield.[459] The latter type of catalyst can also be used for an alternative, though closely related, pyrrolidinone synthesis [Eq. (10.5)], involving reaction of an allylic halide with carbon monoxide and an amine.[459]

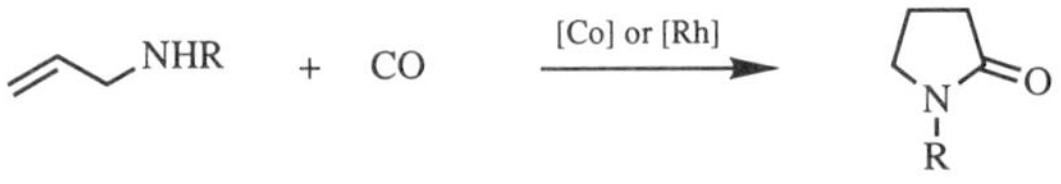

$$\text{(10.4)}$$

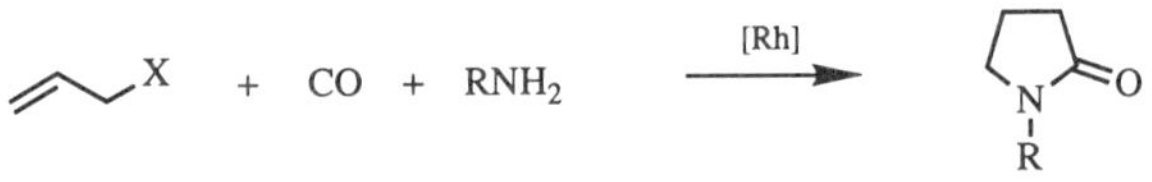

$$\text{(10.5)}$$

Reactions of tetrasubstituted dipropargylamines with carbon monoxide and either water or an alcohol proceed under mild conditions (25°C, 1 bar), in the presence of a palladium-based catalyst, to give carboxylated pyrrolidine derivatives in good yield.[461] The isomeric esters (15) and (16), shown in Eq. (10.6), are formed in a 2:1 ratio.

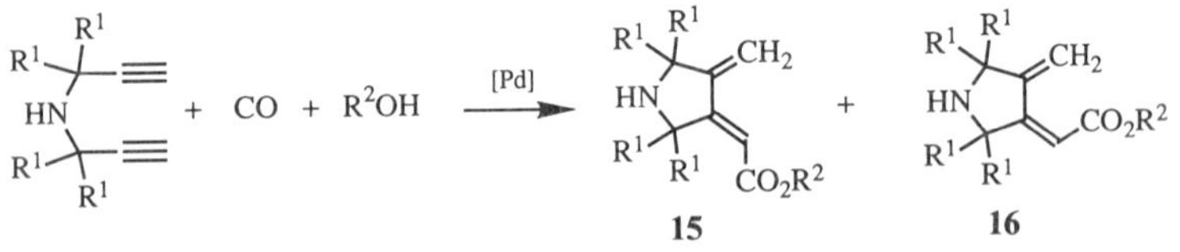

$$\text{(10.6)}$$

The facile *ortho*-palladation reactions of aromatic nitrogen-based ligands such as azobenzenes, tertiary benzylic amines, benzaldimines, and benzaldoximes, provide a group of stable complexes which readily undergo insertion of carbon monoxide into the palladium–carbon bond and thereby give, after workup, a range of heterocyclic carbonyl compounds.[462–464] Azobenzene, for example, reacts with palladium chloride to give a dimeric complex which, on carbonylation,[462] affords 2-phenyl-3-indazolinone (Scheme 10.7). A detailed study of these carbonylation reactions [Eqs. (10.7)–(10.9)] has been carried out by Thompson and Heck,[462] who

Scheme 10.7

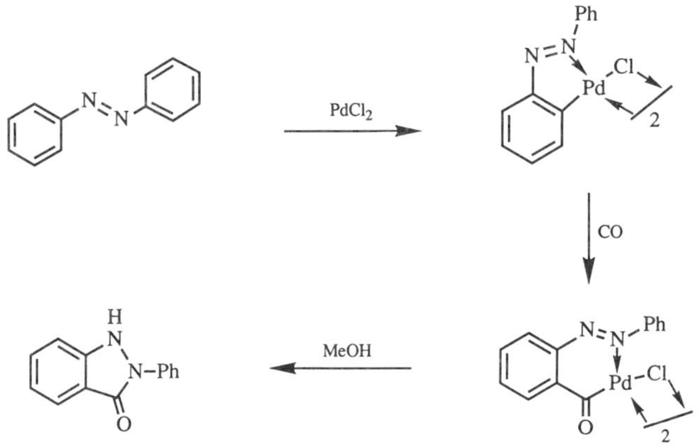

showed that a variety of cyclic products can be obtained in high yield under mild conditions (typically 100°C and 1 bar).

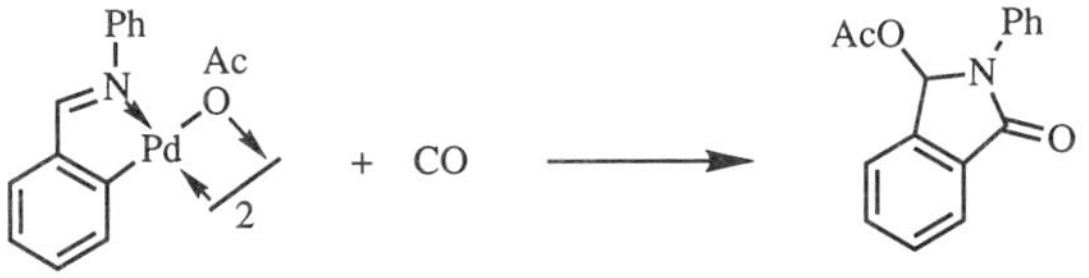

$$(10.7)$$

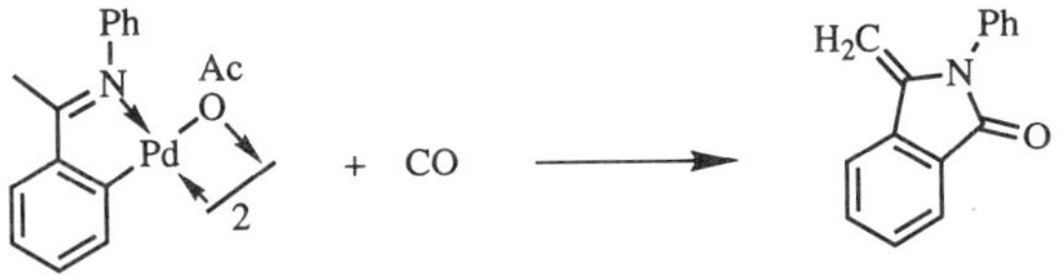

$$(10.8)$$

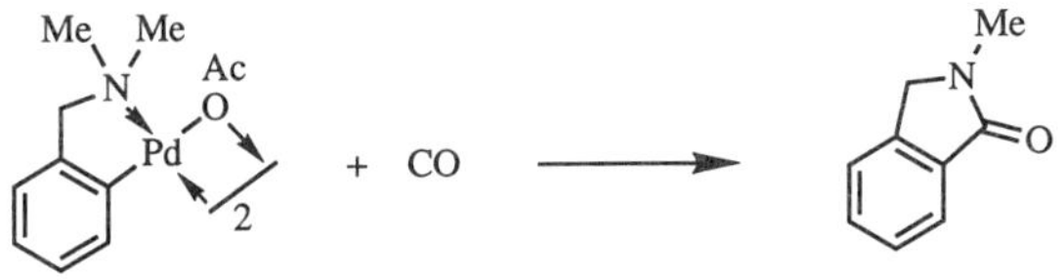

$$(10.9)$$

One obvious drawback with this type of reaction is that it requires a stoichiometric quantity of expensive palladium salt, but, as shown in Eq. (10.10), *catalytic* lactam syntheses *can* be achieved under mild conditions (100°C, 1 bar) if *ortho*-halogenated substrates are available.[463]

$$ \text{(10.10)} $$

17

Synthesis of N-Benzylisoindolin-1-one (17), (R = CH₂Ph)[463]

A mixture of N-benzyl-2-bromobenzylamine (1 mole equivalent), tri-*n*-butylamine (1.1 equivalents), palladium acetate (0.02 equivalents), and triphenylphosphine (0.04 equivalents) is heated at 100°C under carbon monoxide (1 bar) for 26 h. After cooling and extracting with diethyl ether, the product is obtained as colorless crystals (63% yield) by column chromatography on silica gel (1:1 hexane/ether as eluent).

Catalytic reactions can also be achieved using nonhalogenated aromatic substrates, but these generally require much more vigorous conditions [typically 200°C and 200 bar, with $Co_2(CO)_8$ as catalyst]. Under such conditions aromatic Schiff's bases react with carbon monoxide to give isoindolinones in good yield [Eq. (10.11)].[464]

$$ \text{(10.11)} $$

Products from the cobalt-catalyzed reaction of azobenzene with carbon monoxide[465] vary with conditions, as shown in Scheme 10.8. At 190°C and 150 bar, for example, one molecule of carbon monoxide is incorporated to give 2-phenyl-3-indazolinone (**18**) in 55% yield, whereas at 230°C and 150 bar two carbon monoxide molecules react, with cleavage of the N−N bond, to give a dioxo-tetrahydroquinazoline (**19**) in 80% yield.

Palladium-catalyzed cyclization of N-substituted-2-allylanilines in the presence of carbon monoxide and methanol, at ambient temperature and pressure, affords the dihydroindoleacetic ester (**20**) in ca. 70% yield [Eq. (10.12)].[466]

Although much work has been carried out on heterogeneous catalysts derived from mixed-metal clusters, relatively little has been reported on the application of such complexes as homogeneous catalysts or reagents for organic synthesis. An exception is the palladium–iron cluster compound

Scheme 10.8

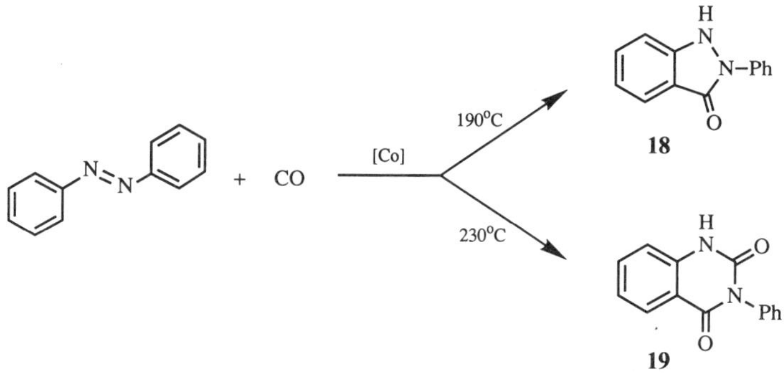

18

19

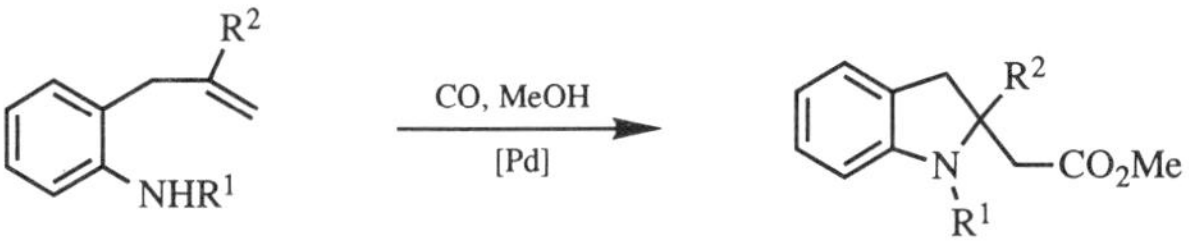

20

(10.12)

(**22**) (derived from the "A-frame" complex (**21**) and $Na_2[Fe(CO)_4]$), which is a much more efficient catalyst for reductive carbonylation of

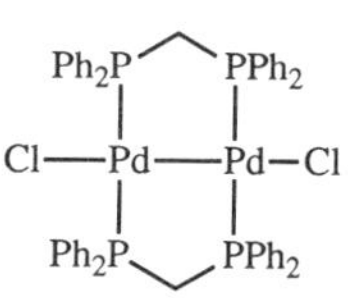

21

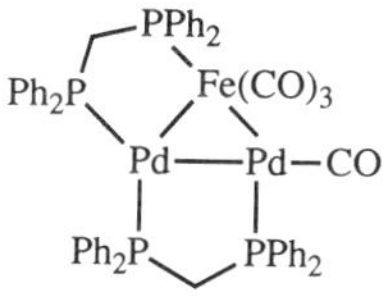

22

2-nitrophenol to benzoxazolinone [Eq. (10.13)] than the conventional mixture of palladium and iron salts.(467)

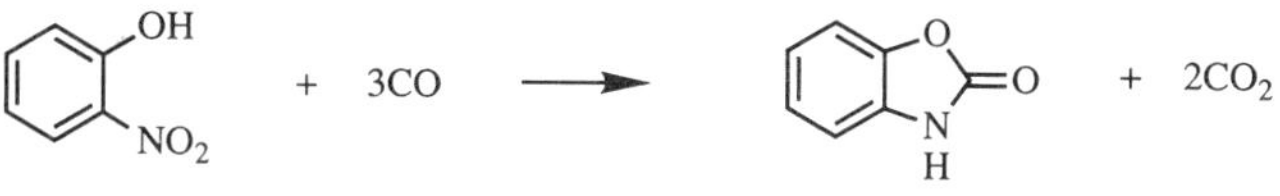

(10.13)

Palladium-catalyzed carbonylation of 2-bromophenyl(alkyl)ketones in the presence of a titanium-isocyanate complex (derived from molecular nitrogen) leads to formation of alkylidenelactams in moderate yield.(468)

The reaction (Scheme 10.9) proceeds via an alkylidenelactone intermediate, which can be isolated if required and converted to the lactam in a separate step. Since enolization of the reactant alkylketone is clearly required yields are, not unexpectedly, enhanced by the presence of electron-withdrawing groups (nitrile, sulfonyl) α to the carbonyl group, but the resulting alkylidene lactams suffer hydration of the exocyclic double bond on aqueous workup.

Scheme 10.9

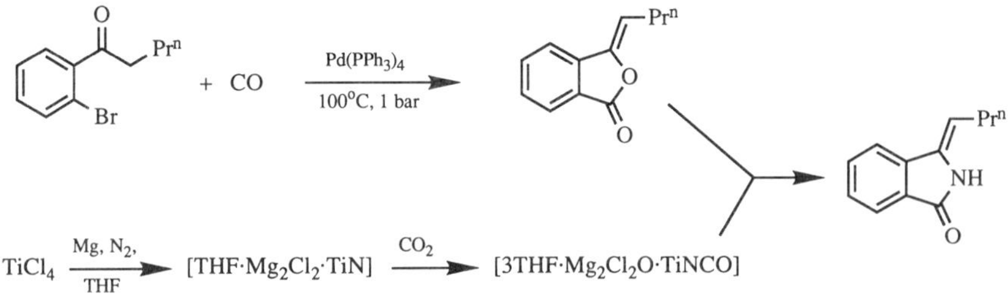

10.4. Six-Membered Rings

Carbonylation of the unsaturated amide (**23**) under vigorous conditions (250°C, 300 bar) in the presence of $Co_2(CO)_8$ as catalyst affords the cyclic imide (**24**) in 41% yield [Eq. (10.14)].[469] With *acyclic* α,β-

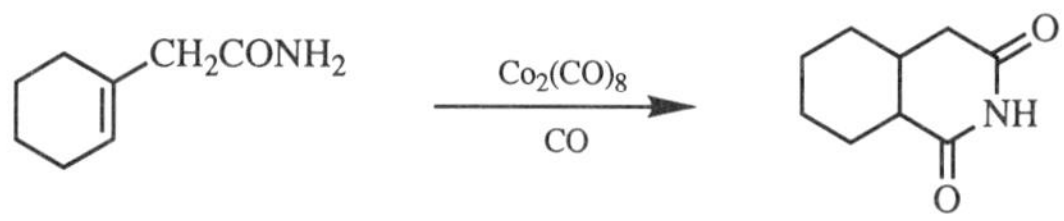

$$(10.14)$$

unsaturated carboxamides as substrates this type of reaction yields a mixture of cyclic five- and six-membered imides,[469] in proportions that depend on the degree of substitution at the acrylic double bond [Eq. (10.15)].

Carbonylative cyclization of N-alkyl-2-bromophenethylamines, in the presence of a palladium acetate/triphenylphosphine catalyst,[463] gives tetrahydroisoquinolin-1-ones in good yield under mild conditions [Eq. (10.16)].

$$(10.15)$$

R = H,	68%	19%
R = Me,	0%	67%

$$(10.16)$$

25

Synthesis of N-Benzyl-1,2,3,4-tetrahydroisoquinolin-1-one (25),
(R = CH$_2$Ph)[463]

A mixture of N-benzyl-2-bromophenethylamine (1 mmol), tri-*n*-butylamine (1.1 mmol), palladium acetate (0.02 mmol), and triphenylphosphine (0.04 mmol) is heated at 100°C under carbon monoxide (1 bar) for 26 h. After cooling, diethyl ether is added to the mixture, and the resulting solution is washed with 10% hydrochloric acid and the organic phase separated and dried over magnesium sulfate. Removal of the solvent under vacuum followed by column chromatography on silica gel (1:1 benzene/ether as eluent) affords the product (25) (R = CH$_2$Ph) in 65% yield.

A range of berbin-8-one derivatives have been synthesized by stoichiometric reactions of metal carbonyl complexes with 2-halogeno-benzyl isoquinolines,[470, 471] as shown in Eq. (10.17). Both Co$_2$(CO)$_8$ and

$$(10.17)$$

Fe$_3$(CO)$_{12}$ are may be used for such reactions, but the optimum reagent for maximizing yield and minimizing competing reactions such as dehalogenation depends very much on the substrate. Thus, for the reaction shown in Eq. (10.17), Co$_2$(CO)$_8$ is the preferred reagent, whereas Fe$_3$(CO)$_{12}$ is more effective in Eq. (10.18). The latter reaction can of course also be achieved by palladium-catalyzed carbonylation in the presence of a tertiary amine.[462]

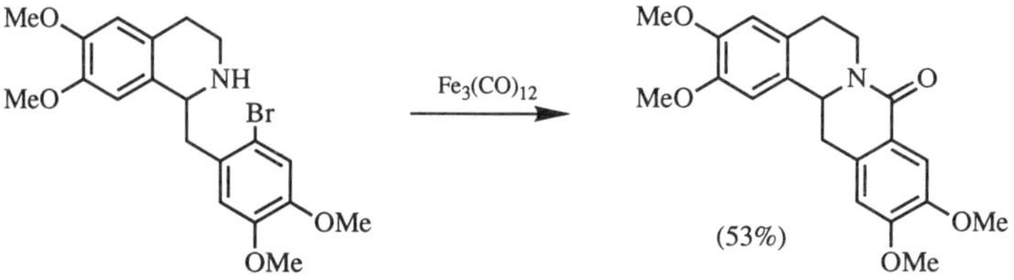

$$(10.18)$$

(53%)

Aniline reacts with carbon tetrachloride and carbon monoxide at 150°C and 150 bar, in the presence of a metal carbonyl catalyst such as $Cr(CO)_6$ or $Mo(CO)_6$, to give an amidine [Eq. (10.19)] and a small amount of 3-phenylquinazolinone. However, under these conditions *substituted* anilines tend to form heterocyclic carbonyl compounds as the major products.[472] 3-Chloroaniline, for example, gives the dichloro-quinazolinone (**26**) in 80% yield [Eq. (10.20)]. 3-Bromo- and 3,4-

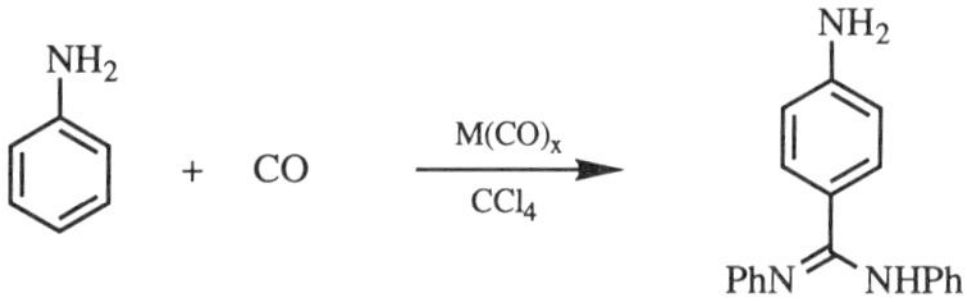

$$(10.19)$$

26

$$(10.20)$$

dichloro-aniline also give quinazolinones, but 3- and 4-methyl anilines give the corresponding quinazolinediones (**27**) [Eq. (10.21)]. The nature of such reactions is at present unclear, but their sensitivity to relatively slight changes of substituent suggests that free-radical chemistry may well be involved.

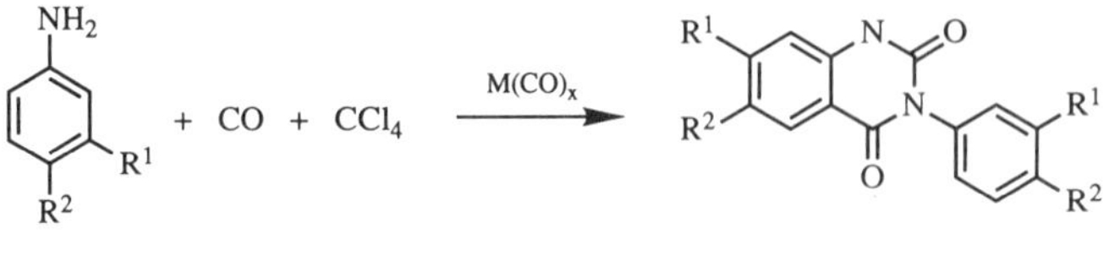

$(R^1, R^2 = H, Me)$ **27**

$$(10.21)$$

10.5. *Seven-Membered Rings*

Benzoazepinone derivatives (**28**) have been synthesized in good yield from 2-bromoaryl alkylamines [Eq. (10.22)] using the same type of palladium catalyst as for the corresponding five- and six-membered benzolactams (Sec. 10.3 and 10.4).[463] This palladium-catalyzed reaction has

$$(10.22)$$

recently been extended to the synthesis of various antibiotics[473, 474]; a key step in the route to anthramycin, for example, involves carbonylation of the intermediate amino-amide (**29**), as shown in Eq. (10.23).

$$(10.23)$$

Tetrahydroazepinones are formed in excellent yield on carbonylation of 2-vinylazetidines, using cobalt carbonyl as catalyst [Eq. (10.24)]. The reaction is tolerant of a number of functional groups including ketone, ester, and nitrile, and proceeds under relatively mild conditions (85–90°C, 3.4 bar).[458]

$$(10.24)$$

Chapter 11

Decarbonylation Reactions

11.1. Introduction

In the absence of a suitable catalyst elimination of carbon monoxide from an organic carbonyl compound is often difficult to accomplish, although gas phase photoinduced decarbonylation of aliphatic aldehydes is relatively well documented [Eq. (11.1)].[475] Related photochemical reactions also

$$RCHO \xrightarrow{\text{hv}} RH + CO \qquad (11.1)$$

result in decarbonylation and ring contraction of some cyclic ketones,[476] as illustrated in Eq. (11.2), but although such reactions may appear to have synthetic potential, in practice they suffer from excessive side reactions when carried out in the liquid phase and often give intractable mixtures of products. Some preparative work has been reported in which decarbonyl-

ation was successfully achieved by a combination of thermal and photochemical techniques.[477] Good yields of heptane, for example, are obtained by photolysis of 2-ethylhexanal at 140°C–145°C in the presence of dibenzyldisulfide [Eq. (11.3)], but such procedures seem unlikely to find more general application.

$$\text{(CHO structure)} \xrightarrow[\text{140°C, 20 h}]{hv, (PhCH_2S)_2} \text{(alkane structure)} + CO \quad (11.3)$$

Some classes of carbonyl compound do, however, undergo relatively facile decarbonylation. Catenated polycarbonyl compounds such as α-keto esters cleanly eliminate carbon monoxide when strongly heated [to about 250°C, Eq. (11.4)], although even here the reaction is facilitated by the presence of powdered glass, suggestive of a surface catalytic process.[478]

$$\text{(cyclohexanone α-keto ester structure)} \xrightarrow{150°C} \text{(cyclohexanone ester structure)} + CO \quad (11.4)$$

Photolysis of α-keto esters also results in decarbonylation, but now with the formation of two carbonyl-containing fragments [Eq. (11.5)].[479]

$$RCOCO_2CHR'R'' \xrightarrow{hv} RCHO + R'COR'' + CO \quad (11.5)$$

Treatment of an α-keto carboxylic acid with hot sulfuric acid leads to decarbonylation and formation of the next lower carboxylic acid, while the central carbonyl group of diphenyl triketone is very readily eliminated on treatment with acidic catalysts such as aluminum trichloride,[480] as shown in Eq. (11.6).

$$PhCOCOCOPh \xrightarrow{AlCl_3} PhCOCOPh + CO \quad (11.6)$$

Of greater synthetic value are the decarbonylations of Diels–Alder adducts derived from substituted cyclopentadienones and either alkenes or alkynes. Thus, the bridging carbonyl groups in bicyclo[2,2,1]hept-2-en-7-ones (1) and bicyclo[2,2,1]hept-2,5-dien-7-ones (2) can be eliminated to give cyclohexadienes and arenes, respectively. Arenes often result even from the former reaction, since the somewhat forcing conditions required for decarbonylation of (1) can also lead to aromatization.[481, 482] The alkyne-derived compounds (2), however, lose carbon monoxide very readily, often at room temperature, to give aromatic compounds directly, and this route has been successfully used to prepare a number of highly substituted benzene derivatives.[483]

Compounds that undergo clean thermal decarbonylation to give useful products are thus relatively few in number, but *transition metal catalyzed* decarbonylation offers very much wider scope. Since many of the elementary reactions involved in catalytic carbonylation are readily reversible, it is not perhaps too surprising that transition metals and their complexes should, under appropriate conditions, also catalyze the elimination of carbon monoxide from organic compounds.

The first reactions of this type to be studied involved decarbonylation of aldehydes in the presence of a heterogeneous palladium catalyst, and although high temperatures are required (typically 200–250°C), reaction times are short and yields good. More recently, however, homogeneous rhodium-based catalysts have been developed that allow aldehyde decarbonylation under much milder conditions (120–150°C), and since these are both convenient and cost effective, they may well become the catalysts of choice. This chapter explores the scope of transition metal mediated decarbonylation, and although emphasis is placed on the newer homogeneous catalysts, heterogeneous systems (particularly palladium on charcoal) are also extensively discussed.

11.2. Decarbonylation of Aldehydes

Although removal of an aldehydic carbonyl group can sometimes be achieved by oxidation to the corresponding carboxylic acid followed by pyrolytic decarboxylation, in practice this approach is limited to some aromatic acids, and a few aliphatic acids such as malonic and trichloroacetic where the carbanion formed on loss of carbon dioxide from the carboxylate salt is stabilized by neighboring electron-withdrawing groups. Even heteroaryl carboxylic acids often give only very low yields of decarboxylated products. In contrast, transition metal catalyzed decarbonylation of aldehydes is a widely applicable, high-yielding reaction, though for aliphatic aldehydes bearing β-hydrogens the decarbonylation product may be an alkene (effectively reversing the hydroformylation process) rather than an alkane.

11.2.1. Aromatic Aldehydes

The absence of β-hydrogens and the stability of the arene product combine to make catalytic decarbonylation of aromatic aldehydes [e.g., Eq. (11.7)] a very straightforward reaction. Hawthorne and Wilt explored the scope of this reaction and found that, using 5% palladium on charcoal as catalyst and working at the boiling point of the aldehyde, reaction times

$$\text{(11.7)}$$

varied from 15 min to 2 h. At least for simple aldehydes, decarbonylation was essentially quantitative.[484] Workup consisted simply of distillation (for volatile liquid products) or dissolution in a suitable solvent, filtration to remove the catalyst, and recrystallization. Table 11.1 illustrates the scope of this reaction.

Table 11.1.   Decarbonylation of Aromatic Aldehydes with Palladium on Charcoal

Substrate	Reaction conditions	Product (yield)	Reference
	179 °C, 1 h	PhH (78%)	a
	205 °C, 0.5 h	PhNO$_2$ (79%)	a
	180 °C, 1 h	(68%)	a
	180 °C, 1 h	(80%)	a, b
	245 °C, 0.5 h	(84%)	a

[a] J. O. Hawthorne and M. H. Wilt, J. Org. Chem., 25, 2215 (1960).
[b] 2-Naphthaldehyde failed to react at 250 °C.
[c] G. W. Kenner, M. A. Murray, and C. M. B. Taylor, Tetrahedron, 1, 265 (1957).

Table 11.1. (Continued)

Substrate	Reaction conditions	Product (yield)		Reference
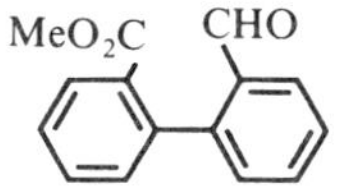MeO$_2$C CHO	250 °C, 0.5 h	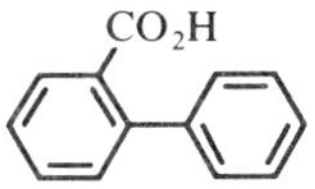CO$_2$H	(20%)	*c, d*
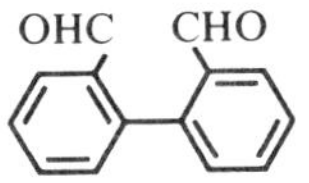OHC CHO	245 °C, 1 h	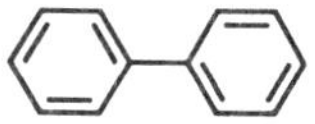	(97%)	*a*
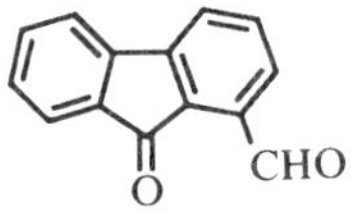O CHO	250 °C, 0.25 h	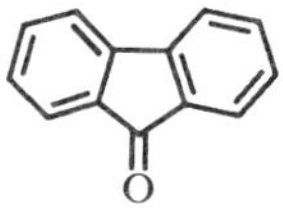O	(82%)	*a*
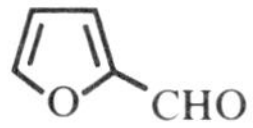O CHO	170 °C, 1 h	O	(90%)	*e*
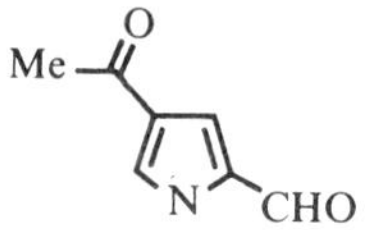Me O N CHO	210 °C, 0.75 h	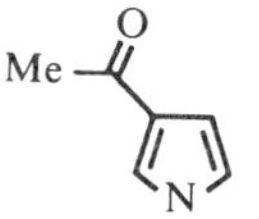Me O N	(67%)	*f*
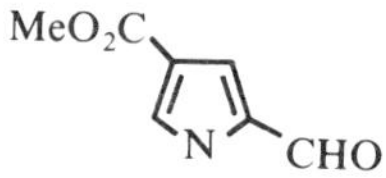MeO$_2$C N CHO	210 °C, 0.75 h	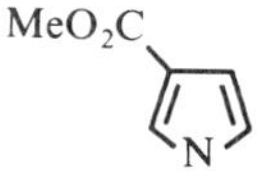MeO$_2$C N	(84%)	*f*
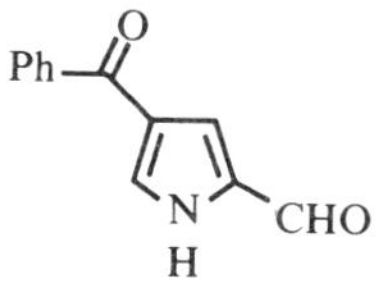Ph O N H CHO	210 °C, 0.75 h	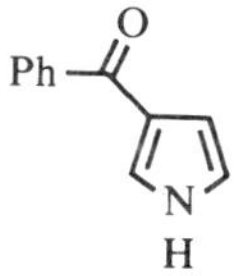Ph O N H	(68%)	*f*
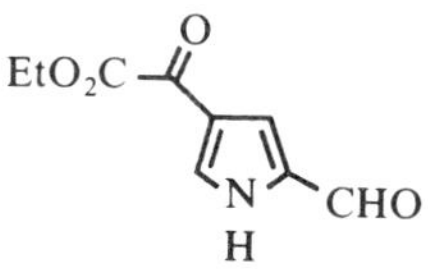EtO$_2$C O N H CHO	Reflux, 6.0 h	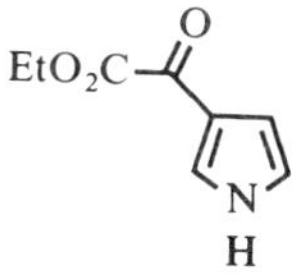EtO$_2$C O N H	(92%)	*g*
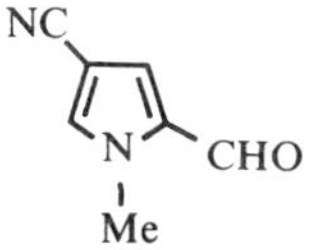NC N Me CHO	210 °C, 0.75 h	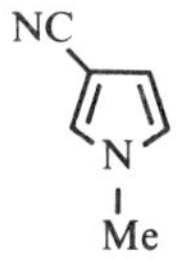NC N Me	(83%)	*h*

[d] Longer reaction times give higher yields. The corresponding acid-aldehyde does not decarbonylate, possibly because 3-hydroxydiphenide is formed.

[e] The presence of K_2CO_3 is beneficial: P. Lejemble, A. Gaset, and P. Kalck, *Biomass*, **4**, 263 (1984); P. Lejemble, Y. Maire, A. Gaset, and P. Kalck, *Chem. Lett.*, 1403 (1983).

[f] H. J. Anderson, C. E. Loader, and A. Foster, *Can. J. Chem.*, **58**, 2527 (1980).

[g] B. J. Demopoulos, H. J. Anderson, C. E. Loader, and K. Faber, *Can. J. Chem.*, **61**, 2415 (1983).

[h] C. E. Loader and H. J. Anderson, *Can. J. Chem.*, **59**, 2673 (1981).

Decarbonylation of furfural (obtained from oat-hull pentosans) is carried out on an industrial scale as a route to furan and thence to THF and butane-1,4-diol. Although, as shown in Table 11.1, palladium catalysis[485–492] allows the reaction to proceed at only 200°C, the commercial process in fact operates in the gas phase at around 400°C and uses an oxide catalyst. Poisoning of palladium by impurities present in commercial furfural appears to be the main reason why palladium catalysis is not used in the industrial process.

2-Formyl pyrroles carrying an electron-withdrawing group in the 4-position are smoothly decarbonylated over palladium on carbon, either in the absence of solvent or using a high-boiling diluent such as mesitylene.[493–495] An advantage of the solvent-free method is that the product can be sublimed directly from the reaction mixture. When an electron-withdrawing 4-substituent is absent, however, decarbonylation by an excess of Raney nickel is preferred over palladium catalysis, perhaps because more basic pyrroles bind irreversibly to the palladium surface.[495] Raney nickel is not generally recommended for decarbonylation because of its adsorbed hydrogen content, which can lead to hydrogenation and by-product formation, so that reaction with compound (**3**), for example, leads to a mixture of products. However, decarbonylation of (**3**) at the 2-position using palladium on charcoal is smooth and essentially quantitative, and subsequent decarbonylation of the product keto-ester with Raney nickel [Eq. (11.8)] can be successfully achieved using a two-phase solvent system (toluene/aqueous ethanol).

The nature of the support on which palladium is dispersed can strongly influence its decarbonylation activity. For example, the higher degree of metal dispersion obtained with activated charcoal as support compared with, say, barium sulfate, calcium carbonate, or silica, is reflected in significantly higher catalytic activity for decarbonylation of furfural.

Generalized Procedure for Decarbonylation of Aromatic Aldehydes[484]
    A mixture of the aldehyde (5–10 g) and 5% palladium on charcoal (0.05–0.1 g) is placed in a small flask fitted with a miniature distillation column. The air in the flask is displaced by a flow of nitrogen or carbon dioxide (this flow is maintained at a low rate during the reaction) and the

mixture is heated to the boiling point. The product is collected as it distils from the reaction mixture.

Although a generally useful technique, heterogeneous palladium-catalyzed aldehyde decarbonylation does require high temperatures and as a result there is significant functional group sensitivity. Double-bond migration and transfer-hydrogenation reactions[496] may occur, for example, and phenolic oxygen can be lost.[497] Homogeneous catalysts often operate under much milder conditions, but those based on palladium and nickel have found only very limited application. *Rhodium* complexes, on the other hand, have emerged as highly efficient decarbonylation reagents and catalysts, and early work (to 1976) on decarbonylation using Wilkinson's complex, $RhCl(PPh_3)_3$, is the subject of an excellent review by Baird.[498] This compound reacts stoichiometrically with aromatic aldehydes, at much lower temperatures than those usually required with palladium on carbon, to form arenes and the yellow, crystalline rhodium carbonyl complex $RhCl(CO)(PPh_3)_2$, as shown in Eq. (11.9).

$$ArCHO + RhCl(PPh_3)_3 \longrightarrow ArH + RhCl(CO)(PPh_3)_2 + PPh_3 \quad (11.9)$$

Despite the high cost of rhodium and its complexes, the effectiveness of $RhCl(PPh_3)_3$ has led to its being used extensively as a decarbonylation reagent for aromatic aldehydes. Some typical examples[499–501] are given in Eqs. (11.10)–(11.12), the last of these being part of a route to 19-nor-steroids.[501]

(11.10)

(11.11)

(11.12)

A number of methods are available$^{(502, 503)}$ for regenerating RhCl(PPh$_3$)$_3$ from the carbonyl complex RhCl(CO)(PPh$_3$)$_2$. Probably the most effective approach is that shown in Scheme 11.1, where reaction with benzyl chloride gives a rhodium(III) benzyl complex which on treatment with triphenylphosphine in ethanol affords RhCl(PPh$_3$)$_3$ in high yield.

Scheme 11.1

RhCl(CO)(PPh$_3$)$_2$ + PhCH$_2$Cl $\longrightarrow$ RhCl$_2$PPh$_3$(η^3-CH$_2$Ph)

RhCl$_2$PPh$_3$(η^3-CH$_2$Ph) $\xrightarrow{\text{PPh}_3,\ \text{EtOH}}$ RhCl(PPh$_3$)$_3$

Rather than recycling stoichiometric quantities of rhodium, it would clearly be more effective if the reaction were genuinely catalytic. Rhodium-catalyzed decarbonylation using Wilkinson's complex can in fact be achieved for some aldehydes and acyl chlorides by operating at or above 200°C, when loss of carbon monoxide occurs from RhCl(CO)(PPh$_3$)$_2$ to give active decarbonylation species. Unfortunately, such forcing conditions offer little advantage over the more conventional palladium catalysts, and Wilkinson's complex is still most often used as a stoichiometric decarbonylation *reagent.*$^{(504–507)}$

Replacing triphenylphosphine by chelating diphosphine ligands, however, leads$^{(508, 509)}$ to cationic rhodium(I) complexes of the type [Rh(Ph$_2$P(CH$_2$)$_n$PPh$_2$)$_2$]$^+$, which are very active decarbonylation catalysts in the temperature range 120–150°C. The most effective compound of this type, [Rh(Ph$_2$P(CH$_2$)$_3$PPh$_2$)$_2$]$^+$, forms only a very weakly bound carbonyl complex,$^{(510, 511)}$ and, since loss of carbon monoxide from the metal is rate determining in this type of decarbonylation, the chelate complex is about two orders of magnitude more active than RhCl(PPh$_3$)$_3$. The reaction mixture must be continuously purged with high-purity nitrogen, not only to remove carbon monoxide from the system and thereby drive the equilibrium in the direction of decarbonylation, but also to prevent ingress of oxygen, which deactivates the catalyst even at

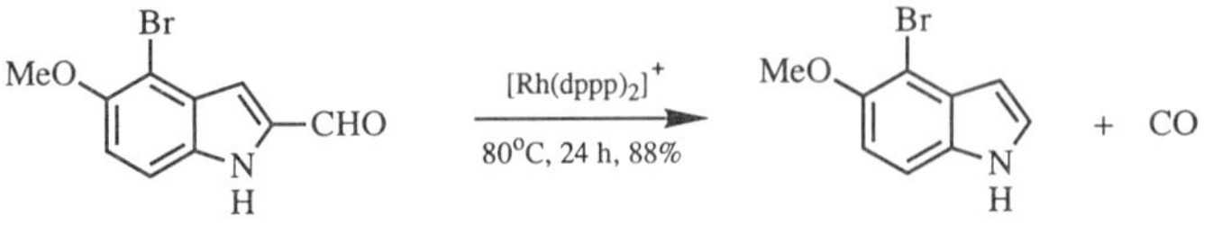

$$\xrightarrow[\text{80°C, 24 h, 88\%}]{\text{[Rh(dppp)}_2]^+}$$

$+ \ CO \quad (11.13)$

quite low levels. In the example given below,[512] the cationic catalyst is generated *in situ* by reaction of the readily available complex $RhCl(CO)(PPh_3)_2$ with commercially supplied $Ph_2P(CH_2)_3PPh_2$.

Decarbonylation of 4-Bromo-5-methoxyindole-2-carboxaldehyde [Eq. (11.13)][512]

To a hot (80°C) solution of $RhCl(CO)(PPh_3)_2$ (0.10 g, 0.145 mmol) in xylene (100 cm³), under a slow purge of nitrogen, is added 1,3-bis(diphenylphosphino)propane (0.13 g, 0.315 mmol), and the reaction mixture is stirred for 30 min. During this time, yellow crystals of the chloride salt of the cationic complex $[Rh(Ph_2P(CH_2)_3PPh_2)_2]^+$ are formed. The aldehyde (3.67 g, 14.5 mmol) is then added, and the mixture is refluxed for 24 h under nitrogen. The solvent is removed, the solid dissolved in 1:1 ethyl acetate/hexane, the solution filtered through a little fine silica gel, and after evaporation of the solvent the product is recrystallized from chloroform/ hexane. Yield 3.03 g, (92%), mp 111–112°C.

11.2.2. Aliphatic Aldehydes

Aliphatic aldehydes may be decarbonylated using the reagents and catalysts described in the previous section for aromatic substrates, although, as shown in Scheme 11.2, the situation is more complicated for aldehydes having β-hydrogens. This is because β-hydrogen elimination from the metal alkyl intermediate (pathway B) can now lead to formation of alkene and hydrogen as well as the parent alkane (pathway A).

Scheme 11.2

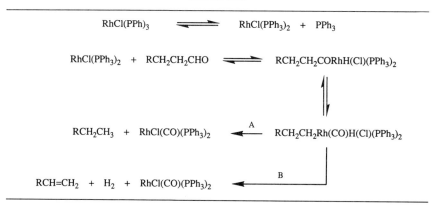

Where β-hydrogens are absent, however, stoichiometric decarbonylation with RhCl(PPh$_3$)$_3$ can proceed with retention of both geometrical and stereochemical configuration,[513–515] as shown in Eqs. (11.14)–(11.16). The

$$\text{(11.14)}$$

retention of aldehydic deuterium in Eq. (11.15) and optical activity in Eq. (11.16) both attest to the high stereoselectivity of the decarbonylation process.

$$\text{(11.15)}$$

$$\text{(11.16)}$$

The tolerance of RhCl(PPh$_3$)$_3$ towards protecting groups such as ester functions is particularly good, as exemplified in Eqs. (11.17)–(11.19). These reactions represent important steps in routes to disaccharides,[516] desoxypodocarpate,[517] and grandisol,[518] respectively.

$$\text{(11.17)}$$

$$\text{(11.18)}$$

$$\text{(11.19)}$$

For sterically hindered aldehydes higher temperatures may be required [Eq. (11.20)]. Here benzonitrile is a useful solvent both because of its high boiling point (160°C) and because of its ability to cleave the chloride-bridged dimer $[RhCl(PPh_3)_2]_2$. This rather insoluble compound is often formed by decomposition of $RhCl(PPh_3)_3$ at high temperatures and is normally inactive for decarbonylation, but in benzonitrile as solvent the dimer itself may even be used instead of Wilkinson's complex.[505]

$$\underset{\text{CHO}}{\text{(structure)}} \xrightarrow[\text{PhCN, 160°C}]{\text{RhCl(PPh}_3)_3} \text{(structure)} \qquad (11.20)$$

Elimination of β-hydrogen does not occur in Eq. (11.20) as this would require double-bond formation at a bridgehead carbon, but more often, as in the synthesis of occidentalol [Eq. (11.21)] decarbonylation of aliphatic aldehydes is accompanied by extensive dehydrogenation.[519] This type of

$$\text{(structure)} \xrightarrow{\text{RhCl(PPh}_3)_3} \text{(structure)} \qquad (11.21)$$

reaction can be used industrially to regenerate propene from 2-methyl-propanal, enabling the less desirable isomeric product of hydroformylation to be decomposed and the propene recycled,[520] as shown in Eq. (11.22).

$$(CH_3)_2CHCHO \longrightarrow CH_3CH{=}CH_2 + CO + H_2 \qquad (11.22)$$

Alkene formation is particularly favored by high temperatures, so that the very active cationic rhodium catalyst $[Rh(Ph_2P(CH_2)_3PPh_2)_2]^+$, which can be used at relatively low temperatures, has only a slight tendency to form alkenes. Decarbonylation of heptanal with this catalyst, for example, gives only hexane,[509] whereas use of $RhCl(PPh_3)_3$ requires higher temperatures and leads[505] to a mixture of hexene (14%) and hexane (86%).

Rhodium-promoted decarbonylation of aldose sugars to the next lower alditol is a very recent discovery.[521] The preferred reagent is $RhCl(PPh_3)_3$, and interestingly, alkene formation does not occur. The dipolar aprotic solvent N-methyl pyrrolidone is favored, perhaps because the sugar has a significant equilibrium concentration of the open-chain

Table 11.2.   Stoichiometric Decarbonylation of Aldoses to Alditols[a]

Aldose	Product	Reaction time (h)	Yield (%)
D-Glycero- D-gulo-Heptose	Glucitol	3.5	88
Glucose	Arabinitol	5.0	88
Arabinose	Erythritol	2.0	84
2-Deoxyribose	1-Deoxyerythritol	0.5	90
Glyceraldehyde	Ethylene glycol	0.5	95

[a] Equimolar amounts of aldose and RhCl(PPh$_3$)$_3$ at 130 °C under nitrogen in N-methylpyrrolidone. Data from M. A. Andrews and S. A. Klaeren, *J. Chem. Soc., Chem. Commun.*, 1266 (1988).

aldehyde form in this solvent, in contrast to an almost negligible amount in aqueous solution. Decarbonylations [e.g., of glucose; Eq. (11.23)] occur smoothly at about 130°C, and proceed with complete retention of stereochemistry. Available data for a range of sugars are summarized in Table 11.2.

$$(11.23)$$

A number of unsaturated aldehydes undergo cyclization rather than decarbonylation on treatment with RhCl(PPh$_3$)$_3$. This type of reaction gives cyclic ketones, effectively by addition of the aldehydic C−H group across the C=C double bond, so that pent-4-enal, for example, yields cyclopentanone.[522]

11.3.   Decarbonylation of Acyl Chlorides

11.3.1.   General Considerations

Acyl chlorides are readily decarbonylated by reaction with a number of transition metal complexes, although the product obtained depends strongly on the type of substrate (aliphatic or aromatic), the transition metal involved, and the reaction conditions. With palladium- and rhodium-based systems chlorocarbons or alkenes (and HCl) are usually obtained, but reactions of acyl halides with nickel complexes often lead to dehalogenated coupling products and only partial elimination of carbon monoxide.

The role of $RhCl(PPh_3)_3$ in decarbonylation of acyl halides has been extensively investigated, and a number of early reports of facile halocarbon formation by stoichiometric, low-temperature decarbonylation of aroyl chlorides (see Sec. 11.3.3) are now known to be erroneous. This complex certainly reacts stoichiometrically with acyl chlorides under mild conditions (below 100°C) to form relatively stable and isolable rhodium(III) chloro–acyl complexes,[523, 524] but at higher temperatures the subsequent fate of such complexes varies widely with the nature of the acyl group.

11.3.2. Aliphatic Acyl Chlorides

Aliphatic acyl halides lacking β-hydrogens react smoothly and stoichiometrically with $RhCl(PPh_3)_3$ to give the corresponding chloro-alkanes,[525] as shown, for example, in Eq. (11.24). Although some

$$CF_3Ph\overset{*}{C}HCOCl + RhCl(PPh_3)_3 \longrightarrow \underset{\text{(racemic)}}{CF_3PhCHCl} + RhCl(CO)(PPh_3)_2 \quad (11.24)$$

reactions of this type proceed with retention of configuration,[526] racemization [as shown in Eq. (11.24) for the decarbonylation of $CF_3PhCHCOCl$] is not uncommon and may reflect the strength of the $Rh-Cl$ bond which results in only slow elimination of RCl. This in turn leads to racemization because the rapid equilibrium between acyl and alkyl–carbonyl complexes shown in Scheme 11.3 is not 100% stereospecific, and multiple interconversions lead to progressive loss of stereochemistry. In contrast, aldehyde decarbonylation appears to involve fast reductive elimination of RH, allowing no time for racemization. The reaction mechanism for decarbonyl-

---

Scheme 11.3

$$RhCl(PPh_3)_3 \; \rightleftharpoons \; RhCl(PPh_3)_2 + PPh_3$$

$$RhCl(PPh_3)_2 + PhCH_2COCl \; \rightleftharpoons \; PhCH_2CORhCl_2(PPh_3)_2$$

$$PhCH_2CORhCl_2(PPh_3)_2 \; \rightleftharpoons \; PhCH_2Rh(CO)Cl_2(PPh_3)_2$$

$$PhCH_2Rh(CO)Cl_2(PPh_3)_2 \; \longrightarrow \; PhCH_2Cl + Rh(CO)Cl(PPh_3)_2$$

---

ation of phenylacetyl chloride to benzyl chloride has been thoroughly investigated[523] and is outlined in Scheme 11.3.

When β-hydrogens *are* present, decarbonylation of aliphatic acyl chlorides generally proceeds with much higher selectivity for alkene formation than does the corresponding decarbonylation of aldehydes. This again reflects the relative slowness of reductive elimination of R−Cl, which, unlike reductive elimination of R−H, cannot compete with the rapidity of β-hydrogen elimination and alkene formation. This means that stereochemical information is generally retained when alkenes are formed, so that decarbonylation of the *eyrthro-* and *threo*-isomers of 2,3-diphenylbutanoyl chloride, for example, leads to selective formation of the two corresponding geometric isomers of methylstilbene [Eqs. (11.25) and (11.26)].[527]

$$\underset{\text{Ph}}{\overset{\text{Ph}}{\underset{\text{H}}{\overset{\text{ClOC}}{\diagup}}}}\quad\longrightarrow\quad\underset{\text{Ph}}{\overset{\text{Me}}{\diagup}}\overset{\text{Ph}}{\underset{\text{H}}{\diagdown}}\qquad(11.25)$$

$$\underset{\text{H}}{\overset{\text{Ph}}{\underset{\text{H}}{\overset{\text{ClOC}}{\diagup}}}}\quad\longrightarrow\quad\underset{\text{Me}}{\overset{\text{Ph}}{\diagup}}\overset{\text{Ph}}{\underset{\text{H}}{\diagdown}}\qquad(11.26)$$

The regiochemistry of alkene formation follows Saytzeff's rule, i.e., carbon atom preference for β-hydrogen elimination follows the order tertiary > secondary > primary, so that branched acyl chlorides give predominantly the most substituted alkene.[527] The first-formed alkene may undergo subsequent rhodium-catalyzed isomerization, but this can generally be prevented by addition of excess triphenylphosphine to the system.[527, 528]

Although *catalytic* decarbonylation of acyl chlorides can be achieved at high temperatures in the presence of $RhCl(CO)(PPh_3)_2$, the cationic complex $[Rh(Ph_2P(CH_2)_3PPh_2)_2]^+$ is surprisingly ineffective considering its very high activity for aldehyde decarbonylation.[509]

Whereas palladium on carbon has been used extensively as a decarbonylation catalyst for aldehydes, it finds much less application in the decarbonylation of acyl chlorides. *Soluble* palladium complexes, on the other hand, probably deserve more detailed investigation, since they appear to show high activity. For example, the decarbonylation of decanoyl chloride with palladium on charcoal affords only a 58% yield of nonenes, while a soluble palladium complex leads to almost quantitative conversion.[529] It is also worth noting that decarbonylated by-products are often formed in the palladium-catalyzed Rosenmund reduction of acyl chlorides to aldehydes.[530–532]

Decarbonylation of α,β-unsaturated acyl chlorides has only recently been studied in any detail. Unexpectedly, and in contrast to reactions of related aldehydes, the products of reaction between α,β-unsaturated acyl chlorides and RhCl(PPh$_3$)$_3$ are vinyl triphenyl phosphonium salts.[533] This type of reaction clearly involves the normal sequence of oxidative addition followed by migration of the vinylic group from carbon monoxide to rhodium, but reductive elimination of a phosphonium cation, rather than of vinylic halide, then occurs (Scheme 11.4). Nucleophilic attack on the vinyl ligand by external PPh$_3$ cannot be ruled out, but unactivated vinylic

Scheme 11.4

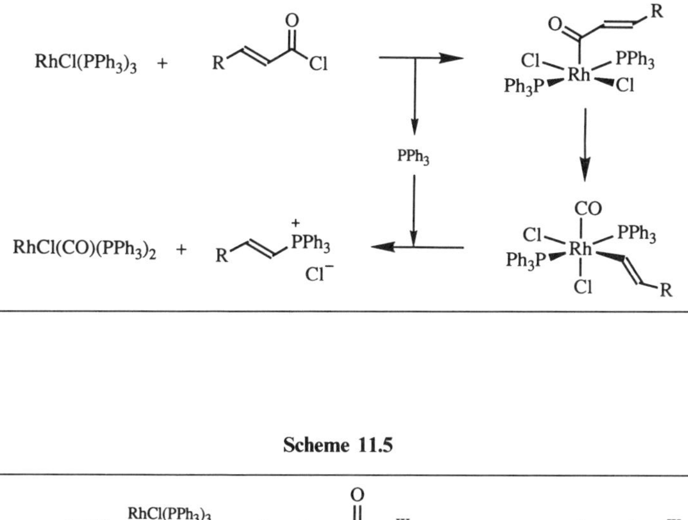

Scheme 11.5

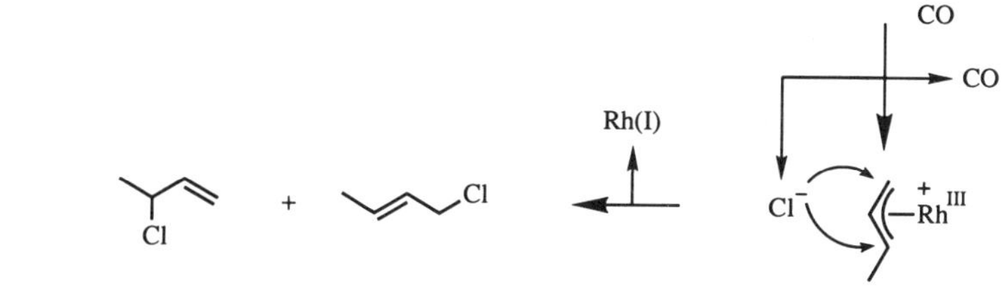

substitution is so uncommon that reductive elimination seems much the more likely pathway.

Allylic (β,γ-unsaturated) acyl chlorides undergo decarbonylation at room temperature when treated with a stoichiometric quantity of RhCl(PPh$_3$)$_3$. The reaction is nonregiospecific, giving mixtures of isomeric allylic halides resulting from attack of chloride at the 1- and 3-positions of an intermediate π-allyl complex (Scheme 11.5).[534]

11.3.3. Aroyl Chlorides

Catalytic decarbonylation of aroyl chlorides to chloroarenes [e.g., Eq. (11.27)] occurs[505] above 200°C in the presence of RhCl(PPh$_3$)$_3$ or RhCl(CO)(PPh$_3$)$_2$.

$$\text{(11.27)}$$

Decarbonylation of 3,4-Dichlorobenzoyl Chloride[505]

In a 35-cm^3 Claisen flask are placed 3,4-dichlorobenzoyl chloride (10.0 g, 47.7 mmol) and RhCl(CO)(PPh$_3$)$_2$ (0.30 g, 0.43 mmol), and the mixture is heated at 250°C under a nitrogen atmosphere for 1.5 h. During this time crude 1,3,4-trichlorobenzene (8.0 g, 92% yield) is collected by distillation.

The older literature contains several indications that aroyl chlorides are *stoichiometrically* decarbonylated to chloroarenes at low temperature by RhCl(PPh$_3$)$_3$, but this is now known to be incorrect.[535, 536] Thus, although oxidative addition of benzoyl chloride occurs, followed by phenyl migration to rhodium to give RhCl$_2$(CO)Ph(PPh$_3$)$_2$, reductive elimination of chlorobenzene is not observed and further heating only converts the latter complex to a mixture of RhCl(CO)(PPh$_3$)$_2$ and RhCl$_2$(Ph)(PPh$_3$)$_2$. As noted above, however, *catalytic* decarbonylation of aroyl chlorides does occur at high temperature, so that other catalytic species must be involved.

One of the most interesting aspects of aroyl chloride decarbonylation is that, in the presence of palladium acetate as catalyst precursor, an arylpalladium intermediate is readily formed and this is sufficiently long-lived to be trapped by insertion of an alkene. β-Hydrogen elimination then liberates an arylated alkene, as shown in Scheme 11.6, so that the "Heck arylation" reaction can be achieved using aroyl chlorides in place of the often less-accessible iodoarenes.[537] The aroyl chloride variant is superior to the original (iodo- or bromo-arene) method for arylation of electron-rich alkenes such as vinyl ethers,[538] and is particularly effective with activated substrates such as acrylate esters, giving good yields with very high selectivity for arylation at the β-position of the alkene.[537]

Scheme 11.6

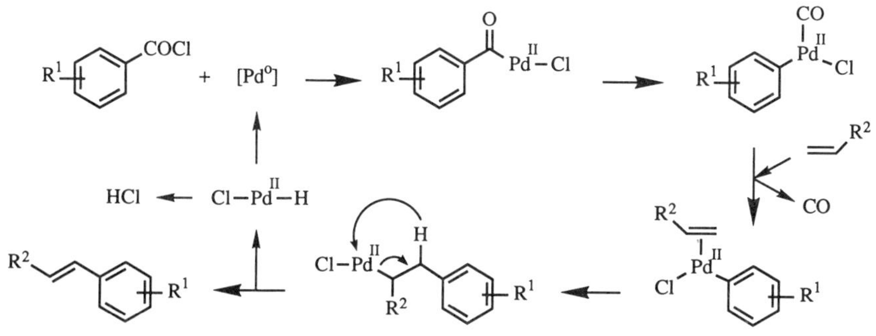

$$R^1\text{-}C_6H_4\text{-}COCl + [Pd^0] \longrightarrow R^1\text{-}C_6H_4\text{-}CO\text{-}Pd^{II}\text{-}Cl \longrightarrow R^1\text{-}C_6H_4\text{-}Pd^{II}(CO)\text{-}Cl$$

Arylation of Activated Alkenes with Aroyl Chlorides[537]

A mixture of *p*-xylene (100 cm^3), palladium acetate (0.112 g, 0.5 mmol), aroyl chloride (50 mmol), activated alkene (62.5 mmol), and tri-*n*-butylamine (62.5 mmol), is stirred at 100–130°C under argon or nitrogen, in a flask fitted with a reflux condenser to avoid loss of alkene. When the aroyl chloride has been consumed the cooled reaction mixture is extracted with 2 M hydrochloric acid (50 then 25 cm^3), 2 M sodium hydroxide (25 cm^3), and water (25 cm^3), and dried over magnesium sulfate. After removal of the solvent under reduced pressure the product is isolated (typically in 50%–80% yield) by recrystallization or distillation.

The facility with which aroylpalladium species undergo decarbonylation has been further exploited in a recently discovered route to arylsilanes.[539] Catalytic reactions of aroyl chlorides with certain disilanes, in the presence of Pd(PhCN)$_2$Cl$_2$ and triphenylphosphine, afford high yields (60%–85%) of arylsilanes, as shown in Eq. (11.28). The procedure is tolerant of a variety of aromatic substituents including alkyl, halo, nitro, cyano, anhydride, and imide, and formation of aryl- versus aroyl-silane is favored by electron-withdrawing groups on the ring. More significantly, the formation of aroyl-silane can be suppressed altogether by use of chlorodisilanes such as ClMe$_2$SiSiMe$_2$Cl. A heterogenized catalyst derived from bis[(diphenylphosphinoethyl)triethoxysilane]palladium dichloride and silica shows almost identical activity and selectivity to the original homogeneous catalyst.[540]

$$
R\text{—}\langle\!\!\!\bigcirc\!\!\!\rangle^{\text{COCl}} + \text{ClMe}_2\text{Si–SiMe}_2\text{Cl} \xrightarrow[\text{PPh}_3]{\text{Pd(PhCN)}_2\text{Cl}_2} R\text{—}\langle\!\!\!\bigcirc\!\!\!\rangle^{\text{SiMe}_2\text{Cl}} + \text{SiMe}_2\text{Cl}_2 + \text{CO} \quad (11.28)
$$

11.4. Decarbonylation of Ketones

Relatively little synthetic work has been reported on transition metal catalyzed decarbonylation of ketones, although a number of α- and β-dicarbonyl compounds are known to decarbonylate[541] when heated in refluxing toluene with catalytic amounts of $RhCl(PPh_3)_3$, $RhCl(CO)(PPh_3)_2$, or even $RhCl_3 \cdot 3H_2O$. The process is not straightforward for β-diketones, since these can form inactive chelate complexes with rhodium, but the available data for ketone decarbonylation summarized in Table 11.3 suggest that this type of reaction may well be worthy of further optimization.

The ketose sugar fructose, when heated with $RhCl(PPh_3)_3$ in N-methylpyrrolidone, gives a surprisingly good yield of furfuryl alcohol.[542] It seems probable that this reaction involves catalytic dehydration of the sugar, to give 5-hydroxymethyl furfural, which then undergoes decarbonylation as shown in Scheme 11.7.

Table 11.3. Decarbonylation of Dicarbonyl Compounds
by $RhCl(PPh_3)_3$[a]

Reactant dicarbonyl	Product	Yield (%)[b]
Acetylacetone	2-Butanone	10.1
Methyl acetoacetate	Methyl propionate	5.0
Ethyl acetoacetate	Ethyl propionate	2.6
Diacetyl[c]	Acetone	10.8
Pyruvic acid	Acetic acid	36.3
Acetylbenzoyl	Acetophenone	20.5

[a] Conditions: diketone (20 mmol), $RhCl(PPh_3)_3$ (0.5 mmol), in toluene (50 ml) under reflux for 6 h.
[b] Calculated from data given by K. Kaneda, H. Azuma, M. Wayaku, and S. Teranishi, *Chem. Lett.*, 215 (1974).
[c] Reaction period 4 h.

Scheme 11.7

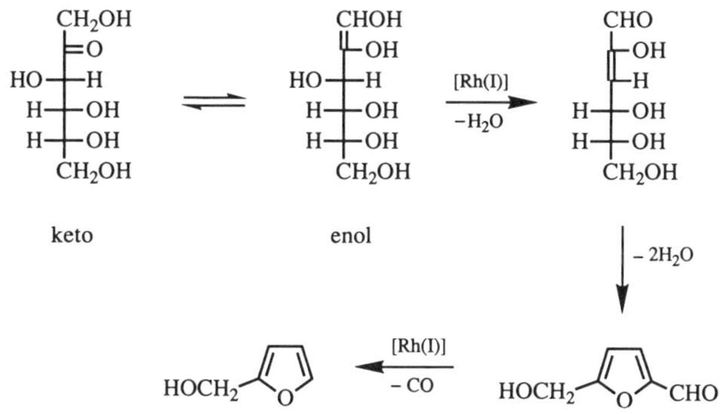

11.5. Miscellaneous Decarbonylations

11.5.1. Acyl Iodides, Bromides, and Fluorides

Despite extensive and successful work on the decarbonylation of acyl chlorides, reactions involving other acyl halides seem to have been somewhat neglected. Aroyl bromides and iodides in fact appear to behave rather like aroyl chlorides (Section 11.3.3),[543, 544] and decarbonylation thus provides a new approach to iodoarenes, which are often relatively inaccessible though valuable synthetic intermediates. Good yields of iodoarenes are obtained by catalytic decarbonylation of aroyl iodides in the presence of $RhCl(PPh_3)_3$, the product distilling from the reaction as it is formed.

Decarbonylation of aroyl fluorides using $RhCl(PPh_3)_3$ was originally reported to give fluoroarenes in essentially quantitative yield.[545] A subsequent reevaluation of this work, however, has shown that the "fluoroarene" products (previously identified only by GLC) are in fact halogen-free, and that this decarbonylation leads to production of aryl radicals which form arenes by abstraction of hydrogen from the solvent.[546]

11.5.2. Acyl Cyanides

Early studies of the decarbonylation of aroyl cyanides to the corresponding nitriles using $RhCl(PPh_3)_3$ as catalyst involved fairly drastic conditions (300°C) and showed the reaction to be limited to activated substrates such as naphthoyl cyanides.[547] More recently, however, the palladium-based catalyst $Pd(PPh_3)_4$ has been found[548] to catalyze decarbonylation of a wide range of aroyl and heteroaroyl cyanides under mild conditions (120°C) in very high yield. The starting materials are readily available by ruthenium-catalyzed oxidation of aldehyde cyanohydrins (Scheme 11.8). As in the decarbonylation of acyl chlorides, aliphatic substrates having β-hydrogens give mainly alkenes.

Scheme 11.8

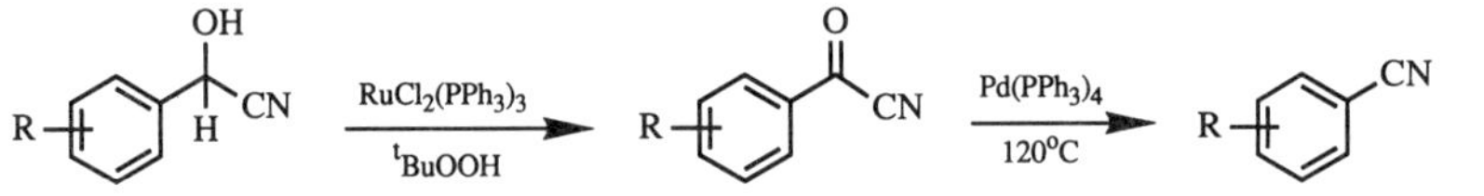

General Procedure for Decarbonylation of Aroyl Cyanides[548]

A mixture of aroyl cyanide (1.0 mmol) and $Pd(PPh_3)_4$ (0.03 mmol) in dry toluene (1.0 cm³) is heated at 120°C for 12 h in a sealed Pyrex tube (150 × 15 mm) under an atmosphere of argon. After evaporation of the solvent the residue is subjected to column chromatography on silica gel (5 g) using diethyl ether/hexane (1:1) as eluent, to give the product nitrile in 90%–98% yield.

11.5.3. Allylic Alcohols

Allylic alcohols are stoichiometrically decarbonylated by $RhCl(PPh_3)_3$ in acetonitrile or benzonitrile as solvent, to give the next lowest alkane as

Scheme 11.9

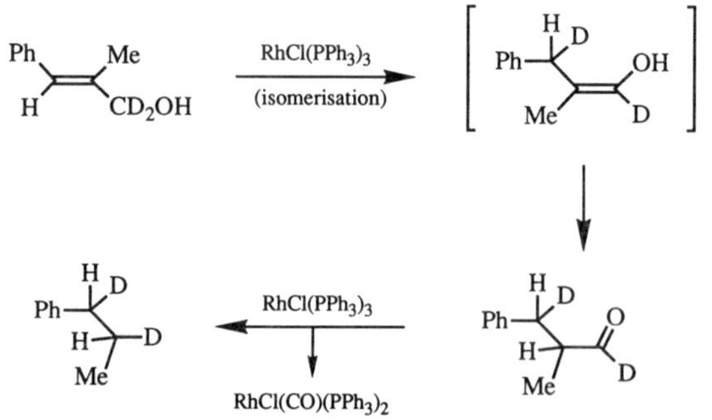

the major product. Cinnamyl alcohol, for example, gives ethyl benzene as the major product, and *cis*-but-2-ene-1,4-diol gives 89% of propan-1-ol.[549] Such reactions probably occur via isomerization to the corresponding aldehyde followed by the familiar steps of aldehyde decarbonylation. Experiments using deuterium-labeled substrates (Scheme 11.9) are entirely consistent with this type of mechanism.[549]

11.5.4. Organo-sulfur and -phosphorus Compounds

Thiol esters are readily converted to sulfides by either stoichiometric or catalytic decarbonylation.[550] Essentially quantitative yields of aryl vinyl sulfides can be obtained using stoichiometric amounts of RhCl(PPh$_3$)$_3$, but efficient catalytic reactions are achievable in the presence of the palladium complex PdCl$_2$(PCy$_3$)$_2$ (Cy = cyclohexyl), as shown in Eq. (11.29).

$$Me\text{-}C_6H_4\text{-}C(O)\text{-}S\text{-}C_6H_4\text{-}Me \xrightarrow[100°C]{PdCl_2(PCy_3)_2} Me\text{-}C_6H_4\text{-}S\text{-}C_6H_4\text{-}Me + CO \quad (11.29)$$

Ketophosphonates, RCOPO(OR′)$_2$, undergo catalytic decarbonylation to alkyl or aryl phosphonates, RPO(OR′)$_2$, in the presence of a range of homogeneous palladium-based complexes.[551] The most active catalyst precursors are those such as PdMe$_2$(PPh$_2$Me)$_2$ and Pd(styrene)(PPh$_2$Me)$_2$, which are able to decompose or dissociate readily to a 14-electron phosphine complex PdL$_2$. In agreement with such a complex being the catalytically active species, addition of excess diphenylmethylphosphine results in a drastic fall-off in reaction rate.

Chapter 12

Catalyst Preparations and Recovery of Precious Metals

12.1. Introduction

Most of the transition metal species discussed in this book are used only in catalytic quantities rather than as stoichiometric reagents. Although the catalyst precursor that is added to the reaction mixture is normally a well-defined compound, the actual species participating in the catalytic cycle are often (though not always) relatively poorly understood. A number of catalyst precursors such as $PdCl_2(PPh_3)_2$ are commercially available, but they can be surprisingly expensive, even for precious metal compounds. When carbonylation reactions are routinely carried out, it is thus often more economical to prepare the necessary catalyst precursors in the laboratory. This chapter gives experimental details for preparing many of the transition metal complexes that find application in carbonylation chemistry. Although most preparations are relatively straightforward, it is strongly recommended that the original literature references be consulted for further information.

Catalysts based on inexpensive metals such as cobalt, iron, or nickel are often used in carbonylation procedures, but more efficient precious metal catalysts (notably complexes of palladium, platinum, and rhodium) are also frequently employed, and by definition, these materials are costly.

Although such catalysts can often be cost effective because of their very high activity, residues containing even small amounts of precious metals should be kept for subsequent work-up and recovery of the metal. The role of precious metal suppliers in recycling such metals is discussed in this chapter, and some practical suggestions are given for recovery of certain of the platinum metals from laboratory residues.

12.2. Preparation of Catalysts

12.2.1. General Considerations

Although many homogeneous catalyst precursors are reasonably stable complexes that are only converted to the active catalytic species under reaction conditions, care is needed in handling the more reactive compounds. The well-known hydrogenation catalyst and decarbonylation reagent $RhCl(PPh_3)_3$ (Wilkinson's catalyst), for example, undergoes ligand dissociation in solution to form a coordinatively unsaturated species, and this takes up oxygen if exposed to air and undergoes oxidative degradation. On the other hand, the palladium(II) complex $PdCl_2(PPh_3)_2$ is stable to oxygen both in the solid state and in solution, and only becomes susceptible to degradation when reduced to the active palladium(0) species under carbonylation conditions. It is, however, generally prudent to store catalyst precursors under an atmosphere of dry nitrogen. Care is also required when handling heterogeneous catalysts such as palladium on carbon, which can cause ignition (explosively in some cases) of flammable solvent vapors in air. Such catalysts are also far more susceptible than homogeneous catalysts to poisoning by minute traces of sulfur, arsenic, trivalent phosphorus compounds, and nitrogen bases, so that fume cupboards previously used for handling sulfur compounds, for example, should be avoided. When aqueous procedures are involved in the preparation of a heterogeneous catalyst, distilled or deionized water should always be used.

The most commonly used homogeneous catalyst precursors are probably $Co_2(CO)_8$, $Pd(CH_3CO_2)_2$ (actually a trimer), $PdCl_2(PPh_3)_2$, and $RhCl(CO)(PPh_3)_2$. Preparative details for these and many other complexes with more specific applications are given here, arranged alphabetically according to the metal concerned.

12.2.2. Cobalt Complexes

Cobalt species have a long history as carbonylation catalysts. The most commonly used in the laboratory is $Co_2(CO)_8$, which is commercially available at moderate cost. Originally this was obtained by decomposition of the toxic, volatile hydride $HCo(CO)_4$, formed by acidification of solutions of the carbonyl anion $[Co(CO)_4]^-$. This anion in turn can be prepared by reduction of a cobalt(II) salt under carbon monoxide at atmospheric pressure.[552] This approach has not proved particularly reliable on a large scale, but good yields of $Co_2(CO)_8$ can be obtained from cobalt(II) acetate using high-pressure hydrogen and carbon monoxide. An atmospheric pressure synthesis starting from CoI_2, zinc metal, and carbon monoxide has also been reported.[553]

Since $Co_2(CO)_8$ is readily available, it is a convenient starting point for the preparation of $[Co(CO)_4]^-$, which like most carbonyl anions is air sensitive. However, the solid $[(Ph_3P)_2N]^+$ salt is surprisingly air stable, and it has considerable solubility in organic solvents. Accordingly procedures for preparing $[(Ph_3P)_2N]^+$ and its $[Co(CO)_4]^-$ salt are given here.

Octacarbonyldicobalt(0)—$Co_2(CO)_8$

Orange crystals of $Co_2(CO)_8$ are soluble in organic solvents, but decomposition takes place quite rapidly in the solid state and in solution, even at room temperature. Storage under carbon monoxide (care should be taken) at 0°C or lower temperature is recommended. However, small amounts can be weighed out quickly in air. When required very pure it can be sublimed at about 25°C onto a cold finger in vacuum (less than 0.1 mbar).

Synthesis of Octacarbonyldicobalt(0) [554, 555]

$$Co(CH_3CO_2)_2 + H_2/CO \rightarrow Co_2(CO)_8 + CH_3CO_2H$$

A rocking high-pressure autoclave (1 liter) is charged with $Co(CH_3CO_2)_2 \cdot 4H_2O$ (200 g, 0.80 mol) and acetic anhydride (330 g, 3.25 mol). After connection to a suitable supply of carbon monoxide and hydrogen the autoclave is flushed with either of these gases at 50 bar before pressurizing with 40 bar of hydrogen and 160 bar of carbon monoxide. Upon heating to 160–180°C a rapid reaction takes place, and after 2 h additional carbon monoxide is added to bring the total pressure back to 200 bar. After a further hour the autoclave is allowed to cool to room temperature when the system is vented. The orange crystals of $Co_2(CO)_8$ are carefully filtered off under nitrogen, washed several times with deionized water and dried *in vacuo*. Yield 82 g (30%).

Bis(triphenylphosphoranylidene)ammonium Tetracarbonylcobaltate(−I)— $[(Ph_3P)_2N][Co(CO)_4]$

The bulky cation $(PPh_3)_2N^+$ is commercially available as the chloride but is relatively expensive. Since it is readily prepared from triphenylphosphine (via chlorination to give Ph_3PCl_2 and reaction of this *in situ* with H_2NOH in the presence of PPh_3) the procedure is given here. Reaction of $Co_2(CO)_8$ with sodium amalgam gives $NaCo(CO)_4$, which, on reaction with $[(Ph_3P)_2N]^+Cl^-$, affords the air-stable salt $[(Ph_3P)_2N]^+[Co(CO)_4]^-$.

Synthesis of Bis(triphenylphosphoranylidene)ammonium Tetracarbonyl-cobaltate(−I) [556]

   Part 1

$$2PPh_3 + Cl_2 \rightarrow 2Ph_3PCl_2$$

$$2Ph_3PCl_2 + Ph_3P + NH_2OH \cdot HCl \rightarrow (Ph_3P)_2NCl + Ph_3PO + 4HCl$$

Into a well-stirred, cold (−20 to −30°C) solution of triphenylphosphine (78.6 g, 0.30 mol) in 1,1,2,2-tetrachloroethylene (100 cm^3) is bubbled chlorine gas from a lecture bottle on a top-loading balance (Care: all operations in a fume cupboard). Solid separates when about 4 g of chlorine has been used, and in order to maintain the low temperature the chlorine rate must be restricted. After 14.2 g of chlorine has been consumed, hydroxylamine hydrochloride (6.9 g, 0.10 mol) is added with stirring and the solution is allowed to warm to room temperature. Then it is boiled for 8 h, during which time hydrogen chloride is evolved. When cool, the resulting solution is poured into ethyl acetate (400 cm^3) and the product allowed to crystallize. The crude material can be recrystallized from water ($\sim$500 cm^3) and vacuum dried. Yield 49 g (84%).

   Part 2

$$Co_2(CO)_8 + 2Na/Hg \rightarrow 2NaCo(CO)_4$$

$$NaCo(CO)_4 + (Ph_3P)_2NCl \rightarrow (Ph_3P)_2NCo(CO)_4 + NaCl$$

This part of the preparation is best conducted in a nitrogen-filled glove bag contained in a fume cupboard. $Co_2(CO)_8$ (3.42 g, 10.0 mol) is added to a stirred mixture of dry tetrahydrofuran (100 cm^3) and sodium amalgam (1%, 50 g) under nitrogen. After 3 h a solution of $(Ph_3P)_2NCl$ (10.0 g, 17.4 mmol) in dichloromethane (100 cm^3) is added. The liquid is decanted from the excess amalgam and filtered. After removal of solvent the residue is dissolved in dichloromethane (100 cm^3), and ether (150–200 cm^3) is slowly added until precipitation begins. The mixture is then chilled (to about 0°C) overnight before filtering off the product and vacuum drying. Yield 9.7 g (78%).

12.2.3. Iron Complexes

   Iron carbonyl anions are easily derived from readily available iron pentacarbonyl. They are used in carbonylation reactions but other species often have practical advantages that outweigh their higher cost. Iron compounds are normally used as reagents rather than catalysts in carbonylation, although under phase transfer conditions catalytic carbonylations are possible. Iron carbonyl anions are frequently generated in *situ* from $Fe(CO)_5$; for example, $K[HFe(CO)_4]$ is readily formed from $Fe(CO)_5$ in ethanolic potassium hydroxide, but cleaner reactions can result from the use of preformed salts of $[Fe(CO)_4]^{2-}$.

Disodium Tetracarbonylferrate(−II)—$Na_2Fe(CO)_4$

The commercially available salt $Na_2Fe(CO)_4$ has been used extensively in aldehyde synthesis, and a procedure for its preparation is given here that involves the reaction of sodium with $Fe(CO)_5$ facilitated by benzophenone. This salt is very air and water sensitive and can be pyrophoric, so it must be prepared, stored, and used under nitrogen. Carbon monoxide is released during this preparation, and since iron pentacarbonyl is involved, all operations must be carried out in a fume cupboard.

Synthesis of Disodium Tetracarbonylferrate(−II) [557]

$$Fe(CO)_5 + 2Na/PhCOPh \rightarrow Na_2Fe(CO)_4 + CO$$

Sodium (4.6 g, 0.2 mol) in small pieces is added to dry dioxane (1000 cm^3) contained in a three-necked flask, followed by benzophenone (3.6 cm^3, 0.02 mol). The blue mixture is refluxed under nitrogen for 15 min, then with rapid stirring $Fe(CO)_5$ (13.4 cm^3, 0.1 mol) is run in over a period of 40 min. A white precipitate forms as carbon monoxide is given off, and the mixture is stirred for a further hour. After this time the colorless mixture is cooled and the precipitate filtered off under nitrogen, washed several times with petroleum ether, and dried *in vacuo*. The product, $Na_2Fe(CO)_4 \cdot 1\frac{1}{2}$ dioxane (30 g, 87%) is a white powder and is sufficiently pure for most purposes.

Tetracarbonylbis(cyclopentadienyl)diiron—$[C_5H_5Fe(CO)_2]_2$

Dark red-purple $[C_5H_5Fe(CO)_2]_2$ is air stable when pure. It is insoluble in water and sparingly soluble in organic solvents, but these solutions gradually oxidize when in contact with air.

Synthesis of Tetracarbonylbis(cyclopentadienyl)diiron [558, 559]

$$2Fe(CO)_5 + C_{10}H_{12} \rightarrow [C_5H_5Fe(CO)_2]_2 + 6CO$$

This preparation, like the previous one, makes use of volatile iron pentacarbonyl, so an efficient fume cupboard must be used.

A three-necked flask equipped with a thermometer, a nitrogen inlet, and an efficient condenser with attached bubbler is used. After flushing the flask, and maintaining a flow of nitrogen, $Fe(CO)_5$ (10 cm^3, 0.07 mol) is added from a syringe, followed by dicyclopentadiene (60 g, 0.455 mol). With minimal nitrogen flow the mixture is heated to $135 \pm 5°C$ for 8–10 h. After this time the nitrogen flow is increased a little, and the mixture allowed to cool. Red-violet crystals of the product are filtered off and washed with several portions of pentane. Additional purification can be effected by addition of an equal volume of hexane to a chloroform solution followed by slow removal of chloroform with a rotary evaporator at room temperature. Yield 47.5 g (89 %).

12.2.4. Nickel Complexes

Although used extensively in the past, nickel tetracarbonyl is employed only occasionally now, as alternative catalysts are available that are often both more efficient and less toxic. Whenever carbon monoxide is used in the presence of nickel compounds, particularly under reducing conditions, the strong possibility of forming $Ni(CO)_4$ should be borne in mind (see Section 3.3).

Nickel Dichlorobis(phosphine) Complexes

Nickel(II) phosphine complexes can be used as catalysts for carbonylation of alkynes, and they are useful in a number of other carbonylation reactions. Dichlorobis(phosphine)nickel(II) complexes are not very soluble in common solvents, which facilitates their preparation and isolation, but they are sufficiently soluble for catalytic applications. Two preparative procedures are given here: $NiCl_2(PPh_3)_2$ and $NiCl_2[Ph_2P(CH_2)_2PPh_2]$. The latter is obtained as a dull red-brown precipitate by mixing acetone or ethanol solutions of hydrated nickel(II) chloride and the diphosphine. The red-violet complex containing the optically active phosphine DIOP is similarly obtained. The blue complex containing triphenylphosphine is best prepared using aqueous acetic acid as solvent.

Synthesis of Dichloro[bis(diphenylphosphino)ethane]nickel(II)[560]

$$NiCl_2 \cdot 6H_2O + Ph_2P(CH_2)_2PPh_2 \rightarrow NiCl_2(Ph_2PCH_2CH_2PPh_2)$$

A solution of 1,2-bis(diphenylphosphino)ethane (4.0 g, 10 mmol) in hot ethanol (400 cm^3) is added to a solution of $NiCl_2 \cdot 6H_2O$ (2.4 g, 10 mmol) in ethanol (20 cm^3). The mixture is allowed to stand overnight, and the red-brown precipitate collected, washed with a small amount of ethanol, and dried *in vacuo*. Yield almost quantitative.

Synthesis of Dichlorobis(triphenylphosphine)nickel(II)[561]

To a solution of nickel(II) chloride hexahydrate (2.38 g, 0.01 mol) in the minimum amount of water (2 cm^3) is slowly added with stirring a solution of triphenylphosphine (5.25 g, 0.02 mol) in glacial acetic acid (75 cm^3). An olive-green precipitate is formed that is allowed to stand in contact with the mother liquor overnight. The resulting dark blue crystals are filtered off, washed with glacial acetic acid, and dried in vacuum over KOH. Yield 5.5 g (84%).

12.2.5. Palladium Complexes

Palladium has become one of the most commonly used transition metals in organic synthesis, and many of its compounds are versatile carbonylation catalysts.

Palladium(II) Acetate—[Pd(CH$_3$CO$_2$)$_2$]$_3$

Palladium acetate is used in some very mild carbonylations and also in a variety of other carbon–carbon bond-forming reactions. It has a trimeric structure in the solid state, with each pair of palladium atoms bridged by two acetate groups. The air stable brown needles are soluble in some organic solvents (e.g., chloroform, dichloromethane, acetone, and diethylether), but insoluble in water and alcohols. In glacial acetic acid it is monomeric, and in benzene trimeric. Palladium acetate is commercially available at reasonable cost so that its preparation is not always worthwhile. However, the following procedure for obtaining the acetate can be valuable in recovery of the metal from palladium residues (see Sec. 12.3).

Synthesis of Palladium(II) Acetate[562]

$$Pd + CH_3CO_2H/HNO_3 \rightarrow [Pd(CH_3CO_2)_2]_3$$

Palladium sponge (10 g), glacial acetic acid (150 cm^3), and concentrated nitric acid (6 cm^3) are boiled under reflux until no further brown fumes are produced. A small amount of metal should remain undissolved. If none remains, some must be added, and refluxing continued till no further brown fumes are evolved. This procedure avoids contamination of the product with the nitro complex Pd(NO$_2$)(CH$_3$CO$_2$). The boiling solution is filtered from excess metal, and the filtrate cooled overnight. The orange-brown crystals are collected and washed with a little acetic acid and water, before air drying. Yield 80%–95%. A small amount of additional product can be obtained by concentrating the pale brown filtrate. Alternatively the filtrate can be reused in subsequent palladium acetate preparations, or added to palladium residues for later recovery.

trans-Bromo(phenyl)bis(triphenylphosphine)palladium(II)— $Pd(Br)(Ph)(PPh_3)_2$

Colorless, air stable $Pd(Br)(Ph)(PPh_3)_2$ (mp $\sim 220°C$) and related alkyl palladium(II) complexes can be obtained by reaction of the appropriate $PdX_2(PR_3)_2$ complex with lithium alkyl, although with two equivalents of lithium alkyl, *trans*-$PdR_2(PR_3)_2$ is formed. The preferred synthesis of the bromo(aryl)palladium complex, however, does not require the use of lithium alkyl, but involves oxidative addition of ArBr to $Pd(PPh_3)_4$.

Synthesis of trans-Bromo(phenyl)bis(triphenylphosphine)palladium(II) [563]

$$Pd(PPh_3)_4 + PhBr \rightarrow trans\text{-}Pd(Br)(Ph)(PPh_3)_2 + 2PPh_3$$

A mixture of $Pd(PPh_3)_4$ (2.2 g, 1.9 mmol; see below for synthesis) and bromobenzene (1.0 g, 6.4 mmol) in degassed benzene (15 cm³) is heated overnight under nitrogen at reflux temperature. After this period the solvent is removed from the cooled mixture by vacuum distillation. After trituration of the residue with two 25 cm³ portions of ether the crude product is recrystallized from dichloromethane/hexane to give air stable colorless crystals. Yield 1.4 g (94%).

Palladium(II) Chloride—$PdCl_2$

Palladium metal dissolves readily in warm aqua regia, and repeated evaporation to dryness with a little hydrochloric acid affords[564] red hygroscopic crystals, of $PdCl_2 \cdot 2H_2O$. Palladium chloride solutions are available commercially (typically about 10 wt% palladium), but the compound is usually purchased in the anhydrous form. There are at least three modifications (α, β, and γ) of anhydrous $PdCl_2$, and the conditions under which each is formed are not yet fully defined.[565] The β-form is hexameric (Pd_6Cl_{12}) and is best prepared by addition of hydrochloric acid to palladium acetate in acetic acid[565]; on heating at $500°C$ it changes to the polymeric α-form.[566] The hexamer is soluble in aromatic solvents, but commercial anhydrous palladium chloride, obtained by heating the hydrate or by direct chlorination of the metal at red heat, is generally in the very much less soluble α- or γ-forms.[567] The formation of $PdCl_2$ from recovered palladium metal makes it an important intermediate in the recycling of palladium residues to the widely used, soluble carbonylation catalysts $PdCl_2(PPh_3)_2$ and $PdCl_2(PhCN)_2$. Anhydrous $PdCl_2$ is also readily converted[568] to K_2PdCl_4, which is sufficiently soluble in water to be a useful starting material for preparing a wide range of palladium(II) complexes.

Synthesis of Palladium(II) Chloride

    1. Polymeric (Insoluble) Form[569]

$$Pd + HCl/HNO_3 \rightarrow PdCl_2$$

Palladium metal is treated with an excess of warm aqua regia (three volumes of concentrated hydrochloric acid to one volume of concentrated nitric acid) until the metal is dissolved. The resulting solution is evaporated to dryness and the residue heated in a small furnace at 500°C for 1 h to remove chlorine and hydrogen chloride. Anhydrous palladium chloride remains as a brown powder in almost quantitative yield.

    2. Hexameric (Soluble) Form[565]

$$Pd(OAc)_2 + 2HCl \rightarrow PdCl_2 + 2AcOH$$

To a 0.05 M solution of palladium acetate in glacial acetic acid is added the theoretical quantity of concentrated hydrochloric acid at room temperature. The resulting fine brown precipitate is separated by centrifugation, washed with acetic acid, and dried *in vacuo* over sodium hydroxide. Yield is essentially quantitative.

Bis(π-allyl)di-μ-chlorodipalladium(II)—[PdCl(C₃H₅)]₂

    Chloro-π-allylic palladium(II) complexes are generally stable compounds with chloro-bridged dimeric structures. In solutions of certain complexes, the π-allylic group is in dynamic equilibrium with the coordinatively unsaturated species having a palladium–carbon σ-bond. A number of routes are available to the parent π-allyl complex, but in practical terms the most attractive involves reduction of $[PdCl_4]^{2-}$ with carbon monoxide, followed by rapid oxidative addition of allyl chloride to the intermediate palladium(0) complex.

Synthesis of Bis(π-allyl)di-μ-chlorodipalladium(II) [570]

$$2PdCl_4^{2-} + 2H_2O + 2CO + 2CH_2{=}CHCH_2Cl \rightarrow$$

$$[(C_3H_5)PdCl]_2 + 2CO_2 + 4HCl + 4Cl^-$$

Palladium chloride (17.7 g, 0.1 mol) and calcium chloride (11.1 g, 0.1 mol) are stirred in methanol (150 cm³) containing a small amount of water (10 cm³), and allyl chloride (27.9 g, 0.36 mol) is added before passing carbon monoxide through the mixture (2.5–3.0 liters/min) for 30 min (care: use an efficient fume cupboard). After this period the reaction mixture is poured into water (1 liter) and the product extracted with chloroform (3×100 cm^3). The combined extracts are dried (MgSO$_4$) and solvent removed under vacuum at room temperature to give air stable yellow crystals of the product. Yield 12.3 g (67%), mp 158°C with decomposition.

Dichlorobis(triphenylphosphine)palladium(II)—PdCl$_2$(PPh$_3$)$_2$

Pale yellow, air stable PdCl$_2$(PPh$_3$)$_2$ is a much used homogeneous catalyst with a very wide range of applicability. Thus in conjunction with tin(II) chloride (which enhances its mild hydrogenation activity) PdCl$_2$(PPh$_3$)$_2$ catalyzes the selective hydrogenation of polyenes to monoenes. It also has useful activity in alkene isomerization, hydrosilyation of alkenes, and in many carbon–carbon bond-forming reactions. Unlike the polymeric form of palladium(II) chloride, PdCl$_2$(PPh$_3$)$_2$ is soluble in chloroform and moderately soluble in aromatic solvents and alcohols. It is insoluble in water, saturated hydrocarbons, ether, and carbon tetrachloride. It can be prepared by dissolving palladium chloride in molten triphenylphosphine, but sometimes this method gives poor yields. A better procedure, described here, uses aqueous ethanol as solvent.

Synthesis of Dichlorobis(triphenylphosphine)palladium(II)[571, 572]

$$PdCl_2 + 2PPh_3 \rightarrow PdCl_2(PPh_3)_2$$

A solution of palladium chloride (3.0 g, 17 mmol) in dilute hydrochloric acid (0.5 cm^3 of concentrated hydrochloric acid in 150 cm^3 water) is slowly added to a stirred, warm solution of triphenylphosphine (9.0 g, 34.5 mmol) in ethanol (300 cm^3). The stirred mixture is maintained at $\simeq 60$°C for 3 h before collecting the product. After washing with 100 cm^3 portions of warm water, ethanol, and ether, the product is sufficiently pure for most purposes. It can, however, be further purified by precipitation from chloroform with hexane. Yield 11.5 g (95%).

Bis(benzonitrile)dichloropalladium(II)—PdCl$_2$(PhCN)$_2$

The particularly useful carbonylation catalyst PdCl$_2$(PhCN)$_2$ is an air stable, light yellow solid, soluble in aromatic solvents but insoluble in saturated hydrocarbons. It is readily prepared from anhydrous PdCl$_2$, which can itself be difficult to dissolve in most solvents used for carbonylations. The equally useful PdCl$_2$(CH$_3$CN)$_2$ can be obtained by using acetonitrile rather than benzonitrile in the following procedure.

Synthesis of Bis(benzonitrile)dichloropalladium(II) [573]

$$PdCl_2 + 2PhCN \rightarrow PdCl_2(PhCN)_2$$

A suspension of $PdCl_2$ (2.0 g, 11 mmol) in benzonitrile (50 cm^3) is heated to 100°C to form a red solution (20 min). Undissolved material is removed by filtering the hot solution, and the filtrate is poured into low-boiling petroleum (300 cm^3) to precipitate the yellow product. This is filtered off, and washed with low-boiling petroleum. Yield 4.0 g (93 %).

Tetrakis(triphenylphosphine)palladium(0)—Pd(PPh$_3$)$_4$

Many procedures are available for the preparation of this catalyst. The procedure given here uses readily available reagents, and is known to work well in different laboratories. Pd(PPh$_3$)$_4$ is a yellow solid, mp ~ 115°C, which is soluble in most organic solvents except saturated hydrocarbons. In the solid it is air stable, but in solution it is sensitive to oxygen, initially forming [Pd(PPh$_3$)$_2 \cdot O_2$]. It is recommended that the solid be stored under nitrogen.

Synthesis of Tetrakis(triphenylphosphine)palladium(0) [574]

$$PdCl_2 + 4PPh_3 + N_2H_4 \cdot H_2O \rightarrow Pd(PPh_3)_4 + N_2$$

Palladium(II) chloride (17.7 g, 0.1 mol), triphenylphosphine (131 g, 0.5 mol) and dimethysulfoxide (1.2 liters) are heated with stirring to about 140°C under nitrogen. Once a clear solution is obtained, hydrazine hydrate (19 cm^3, 0.4 mol) is added in one portion over a period of 1 min (with exclusion of air). Nitrogen is evolved in the *vigorous* ensuing reaction. The reaction flask is then cooled immediately (water bath) to reduce the temperature of the mixture to about 100°C. After crystallizing overnight, the product is filtered off under nitrogen on a glass sinter, and washed (under nitrogen) with ethanol and ether (2×50 cm^3 portions). Yield 103–113 g (90 %–98 %).

Palladium on Charcoal

Palladium on charcoal, being a heterogeneous catalyst, is readily removed from reaction mixtures, greatly facilitating product work-up. It has significant activity in carbonylation and decarbonylation reactions, albeit under more vigorous conditions than with homogeneous palladium-based catalysts, and its use in these roles will probably increase in the future. It is commercially available, but its preparation is very straightforward. Like many other heterogeneous catalysts, it can be poisoned by impurities adsorbed during preparation or use. Once prepared, catalyst should be dried only at room temperature, since *at high temperatures it may ignite.* Care should be taken to keep heterogeneous catalysts of this type out of contact with combustible vapors. In use, it is advisable that reaction vessels first be blanketed with nitrogen, and the organic solvent (particularly when using low-boiling alcohols) added in large portions to the catalyst, because once wet the catalyst is less likely to cause ignition. In some situations it is more convenient to use carbon impregnated with a palladium salt (e.g., $PdCl_2/HCl$), and to activate this by reduction *in situ*, than to handle the more active prereduced catalyst. Reducing agents used to convert the supported chloride to the metal include hydrazine, formates, formaldehyde, hydrogen, and sodium borohydride. The procedure given here uses formaldehyde. Almost any high surface area carbon can be used, after treatment with nitric acid (10%) for 2–3 h, followed by washing free of acid with water and drying at 100°C.

Preparation of 5% Palladium on Charcoal[575, 576]

A solution of anhydrous palladium chloride (8.2 g) in hydrochloric acid ($20 cm^3$ concentrated acid and $50 cm^3$ water) is obtained by warming for about 2 h. This is added to a stirred, hot (80°C) suspension of nitric acid washed charcoal (93 g) in water (1.2 liters). Formaldehyde ($8 cm^3$ of a 37% solution) is then added, followed by sufficient 30% sodium hydroxide solution to make the suspension strongly alkaline. After 10 min the catalyst is filtered off, washed with water ($10 \times 250 cm^3$), and dried *in vacuo* over calcium chloride, before storing in a tightly closed bottle. Yield 93–98 g.

12.2.6. Platinum Complexes

Although of overall less importance than catalysts based on cobalt or palladium, platinum compounds are useful in a number of carbonylation reactions. Section 12.3.2 should be consulted for safety precautions associated with the use of platinum compounds before work with them is undertaken.

Platinum(II) Chloride—$PtCl_2$

Platinum(II) chloride exists in two forms. The reddish-black β-form is hexameric (Pt_6Cl_{12}), but can only be obtained under carefully controlled conditions.[577] The more common α-form is brownish-green, polymeric, and practically insoluble in water. It does, however, dissolve in aqueous ammonia, and in hydrochloric acid, to form the complex ions

$[Pt(NH_3)_4]^{2+}$ and $[PtCl_4]^{2-}$, respectively. It can be obtained from chloroplatinic acid, H_2PtCl_6, which decomposes on heating in chlorine to give first $PtCl_4$ (at 370°C) and then $PtCl_2$ (at 580°C).[578] In the absence of chlorine, platinum metal is formed. More convenient, however, is the controlled hydrazine reduction of aqueous H_2PtCl_6 described here.

Synthesis of Platinum(II) Chloride[579]

$$2H_2PtCl_6 + N_2H_4 \cdot 2HCl \rightarrow 2H_2PtCl_4 + N_2 + 6HCl$$

$$2H_2PtCl_4 \rightarrow 2PtCl_2 + 4HCl$$

To a stirred solution of chloroplatinic acid (50 cm^3 containing 20.5 mg atom of platinum) are added small portions of solid $N_2H_4 \cdot 2HCl$ (1.07 g, 10.3 mmol) over about 10 min. The cherry-red solution is warmed on a water bath, and filtered to remove any metal that may have formed. The filtrate is evaporated to dryness, and dried at about 120°C. The dry material is powdered, and heated at 150°C for 4 h to remove hydrogen chloride (care should be taken). Soluble impurities are removed with hot water (5 × 10 cm^3) and the product is air-dried at 110°C. Yield 4.6 g (84%).

Potassium Tetrachloroplatinate(II)—K_2PtCl_4

Although only slightly soluble in water (about 1 wt% at 25°C), red crystalline K_2PtCl_4 is a useful starting material for the preparation of other platinum compounds. It is commercially available, but it can also be recovered without too much difficulty from platinum residues. Thus, as indicated in Sec. 12.3.4, a solution of chloroplatinic acid (H_2PtCl_6) can be readily obtained, and on addition of NH_4Cl or KCl the corresponding salt of $[PtCl_6]^{2-}$ is precipitated. Careful reduction then gives the $[PtCl_4]^{2-}$ salt, but overreduction to platinum metal must be avoided. A number of reducing agents have been described; in the procedure described here, hydrazine hydrochloride is used.

Synthesis of Potassium Tetrachloroplatinate(II)[580]

$$2K_2PtCl_6 + N_2H_4 \cdot 2HCl \rightarrow K_2PtCl_4 + 6HCl + N_2$$

A well-stirred suspension of pure K_2PtCl_6 in water (10–12 times the weight of the solid) is warmed on a water bath, and the stoichiometric amount of $N_2H_4 \cdot 2HCl$ is added in small portions. Nitrogen is evolved as the K_2PtCl_6 dissolves, and a clear solution is formed. A deficiency of hydrazine leaves some undissolved K_2PtCl_6 while an excess leads to formation of platinum metal, so the solution is filtered before evaporating to the point of crystallization. After cooling, the product is filtered off and washed with a small volume of water. Typical yield 85%.

Dichlorobis(triphenylphosphine)platinum(II)—PtCl$_2$(PPh$_3$)$_2$

This compound exists in *cis* and *trans* forms. White crystals of the *cis*-compound (mp 310–312°C) are formed from the reaction of triphenylphosphine with K$_2$PtCl$_4$ in aqueous ethanol. The light yellow *trans*-compound is obtained when K[PtCl$_3$(C$_2$H$_4$)] reacts with triphenylphosphine, but this isomer readily isomerizes to the thermodynamically stable *cis* isomer, a process catalyzed by excess triphenylphosphine. Since either isomer may be used as a carbonylation catalyst, only a procedure for *cis*-PtCl$_2$(PPh$_3$)$_2$ is given here. This complex is virtually insoluble in water and ethanol, but moderately soluble in chloroform.

Synthesis of cis-Dichlorobis(triphenylphosphine)platinum(II) [581]

A solution of K$_2$PtCl$_4$ (4.0 g, 9.6 mmol, see previous preparation) in water (50 cm^3) is added dropwise with rapid stirring to a boiling solution of triphenylphosphine (5.1 g, 19.4 mmol) in ethanol (60 cm^3). White crystals separate and the stirred mixture is maintained at 60°C for two hours before filtering off the product and washing it with hot water, hot ethanol, and finally ether. Yield 6.4 g (84%).

Tetrakis- and Tris-(triphenylphosphine)platinum(0)—Pt(PPh$_3$)$_4$ and Pt(PPh$_3$)$_3$

Pale yellow Pt(PPh$_3$)$_4$ melts in air, with decomposition, at ~120°C. Being moderately air sensitive it must be kept under nitrogen. It dissolves in benzene with dissociation of triphenylphosphine, and reacts readily with certain chlorinated solvents; treatment with CCl$_4$, for example, affords good yields of *cis*-PtCl$_2$(PPh$_3$)$_2$. Heating an ethanolic suspension of Pt(PPh$_3$)$_4$ at reflux under nitrogen affords Pt(PPh$_3$)$_3$ in acceptable yield. The properties of this yellow complex are similar to those of Pt(PPh$_3$)$_4$. In solution Pt(PPh$_3$)$_3$ reacts rapidly with carbon monoxide at atmospheric pressure to give Pt(PPh$_3$)$_3$CO, and at higher pressures Pt(PPh$_3$)$_2$(CO)$_2$ is formed.

Synthesis of Tetrakis- and Tris-(triphenylphosphine)platinum(0) [582]

    Part 1

$$K_2PtCl_4 + 4PPh_3 + 2KOH + C_2H_5OH$$

$$\rightarrow Pt(PPh_3)_4 + 4KCl + CH_3CHO + 2H_2O$$

Potassium hydroxide (1.4 g, 25 mmol) in aqueous ethanol (32 cm^3 of ethanol and 8 cm^3 of water) is added to a hot (65°C) stirred solution of PPh$_3$ (1.5 g, 60 mmol) in ethanol (200 cm^3), followed by addition over 20 min of a solution of K_2PtCl_4 (5.2 g, 12.5 mmol, see above) in water (50 cm³). During this period the temperature of the stirred solution is maintained at 65°C, and the yellow product begins to form shortly after the first addition of K_2PtCl_4. The product is separated from the cooled solution and rapidly washed with warm (35°C) ethanol (100 cm³), cold water (50 cm³), and cold ethanol (50 cm³), before drying *in vacuo* and storing under nitrogen. Yield of $Pt(PPh_3)_4$, 12.4 g (79%).

Part 2

$$Pt(PPh_3)_4 \rightarrow Pt(PPh_3)_3 + PPh_3$$

A stirred suspension of $Pt(PPh_3)_4$ (5.8 g, 4.7 mmol) in ethanol (250 cm³) is refluxed under nitrogen for 2 h. The hot solution is quickly filtered and the yellow product washed with cold ethanol (30 cm³), before drying *in vacuo* and storing under nitrogen. Yield of $Pt(PPh_3)_3$, 3.0 g (66%).

cis-Dichlorobis(triphenylphosphite)platinum(II)—$PtCl_2[P(OPh)_3]_2$

Unlike their triphenylphosphine counterparts, the triarylphosphite compounds *cis*-$PtX_2\{P(OPh)_3\}_2$ (X = halide) have attracted little attention in their own right, although interesting *ortho*-metallation reactions take place when they are heated to relatively high temperatures. The air stable crystalline chloro-complex has been used as carbonylation catalyst.

Synthesis of cis-Dichlorobis(triphenylphosphite)platinum(II) [583]

Ethanolic solutions of sodium tetrachloroplatinate(II) tetrahydrate (0.4 g, 1.0 mol in 8 cm³ of ethanol) and triphenylphosphite (0.62 g, 2.0 mmol in 4 cm³ of ethanol) are mixed, before warming for 2 min. After standing for a further 10 min the mixture is cooled (0°C) and the colorless precipitate filtered off. After washing with aqueous ethanol and hexane, the product is dried under vacuum. Yield 0.7 g (80%). Recrystallization from dichloromethane/methanol affords white crystals, mp 187–189°C.

Bis(benzonitrile)dichloroplatinum(II)—$PtCl_2(PhCN)_2$

Because of the lability of the nitrile ligands, this compound is frequently used as a starting material for the preparation of other platinum complexes. The original route from K_2PtCl_4 involved very long reaction times,[584] as the displacement of chloride ligands from platinum is very slow. It is therefore more convenient to start from $PtCl_2$. The product is a mixture of *cis* and *trans* isomers with the relative amounts depending on the reaction temperature. The relative thermodynamic stability of these isomers is solvent dependent and is not a major concern in catalytic applications.

Synthesis of Bis(benzonitrile)dichloroplatinum(II) [585]

A suspension of anhydrous $PtCl_2$ (0.26 g, 1.0 mmol) in benzonitrile (20 cm^3) is stirred at room temperature until a clear solution is obtained (about 8 h). The solution is then filtered and petroleum ether is added to the filtrate to precipitate the product as a yellow powder which is washed with petroleum ether and dried in vacuum. Yield 0.42 g (92%) of an approximately 1:2 *trans/cis* mixture of isomers.

12.2.7. Rhodium Complexes

Rhodium compounds are active hydroformylation and carbonylation catalysts, and are used industrially on a large scale. Selectivity is often better than with cobalt catalysts, but the cost of rhodium is a disincentive unless efficient recovery of metal is possible.

Di-μ-chlorotetracarbonyldirhodium(I)—[RhCl(CO)$_2$]$_2$

Dimeric rhodium carbonyl chloride is a useful catalyst, and one that is soluble in most organic solvents. It can be synthesized by direct reaction of carbon monoxide with $RhCl_3 \cdot 3H_2O$ at about 100°C, but this reaction has not proved reliable and $[RhCl(CO)_2]_2$ is best obtained by displacement of ethylene from $[RhCl(C_2H_4)_2]_2$ by carbon monoxide. The intermediate ethylene complex can be readily obtained at room temperature. The product $[RhCl(C_2H_4)_2]_2$ precipitates in good yield over a period of about 8 h. This ethylene complex is also a useful intermediate in the preparation of other rhodium(I) compounds; Wilkinson's catalyst for example can be prepared *in situ* by reaction with triphenylphosphine.

Synthesis of Di-μ-chlorotetracarbonyldirhodium(I) [586, 587]

$$2RhCl_3 \cdot 3H_2O + 6C_2H_4 \rightarrow [RhCl(C_2H_4)_2]_2 + 2CH_3CHO + 4HCl + 4H_2O$$

$$[RhCl(C_2H_4)_2]_2 + 4CO \rightarrow [RhCl(CO)_2]_2 + 4C_2H_4$$

A solution of $RhCl_3 \cdot 3H_2O$ (10 g, 37 mmol) in water (15 cm^3) is added to methanol (250 cm^3), and a stream of ethylene is bubbled through the solution. The product precipitates over about 8 h, and is filtered off taking care to minimize exposure to air. The red-brown product is washed once with methanol (30 cm^3) and dried *in vacuo*. Yield 5.0 g (65%). This material is best stored under ethylene at about 0°C if it is not used immediately. To convert it to $[RhCl(CO)_2]_2$, carbon monoxide (care: use an efficient fume cupboard) is bubbled through a stirred suspension of $[RhCl(C_2H_4)_2]_2$ (1 g, 2.6 mmol) in diethyl ether (30 cm^3) at room temperature for about an hour. The mixture is then filtered, the filtrate concentrated to about 10 cm^3 on a rotary evaporator, and the deep-red product filtered from the ice-cold supernatant liquor. Yield 0.6 g (60%). Storage under carbon monoxide in a refrigerator is recommended.

trans-Carbonylchlorobis(triphenylphosphine)rhodium(I)—
$RhCl(CO)(PPh_3)_2$

This complex is a particularly active hydroformylation catalyst, and it has also been used in many other carbonylation reactions. The rapid, one-step preparation described here uses formaldehyde as the source of carbon monoxide, and gives good yields. *Trans*-$RhCl(CO)(PPh_3)_2$ is a yellow air stable crystalline material (mp 195–197°C) with moderate solubility in benzene.

Synthesis of trans-Carbonylchlorobis(triphenylphosphine)rhodium(I)[(588)]

$$RhCl_3 + PPh_3 + H_2CO \rightarrow \textit{trans-}RhCl(CO)(PPh_3)_2$$

A solution of $RhCl_3 \cdot 3H_2O$ (2.0 g, 8 mmol) in ethanol (70 cm^3) is slowly added to a refluxing solution of triphenylphosphine (7.2 g, 28 mmol) in ethanol (300 cm^3). When the resulting solution becomes clear, sufficient formaldehyde (10–20 cm^3 of a 40% solution) is added to make the red solution yellow in about a minute. Yellow crystals of the product are formed. When cool these are filtered off and washed with ethanol and ether (two 50 cm^3 portions of each). Yield 4.5 g (85%).

Chlorotris(triphenylphosphine)rhodium(I)—$RhCl(PPh_3)_3$

Wilkinson's catalyst, $RhCl(PPh_3)_3$, was the first effective homogeneous catalyst for the hydrogenation of alkenes at room temperature and atmospheric pressure.[(589)] Only unhindered double bonds undergo reaction, so polyenes may be selectively hydrogenated. It has moderate carbonylation activity, but it has been more widely used for the decarbonylation of aldehydes. It is dimorphic in the solid state, crystallizing in both orange and red-violet modifications. Both are soluble in benzene and toluene—solvents in which it is commonly used. It should be noted that $RhCl(PPh_3)_3$ slowly reacts with oxygen, and erratic results are sometimes obtained if the catalyst has been stored for a long period.

Synthesis of Chlorotris(triphenylphosphine)rhodium(I)[(590)]

$$RhCl_3 \cdot 3H_2O + 4PPh_3 \rightarrow RhCl(PPh_3)_3 + Ph_3PO$$

A solution of triphenylphosphine (12 g, 46 mmol) in hot ethanol (350 cm^3) is added to a solution of $RhCl_3 \cdot 3H_2O$ (2 g, 8 mmol) in hot ethanol (70 cm^3). The mixture is refluxed for 30 min under nitrogen, and red crystals of product separate. These are quickly filtered off, washed with ether (50 cm^3), and dried *in vacuo*. Yield 6.3 g (86%), mp 157–158°C. Addition of water to the filtrate precipitates most of the excess triphenylphosphine, and this can be collected and reused after recrystallization from ethanol.

Tetrakis(acetato)dirhodium(II)—[Rh(O$_2$CCH$_3$)$_2$]$_2$

This emerald-green diamagnetic complex has the dimeric, bridged copper acetate type structure. It has only slight solubility in water and alcohols, and has uses as a hydroformylation catalyst. It is easily prepared as a precipitate by reduction of rhodium trichloride (RhCl$_3 \cdot$3H$_2$O) with ethanol in the presence of acetic acid. The dimeric acetate can be recrystallized from methanol to give a weakly bound methanol adduct.

Synthesis of Tetrakis(acetato)dirhodium(II)[591]

Rhodium trichloride (5.0 g), sodium acetate trihydrate (10.0 g), glacial acetic acid (100 cm^3), and ethanol (100 cm^3) are refluxed under an atmosphere of nitrogen for 1 h. During this period the deep red solution turns green and [Rh(O$_2$CCH$_3$)$_2$]$_2$ is precipitated. The cold mixture is filtered and the precipitate recrystallized from methanol (650 cm^3—with cooling to 10°C overnight) to give the bis-methanol adduct [Rh(O$_2$CCH$_3$)$_2$]$_2 \cdot$2CH$_3$OH. Heating this adduct at 45°C for 20 h under vacuum gives [Rh(O$_2$CCH$_3$)$_2$]$_2$ in 75% yield (3.2 g).

Bis(diphosphine)rhodium(I) Salts—[Rh(Ph$_2$P(CH$_2$)$_n$PPh$_2$)$_2$]$^+$

A number of bis(diphosphine)rhodium(I) salts have very high catalytic activity in decarbonylation reactions. The chemistry of [Rh(Ph$_2$P(CH$_2$)$_2$PPh$_2$)$_2$]$^+$ has been the most extensively studied, but its bis(diphenylphosphino)propane analogue, [Rh(Ph$_2$P(CH$_2$)$_3$PPh$_2$)$_2$]$^+$, is a more active decarbonylation catalyst. Complexes of this type are formed by reaction of the diphosphine with a variety of rhodium compounds, for example, with [RhCl(CO)$_2$]$_2$, [RhCl(diene)]$_2$, or RhCl(PPh$_3$)$_3$. Some procedures[592, 593] require long reaction times compared with that given here. This high-yield, two-stage reaction involves first the formation of a 1,5-cyclo-octadiene complex, [RhCl(COD)]$_2$, and second its reaction with Ph$_2$P(CH$_2$)$_3$PPh$_2$.

Synthesis of Bis[1,3-bis(diphenylphosphino)propane]rhodium(I)
Chloride[594, 595]

Part 1

$$2RhCl_3 \cdot 3H_2O + 4C_8H_{12} + EtOH \rightarrow [RhCl(C_8H_{12})_2]_2$$

A stirred mixture of RhCl$_3 \cdot$3H$_2$O (2.6 g, 10.0 mmol), 1,5-cyclooctadiene (COD) (4.42 g, 41 mmol), and ethanol (50 cm^3) is heated under reflux for 3 h. After cooling to room temperature the yellow product is filtered off, washed with ethanol (2 × 25 cm^3) and hexane (2 × 25 cm^3), and air-dried. Yield 2.0 g (81%), mp 258°C.

Part 2

$$[RhCl(C_8H_{12})_2]_2 + 2Ph_2P(CH_2)_3PPh_2 \rightarrow [Rh(Ph_2P(CH_2)_3PPh_2)_2]^+Cl^-$$

To a stirred solution of $Ph_2P(CH_2)_3PPh_2$ (0.82 g, 2.0 mmol) in acetone (20 cm^3) under nitrogen is added $[RhCl(C_8H_{12})_2]_2$ (0.25 g, 0.50 mmol). The desired product precipitates, and after stirring for $1\frac{1}{2}$ h it is filtered off, washed with acetone and hexane, and dried in vacuum. Yield almost quantitative.

Carbonylhydridotris(triphenylphosphine)rhodium(I)—RhH(CO)(PPh$_3$)$_3$

Bright yellow $RhH(CO)(PPh_3)_3$ is a very active hydroformylation catalyst. It is moderately soluble in aromatic solvents, but insoluble in water, ethanol, and saturated hydrocarbons. In air it melts with decomposition at about 120°C. It can be obtained by the reaction of *trans*-$RhCl(CO)(PPh_3)_2$ with sodium borohydride in the presence of excess triphenylphosphine, but is more conveniently prepared directly from $RhCl_3 \cdot 3H_2O$ by reaction with triphenylphosphine and formaldehyde in refluxing ethanolic potassium hydroxide.

Synthesis of Carbonylhydridotris(triphenylphosphine)rhodium(I)[596]

$$RhCl_3 \cdot 3H_2O + PPh_3 + CH_2O \rightarrow RhH(CO)(PPh_3)_3$$

To a refluxing solution of triphenylphosphine (2.64 g, 10 mmol) in ethanol (100 cm^3) is added $RhCl_3 \cdot 3H_2O$ (0.26 g, 1.0 mmol) dissolved in ethanol (20 cm^3). After a very short delay (~ 15 s), aqueous formaldehyde (10 cm^3 of 40% solution) and hot ethanolic potassium hydroxide (0.8 g, 14 mmol, in 20 cm^3 of ethanol) are added. After refluxing for 10–20 min the mixture is allowed to cool and the product crystallizes. Yield 0.8–0.9 g ($\sim 95\%$ based on rhodium).

Dodecacarbonyltetrarhodium(0)—Rh$_4$(CO)$_{12}$

Dark red crystalline $Rh_4(CO)_{12}$ is commercially available. It is soluble in most organic solvents but readily decomposes to $Rh_6(CO)_{16}$ in the presence of water. At 100°C this reaction takes place smoothly in the solid state. $Rh_4(CO)_{12}$ has been used as a source of soluble rhodium for the catalysis of carbonylation reactions. The preparation described here makes use of the reaction of $[Rh(CO)_2Cl]_2$ with carbon monoxide at atmospheric pressure, in the presence of a carefully controlled amount of water.

Synthesis of Dodecacarbonyltetrarhodium(0)[597]

$$[Rh(CO)_2Cl]_2 + CO/NaHCO_3 \rightarrow Rh_4(CO)_{12}$$

Into a carefully dried three-neck round bottom flask (500 cm^3) is placed $[Rh(CO)_2Cl]_2$ (1.0 g, 2.57 mmol), dried sodium bicarbonate (2.86 g, 34.0 mmol), and dry degassed *n*-hexane (425 cm^3). Carbon monoxide, dried over P_4O_{10}, is flushed through the flask for 5 min before bubbling it through the magnetically stirred reaction mixture for 2 h. Addition of water (0.03 cm^3) causes the rapid formation of $Rh_4(CO)_{12}$ (red coloration). After stirring for another $1\frac{1}{2}$ h under carbon monoxide, the solution is filtered and the liquid concentrated under reduced pressure to about 140 cm^3. When the concentrate is cooled ($-70°C$) the product crystallizes out and is filtered off at $-70°C$. Yield 0.5 g (52%).

12.3. Recovery of Precious Metals

12.3.1. Introduction

In reactions where precious metal compounds are used as stoichiometric reagents, considerable amounts of metal can be involved, and it is then economically sensible to recover the metal from the reaction mixture after removal of the organic product. Even when only small amounts of precious metal *catalysts* are used, residues can be collected together and retained for subsequent work-up. It is important that residues containing different precious metals be kept separate, since although recovery of a single metal can be straightforward, the separation of metals from mixed residues by selective precipitation,[598] or liquid–liquid extraction procedures,[599] would be so time consuming as not to be worthwhile in the research laboratory.

In general, if arrangements can be made for returning precious metal residues to a refiner for credit, it is recommended that this be done, since it is the most straightforward means of recycling the metal. However, occasionally it *is* useful to be able to obtain specific compounds from precious metal residues. Although there is relatively little work published in this area, only simple reactions are needed to recover some metals.[600-602] It is fortunate that palladium is one of the easiest metals to recover, since it is probably now the most widely used precious metal in laboratory-scale catalytic carbonylation.

12.3.2. Safety Aspects

In general precious metal compounds should be handled carefully, and the use of a fume cupboard is recommended. Ruthenium and osmium form extremely toxic volatile tetraoxides (RuO_4, bp $100°C$; OsO_4, bp $131°C$), which are injurious to both eyes and lungs, but compounds of these metals are seldom used in carbonylation reactions. Platinum compounds can cause dermatitis, and they present an additional potential hazard for workers who become sensitized to them.[603] For such individuals, exposure to even very low levels of soluble platinum compounds causes an allergic reaction known as platinosis (see also Sec. 3.4). Typical symptoms are repeated sneezing and profuse running of the nose, followed by tightness of the chest, shortness of breath and wheezing. Accordingly, those suffering from respiratory disease should not be subjected to further risk by working with platinum compounds, nor should sensitized workers be exposed to them.

When working with platinum compounds an efficient fume cupboard should always be used, and gloves must be worn. If symptoms of platinosis appear, exposure of the individual to platinum compounds must cease, and this normally causes the symptoms to disappear. Medical advice must however be obtained.

Compounds of other precious metals do not appear to produce the same effects as those of platinum.

12.3.3. Chemical Aspects

The procedures used for recovery of any single precious metal usually involve careful evaporation of the residue to dryness, followed by ignition at red heat in air to remove organic materials and to leave the metal behind. This is then treated with appropriate reagents to produce a pure salt or complex.

CAUTION: Before heating to dryness, a very small quantity of the residue should be tested before working on a larger scale. Mixtures containing organic material and oxidizing agents such as nitrates or perchlorates can explode violently when heated! Ignitions must always be carried out in an efficient fume cupboard behind a suitable shield, and it is suggested that large batches of material be divided into small amounts so that a maximum of 20 g of material is ignited at any one time.

The practical difficulties in obtaining precious metal salts from ignited residues stem largely from the inertness of the metals themselves. Their reactivities are detailed in Table 12.1. Palladium metal is relatively reactive and dissolves in hot concentrated nitric acid. A more useful reagent,

Table 12.1. Chemical Reactivity of Selected Precious Metals and Procedures for their Recovery from Ignited Residues

Metal	Reactivity	Conversion of metal to useful salt
Palladium	Dissolves in concentrated nitric acid, and even in aerated hydrochloric acid.	Refluxing in acetic/nitric acid affords palladium acetate (see Ref. 562), and the chloride via treatment with aqua regia.
Platinum	Almost completely inert to mineral acids other than aqua regia. Reacts with chlorine at red heat, and rapidly with fused alkali.	Dissolving in aqua regia, followed by repeated evaporation with hydrochloric acid, gives chloroplatinic acid (see Ref. 604), which is easily converted to platinum(II) chloride (see Ref. 605).
Rhodium	Reacts at red heat with chlorine to form $RhCl_3$, and in the presence of potassium chloride, K_3RhCl_6. Dissolves in fused alkali containing oxidizing agents. Insoluble in mineral acids, but when very finely divided will dissolve slowly in aqua regia.	Chlorination of metal mixed with potassium chloride at 575 °C produces K_3RhCl_6, which can be converted to $RhCl_3 \cdot 3H_2O$ (see Ref. 606).
Iridum	Very inert and insoluble in aqua regia. Dissolves in fused alkali nitrates. At red heat reacts with chlorine to form $IrCl_3$.	Chlorination of metal mixed with sodium chloride at 625 °C. Work-up gives $(NH_4)_2IrCl_6$ (see Ref. 607).
Ruthenium	Ruthenium dioxide is slowly formed when heated in air. Attacked by chlorine, and dissolves in fused alkali. Insoluble in mineral acids including aqua regia, but can be dissolved in aqua regia containing $KClO_3$ (caution!)	Fusion with sodium peroxide gives Na_2RuO_4, which in aqueous solution is converted by chlorine to highly toxic RuO_4—not a procedure recommended for general use.

however, is hot acetic acid containing about 5% concentrated nitric acid, which provides a convenient means of obtaining the versatile carbonylation catalyst palladium(II) acetate (see Sec. 12.2.5).[562] Although of only limited application in carbonylation chemistry, silver is also readily converted to the nitrate by nitric acid, but the other precious metals are considerably more inert to attack by mineral acids.

Platinum dissolves in hot aqua regia (three volumes of concentrated hydrochloric acid to one volume of concentrated nitric acid). Repeated evaporation of the solution so formed with hydrochloric acid removes the nitric acid, and affords chloroplatinic acid,[604, 605] which crystallizes in brownish-red very deliquescent prisms ($H_2PtCl_6 \cdot 6H_2O$). In practice, the chloroplatinic acid solution is more often used directly as a source of other platinum compounds. Treatment of palladium with aqua regia gives H_2PdCl_4 on evaporation, and this, when heated to 500°C, gives[569] polymeric $PdCl_2$ (see Sec. 12.2.5). The hydrated chloride, $PdCl_2 \cdot 2H_2O$, can be obtained from H_2PdCl_4 by repeated evaporation to near dryness with a little hydrochloric acid.

The remaining precious metals (rhodium, iridium, ruthenium, and osmium) are more inert than platinum, and they are very reluctant to dissolve in any acid. As a result of their inert character these metals are difficult to recycle in the laboratory. It is possible to convert them to their chlorides by heating strongly in chlorine, but although this is the recommended procedure for recycling rhodium[606] and iridium,[607] it is not an attractive proposition. Even the most inert of the precious metals can (with some difficulty) be dissolved in fused potassium hydroxide containing potassium nitrate, to form oxide species, which can subsequently be dissolved in acid.[608] However, although this approach works well for ruthenium and osmium, subsequent purification steps invariably involve distillation of the highly toxic tetraoxides,[609–611] an operation not to be undertaken without serious consideration of the hazards involved.

12.3.4. Recovery of Palladium

Volatile solvents are carefully evaporated, and the residue is dried before it is strongly ignited in air at the maximum available temperature for 2 h (caution: a small sample must first be examined for explosive behavior). When cold the remaining black material is finely ground and soluble impurities (salts) are removed by washing with dilute hydrochloric acid, before extraction with two portions of hot aqua regia. The decanted aqua regia is filtered and concentrated by evaporation to small volume. Dilution of this solution with dilute hydrochloric acid followed by repeated evaporation to dryness removes residual nitrate. Dilution with water and addition of excess sodium hydroxide then enables the metal to be precipitated by addition of excess hydrazine hydrate. The black metal powder can be converted to palladium acetate or chloride as required using the procedures described in Sec. 12.2.5.

12.3.5. Recovery of Platinum

See Secs. 3.4 and 12.3.2 for potential hazards associated with the use of platinum compounds. Treatment of crude residues is carried out as described above for recovery of palladium—they are dried and strongly ignited in air to form finely divided platinum (caution: a small sample must first be examined for possible explosive behavior). Soluble impurities (salts) are removed by washing with hydrochloric acid and water, before dissolving the metal in aqua regia. After destroying any remaining nitro complexes by repeated evaporation to dryness with hydrochloric acid, the resulting chloroplatinic acid can be used as such, or converted to its sparingly soluble ammonium salt. This involves reaction with aqueous ammonium chloride followed by addition of alcohol to precipitate yellow $(NH_4)_2PtCl_6$. The solid is collected and washed with portions of ice-cold ammonium chloride solution, ethanol, and finally ether. Careful thermal decomposition of $(NH_4)_2PtCl_6$ affords $PtCl_2$.

12.3.6. Conclusions

It is not difficult to recover palladium and platinum from residues in which they are the only metal, and when their compounds are used as carbonylation catalysts they can be fairly readily recycled in the laboratory. On the other hand, rhodium, iridium, ruthenium, and osmium can be very difficult to recover without special facilities, and it is recommended that residues containing these metals be kept separate and always returned to refiners for metal recovery. In the absence of severe economic constraints, it is clearly more straightforward for research laboratories to recycle all precious metal residues (including platinum and palladium) via a commercial refiner.

Appendix 1

Suppliers of Transition Metal Catalysts and Reagents

Many of the catalysts and reagents referred to in this book are available through the chemical supply houses (or their local agencies) whose addresses are given below. Starting materials for catalyst preparation are also available from these sources, and from a number of refiners of precious metals whose addresses are listed; such refiners may arrange to reprocess reaction residues containing significant amounts of the platinum group metals.

Strem Chemicals Inc.
7 Mulliken Way
Dexter Industrial Park
Newburyport
Massachusetts 01950
USA

Available through:

Strem Chemicals GmbH
Querstrasse 2
7640 Kehl
Germany

Strem Chemicals Inc.
15 Rue de L'Atome
Zone Industrielle
67800 Bischheim
France

Fluorochem Ltd.
Wesley Street
Old Glossop
Derbyshire SK13 9RY
England

Strem Chemicals Inc.
7 Mulliken Way
Dexter Industrial Park
Newburyport
Massachusetts 01950
USA

Available through:

Kokusai Kinzoku Yakuhin Company
8F Star Plaza
Aoyama Building
10-3 Shibuya 1-Chome
Shibuya-ku Box 147
Tokyo
Japan

Aldrich Chemical Company Inc.
940 West Saint Paul Avenue
Milwaukee
Wisconsin 53233
USA

Available through:

Aldrich Chemie N.V./S.A.
Boulevard Lambermontlaan 140
B-1030 Brussels
Belgium

Aldrich-Chemie S.a.r.l.
27 Fosse des Treize
F-67000 Strasbourg

Aldrich Japan
Kyodo Building Shinkanda
10 Kanda-Mikuracho
Chiyoda-ku
Tokyo
Japan

Aldrich Chemical Company Ltd.
The Old Brickyard
New Road
Gillingham
Dorset SP8 4JL
England

Aldrich-Chemie GmbH & Co. KG
D-7924 Steinheim
Germany

Johnson Matthey/Alpha Products
152 Andover Street
Danvers
Massachusetts 01923
USA

Available through:

Johnson Matthey Gmbh.
Alpha Products
Zeppelinstrasse 7
Postfach 6540
D-7500 Karlsruhe 1
Germany

Maagar-Scientific Services Ltd.
Kiryat Weizman
Rohovot 70400
Israel

Johnson Matthey/Alpha Products
152 Andover Street
Danvers
Massachusetts 01923
USA

Available through:

Johnson Matthey Ltd.
Toronto Postal Station W
Toronto, Ontario M6M 5C2
Canada

Johnson Matthey Australia
160 Rocky Point Road
Kogarah
New South Wales 2217
Australia

Organometallics Inc.
PO Box 287
East Hampstead
New Hampshire 03826
USA

Pressure Chemical Company
3419 Smallman Street
Pittsburgh
Pennsylvania 15201
USA

Grilyt Emser Werker AG
Market Development Department
CH-8039 Zurich
Switzerland

The following precious metal refiners supply compounds and catalysts, and also reprocess precious metal residues:

Johnson Matthey Chemicals Ltd.
Orchard Road
Royston
Hertfordshire SG8 5HE
England

Johnson Matthey Inc.
Malvern
Pennsylvania 19355
USA

Engelhard Industries
529 Delancy Street
Newark
New Jersey 07105
USA

Engelhard Sales Ltd.
Chemical Group
Valley Road
Cinderford
Gloucester GL14 2PB
England

Appendix 2
Carbon Monoxide Suppliers

Carbon monoxide can be prepared by dehydration of formic or oxalic acid with sulfuric acid,* but this is seldom done since the gas is readily available in cylinders at moderate cost. The following are some of the companies that supply carbon monoxide in small cylinders, together with suitable pressure regulators and associated equipment:

Aldrich Chemical Company Inc.
940 West Saint Paul Avenue
Milwaukee
Wisconsin 53233, USA

Air Products Ltd.
Speciality Gases
Western Road
Crewe CW1 1DF, UK

BOC Special Gases
24 Deer Park Road
London SW19 3UF, UK

Airco Industrial Gases
Murry Hill
New Jersey 07974, USA

MG Industries, Gas Products
2460 Boulevard of the Generals
Valley Forge
Pennsylvania 19482, USA

Aldrich Chemical Company Ltd.
The Old Brickyard
New Road
Gillingham
Dorset SP8 4JL, UK

Air Products & Chemical Inc.
Allentown
Pennsylvania 18105, USA

Union Carbide Corporation
Linde Division
National Speciality Gases Office
100 Davidson Avenue
Somerset
New Jersey 08873, USA

Matheson Gas Products
932 Paterson Plank Road
P.O. Box 85
East Rutherford
New Jersey 07073, USA

* W. L. Gilliland and A. A. Blanchard, *Inorg. Synth.*, **2**, 81 (1946); W. Rudorff, in *Handbook of Preparative Inorganic Chemistry* (G. Brauer, ed.), Academic Press, New York, pp. 645–647 (1963).

Appendix 3
Gas Monitors

Detectors used for monitoring carbon monoxide may be divided into two classes: those that rely on some color change associated with a specific chemical reaction, and those that make use of electronic sensing. Systems involving chromatographic separation prior to detection are included in the second class. Examples of the former includes the black coloration due to iodine when iodine pentoxide quantitatively oxidizes carbon monoxide to carbon dioxide, and examples of electronic sensors include infrared cells, and those that have catalytic oxidation elements (usually based on platinum) or electrochemical elements.

Information contained in this appendix was obtained from various suppliers' catalogs, and is intended to be representative of what is available at the time of writing, rather than any recommendation or endorsement of particular products. Since products and specifications constantly change, interested readers should contact a number of suppliers for the latest information and recommendations.

Chemical Tube Detectors

These are easy to operate, incorporating a simple hand pump to pass a measured volume of air through a short tube containing the indicator reagent (typically I_2O_5 for CO detection). The concentration of carbon monoxide is indicated by the length of darkened zone in the tube. Detectors of this type are obtainable from a number of suppliers including the following:

Dragerwerk AG   Draeger Ltd.
Lubeck   The Willows
Germany   Mark Road
                                 Hemel Hempstead
                                 Herfordshire, HP2 7BW
                                 England

Dragerwerk's authoritative *Detector Tube Handbook* details the range of gases that can be detected by their products, and also provides information about the underlying chemistry and physics involved. "Sensidyne" toxic gas detection kits, are available from:

MG Industries, Gas Products   Union Carbide Corporation
2460 Boulevard of the Generals   Linde Division
Valley Forge   National Speciality Gases Office
Pennsylvania 19482, USA   100 Davidson Avenue
                                     Somerset
                                     New Jersey 08873, USA

The related chemical-based "Matheson-Kitagawa Toxic Gas Detector System" is available from:

Matheson Gas Products
932 Paterson Plank Road
P.O. Box 85
East Rutherford
New Jersey 07073, USA

Electronic Detectors

Many companies supply fixed laboratory electronic carbon monoxide detection systems that provide continuous monitoring of the laboratory atmosphere, or a specific area. Special mention should also be made of the increasing number of small portable electronic hazardous gas detectors that are useful in many situations. For example, Matheson Gas Products (see above for address) supply one (model No. 8057) that uses a solid state/thermal conductivity sensor incorporating a platinum filament. This is sensitive to a wide range of gases and vapors, and has a carbon monoxide detection limit of 1 ppm, with a sensitivity control that permits the alarm to be set above the background level of any solvent vapor present. This can be used as a personal unit clipped to a belt, and it also doubles as a leak detector for checking couplings, etc. Further information is contained in "TechBrief" No. TB-218 available from Matheson. Other examples of portable instruments include one made by Sieger (address is given below) that was originally designed for use in mining, steel, and petrochemical industries. It employs an electrochemical sensing element. GasTech Inc. (address below) supply small personal monitor units containing three or four sensors (e.g., combustibles, oxygen, hydrogen sulfide, and carbon monoxide) that give a continuous display of concentration levels on a LCD display. International Sensor Technology (adress below) supply a small portable detector for a single gas. It is provided with a meter and an audible alarm. Like other models it operates from rechargeable batteries.

The names and addresses of some representative suppliers of electronic carbon monoxide detection systems are given below, and it should be noted that some companies rent gas detection systems.

Neotronics Ltd.
Parsonage Road
Takeley
Bishop's Stortford
Herts CM22 6PU, UK

Complete monitoring systems incorporating electrochemical sensing elements.

Sieger Ltd.
Fulwood Close
Fulwood Industrial Estate
Sutton-in-Ashfield
Notts NG17 2JZ, UK

Complete monitoring systems with infrared, catalytic or electrochemical sensors.

The Foxboro Company
151 Woodward Avenue
South Norwalk
Connecticut 06856-5449, USA

Complete monitoring systems. Rugged industrial portable units.

Microsensor Technology Inc.
41762 Christy Street
Fremont
California 94538, USA

Compact chromatographic based portable systems.

Astro Safety Products Inc.
100 Park Avenue
League City
Texas 77573, USA

Novel infrared sensor-based systems.

GasTech Inc.
8445 Central Avenue
Newark
California 94560, USA

Range of systems mainly electrochemical sensors. Variety of portable instruments with multiple sensors.

International Sensor Technology
17771 Fitch Street
Irvine
California 92714, USA

Complete systems, specialize in sensor elements and wireless links. Small portable single gas instruments.

MSA (Britain) Ltd.
East Shawhead
Coatbridge ML5 4TD
Scotland

Complete systems. Electrochemical sensors.

Appendix 4

Suppliers of Autoclave Equipment

Autoclaves are available in a wide range of reactor sizes and pressure ratings, from small bench-top models ($25\,cm^3$) to those with a capacity of several liters. The normal material of construction for autoclave equipment is stainless steel, but as noted in Chapter 3, glass reactors rated at up to 12 bar are available from Buchi A.G. of Switzerland.

Names and addresses of companies who kindly provided the illustrations of autoclave equipment used in Chapter 3 are given below.

Autoclave Engineers Group
2930 West 22nd Street
P.O. Box 4007
Erie
Pennsylvania
16512 USA

Buchi AG, Technical Glass
Gschwaderstrasse 12
CH-8610 Uster
Switzerland

Parr Instrument Company
211 53rd Street
Moline
Illinois
61265 USA

Available through:

Scientific and Medical Products Ltd.
Shirley Institute
Didsbury
Manchester
M20 8RX, UK

References

1. See for example S. Patai (ed.), *The Chemistry of the Carbonyl Group*, Wiley-Interscience, New York (1966), and S. Warren, *Chemistry of the Carbonyl Group*, Wiley, London (1974).
2. H. M. Colquhoun, J. Holton, D. J. Thompson, and M. V. Twigg, *New Pathways for Organic Synthesis*, Plenum Press, New York (1984).
3. (a) J. Falbe (ed.), *New Syntheses with Carbon Monoxide*, Springer-Verlag, Berlin (1980). (b) M. V. Twigg (ed.), *Catalyst Handbook*, 2nd Edn., Wolfe, London (1989).
4. H. C. Brown, *Acc. Chem. Res.*, **2**, 65 (1969).
5. (a) I. Wender and P. Pino (eds.), *Organic Synthesis via Metal Carbonyls*, Vol. 2, John Wiley and Sons, New York (1977). R. L. Pruett, *Adv. Organomet. Chem.*, **17**, 1 (1979). (b) J. Falbe, *Carbon Monoxide in Organic Synthesis*, Springer-Verlag, Berlin (1970).
6. (a) O. Roelen (to Rührchemie AG), German Patent No. 849,548 (1938). (b) W. A. Herrmann, *J. Organomet. Chem.*, **383**, 21 (1990).
7. M. Orchin, *Acc. Chem. Res.*, **14**, 259 (1981).
8. W. Reppe (to IG Farben), German Patent No. 855,110 (1939).
9. *Hydrocarbon Processing*, 120 (Nov. 1974).
10. E. G. Hancock (ed.), *Propylene and its Industrial Derivatives*, Benn, London (1973).
11. C. W. Bird, *Chem. Rev.*, **62**, 283 (1962).
12. J. A. Osborn, J. F. Young, and G. Wilkinson, *J. Chem. Soc., Chem. Commun.*, 17 (1965).
13. (a) R. F. Heck, *Palladium Reagents in Organic Synthesis*, Academic Press, New York (1985). (b) J. Tsuji, *Organic Synthesis with Palladium Compounds*, Springer-Verlag, Berlin (1980).
14. H. Alper and H. des Abbayes, *J. Organomet. Chem.*, **134**, C11 (1977). L. Cassar and M. Foà, *J. Organomet. Chem.*, **134**, C15 (1977).
15. L. de Picciotto, P. A. Carrupt, and P. Vogel, *J. Org. Chem.*, **47**, 3796 (1982).
16. J. K. Stille and R. Divakaruni, *J. Org. Chem.*, **44**, 3474 (1979).
17. E. Dalcanale and M. Foà, *Synthesis*, 492 (1986).
18. J. L. Eisenmann, R. L. Yamartino, and J. F. Howard Jr., *J. Org. Chem.*, **26**, 2102 (1961).
19. F. Ozawa, N. Kawasaki, H. Okamoto, T. Yamamoto, and A. Yamamoto, *Organometallics*, **6**, 1640 (1987), and references therein.
20. M. Foà and F. Francalanci, *J. Mol. Catal.*, **41**, 89 (1987).
21. S. Cacchi, P. G. Ciattini, E. Morera, and G. Ortar, *Tetrahedron Lett.*, 3931 (1986).

22. U. Gelius, E. Basilier, S. Svensson, T. Bergmark, and K. Siegbahn, *J. Electron Spectrosc.*, **2**, 405 (1973).

23. V. Rautenstrauch and M. Joyeux, *Angew. Chem., Int. Ed. Engl.*, **18**, 83 (1979).

24. C. Narayana and M. Periasamy, *Synthesis*, 253 (1985), and references therein.

25. R. West, *Oxocarbons*, Academic Press, New York (1980).

26. H. Bahrmann, in *New Syntheses with Carbon Monoxide* (J. Falbe, ed.), p. 372, Springer-Verlag, Berlin (1980).

27. A. B. Burg and H. I. Schlesinger, *J. Am. Chem. Soc.*, **59**, 780 (1937).

28. G. A. Olah, F. Pelizza, S. Kobayashi, and J. A. Olah, *J. Am. Chem. Soc.*, **98**, 296 (1976).

29. J. E. Ellis and R. A. Faltynek, *J. Chem. Soc., Chem. Commun.*, 966 (1975).

30. H. C. Clark, K. R. Dixon, and W. J. Jacobs, *J. Am. Chem. Soc.*, **91**, 1346 (1969).

31. K. M. Doxsee and R. H. Grubbs, *J. Am. Chem. Soc.*, **103**, 7696 (1981).

32. G. K. Anderson and R. J. Cross, *Acc. Chem. Res.*, **17**, 67 (1984).

33. H. C. Brown, *Acc. Chem. Res.*, **2**, 65 (1969).

34. J. K. Kochi, *Organometallic Mechanisms and Catalysis*, Academic Press, New York (1978).

35. C. E. Hickey and P. M. Maitlis, *J. Chem. Soc., Chem. Commun.*, 1609 (1984).

36. T. T. Tsou and J. K. Kochi, *J. Am. Chem. Soc.*, **101**, 6319 (1979).

37. G. J. Kubas, R. R. Ryan, B. I. Swanson, P. J. Vergamini, and H. J. Wasserman, *J. Am. Chem. Soc.*, **106**, 451 (1984).

38. D. Evans, J. A. Osborn, and G. Wilkinson, *J. Chem. Soc. (A)*, 3133 (1968).

39. I. Kovacs, F. Ungvary, and L. Marko, *Organometallics*, **5**, 209 (1986).

40. A. Schoenberg and R. F. Heck, *J. Am. Chem. Soc.*, **96**, 7761 (1974).

41. J. F. Fauvarque, F. Pflüger, and M. Troupel, *J. Organomet. Chem.*, **208**, 419 (1981).

42. (a) F. Ozawa, N. Kawasaki, H. Okamoto, T. Yamamoto, and A. Yamamoto, *Organometallics*, **6**, 1640 (1987). (b) F. Ozawa, H. Soyoma, H. Yanagihara, I. Aoyama, H. Takino, K. Izawa, T. Yamamoto, and A. Yamamoto, *J. Am. Chem. Soc.*, **107**, 3235 (1987). (c) H. Yamashita, T. Sakakura, T. A. Kobayashi, and M. Tanaka, *J. Mol. Catal.*, **48**, 69 (1988).

43. J. Mantzaris and E. Weissberger, *J. Am. Chem. Soc.*, **96**, 1880 (1974).

44. F. Calderazzo and K. Noack, *Coord. Chem. Rev.*, **1**, 118 (1966).

45. P. M. Henry and G. A. Ward, *J. Am. Chem. Soc.*, **94**, 673 (1972).

46. K. C. Brinkman and J. A. Gladysz, *Organometallics*, **3**, 147 (1984).

47. R. R. Schrock, *Acc. Chem. Res.*, **12**, 98 (1979).

48. D. Forster, *J. Am. Chem. Soc.*, **98**, 846 (1976).

49. J. F. Roth, J. H. Craddock, A. Hershman, and F. E. Paulik, *Chem. Technol.*, 600 (1971).

50. F. Ozawa, T. Sugimoto, Y. Yuasa, M. Santra, T. Yamamoto, and A. Yamamoto, *Organometallics*, **3**, 683 (1984).

51. R. Takeuchi, Y. Tsuji, and Y. Watanabe, *J. Chem. Soc., Chem. Commun.*, 351 (1986).

52. P. Pyykkö and J. P. Desclaux, *Acc. Chem. Res.*, **12**, 276 (1979).

53. L. Cassar and M. Foà, *J. Organomet. Chem.*, **134**, C15 (1977).

54. R. F. Heck and D. S. Breslow, *J. Am. Chem. Soc.*, **83**, 4023 (1961).

55. Y. Souma, H. Sano, and J. Iyoda, *J. Org. Chem.*, **38**, 2016 (1973).

56. H. Alper, J. K. Currie, and H. des Abbayes, *J. Chem. Soc., Chem. Commun.*, 311 (1978).

57. D. H. Doughty, in *Homogeneous Catalysis with Metal–Phosphine Complexes* (L. H. Pignolet, ed.), p. 343, Plenum, New York (1983).

58. N. I. Sax and R. J. Lewis, *Dangerous Properties of Industrial Materials*, Van Nostrand Reinhold, New York (1989).

59. *Catalyst Handbook*, 2nd Edition (M. V. Twigg, ed.), Chap. 5, Wolfe, London (1989).

60. D. F. Shriver and M. A. Drezdzon, *The Manipulation of Air-Sensitive Compounds*, 2nd Edition, J. Wiley, New York (1986).

61. A. I. Vogel, *Textbook of Practical Organic Chemistry*, 4th Ed., pp. 65–67, Longmans, London (1978).

62. *The Volatile Metal Carbonyls*, Pamphlet concerned with handling and other practical aspects of metal carbonyls, formerly published by the International Nickel Company, Clydach, Swansea, Wales.

63. P. W. Jolly, in *Comprehensive Organometallic Chemistry*, (G. Wilkinson, ed.), Vol. 6, pp. 7–8, Pergamon, Oxford (1982).

64. H. Remy, *Treatise on Inorganic Chemistry*, (J. S. Anderson, trans., J. Kleinberg, ed.), Vol. 2, p. 7, Elsevier, Amsterdam (1956).

65. W. Hieber and E. O. Fischer, *Z. Anorg. Allg. Chem.*, **269**, 292 (1952); **271**, 229 (1953).

66. (a) F. Steel, in *Handbook of Preparative Inorganic Chemistry*, 2nd Ed., (G. Brauer, ed.), Vol. 2, pp. 1747–1749, Academic Press, New York (1965). (b) W. Hieber, E. O. Fischer, and E. Bockly, *Z. Anorg. Allg. Chem.*, **269**, 308 (1952).

67. F. W. Laird, *Recl. Trav. Chim. Pays-Bas*, **46**, 177 (1927).

68. C. F. van Duin, *Recl. Trav. Chim. Pays-Bas*, **46**, 381 (1927).

69. W. L. Gilliland and A. A. Blanchard, *Inorg. Synth.*, **2**, 234 (1946).

70. (a) E. Hirsch and E. Peters, *Can. Metall. Quart.*, **3**, 137 (1964). (b) G. N. Dobrokhotov, *Zh. Prikl. Khim.*, **32**, 757 (1959). (c) G. N. Dobrokhotov, N. I. Onuchkina, and P. S. Kudryavtsev, U.S.S.R. Patent No. 114,061 (1958).

71. P. Gilmont and A. A. Blanchard, *Inorg. Synth.*, **2**, 238 (1946).

72. R. J. Clark, S. E. Whiddon, and R. E. Serfass, *J. Organomet. Chem.*, **11**, 637 (1968).

73. J. A. Roth and M. Orchin, *J. Organomet. Chem.*, **187**, 103 (1980).

74. P. M. Treichel, in *Comprehensive Organometallic Chemistry*, (G. Wilkinson, ed.), Vol. 4, p. 67, Pergamon, Oxford (1982).

75. M. J. Cleare, E. G. Hughes, B. Jacoby, and J. Pepys, *Clin. Allergy*, **6**, 183 (1976).

76. W. L. Jolly, *Synthetic Inorganic Chemistry*, pp. 129–135, Prentice Hall, Englewood Cliffs, New Jersey (1960).

77. Y. Takegami, Y. Watanabe, H. Masada, and I. Kanaya, *Bull. Chem. Soc. Jpn.*, **40**, 1456 (1967).

78. M. P. Cooke Jr., *J. Am. Chem. Soc.*, **92**, 6080 (1970).

79. M. Ryang, I. Rhee, and S. Tsutsumi, *Bull. Chem. Soc. Jpn.*, **37**, 341 (1964).

80. A. Schoenberg and R. F. Heck, *J. Am. Chem. Soc.*, **96**, 7761 (1974).

81. R. Mutin, C. Lucas, J. Thivolle-Cazat, V. Dufaud, F. Dany, and J. M. Basset, *J. Chem. Soc., Chem. Commun.*, 896 (1988).

82. Y. Ben-David, M. Portnoy, and D. Milstein, *J. Chem. Soc., Chem. Commun.*, 1816 (1989).

83. (a) R. Takeuchi, Y. Tsuji, and Y. Watanabe, *J. Chem. Soc., Chem. Commun.*, 351 (1986). (b) R. Takeuchi, Y. Tsuji, M. Fujita, T. Kondo, and Y. Watanabe, *J. Org. Chem.*, **54**, 1831 (1989).

84. A. Kasahara, T. Izumi, and H. Yanai, *Chem. Ind. (London)*, 898 (1983).

85. I. Pri-Bar and O. Buchman, *J. Org. Chem.*, **49**, 4009 (1984).

86. V. P. Baillargeon and J. K. Stille, *J. Am. Chem. Soc.*, **108**, 452 (1986).

87. K. Kikukawa, T. Totoki, F. Wada, and T. Matsuda, *J. Organomet. Chem.*, **270**, 283 (1984).

88. O. Roelen (to Rührchemie A.G.), German Patent No. 849,548 (1938).

89. (a) R. L. Pruett, *Adv. Organomet. Chem.*, **17**, 1 (1979). (b) B. L. Cornils, in *New Syntheses with Carbon Monoxide*, (J. Falbe, ed.), p. 1, Springer-Verlag, Berlin (1980). (c) M. Orchin, *Acc. Chem. Res.*, **14**, 259 (1981).

90. (a) R. F. Heck and D. S. Breslow, *J. Am. Chem. Soc.*, **83**, 4023 (1961). (b) L. Versluis, T. Zeigler, E. J. Baerends, and W. Ravenek, *J. Am. Chem. Soc.*, **111**, 2018 (1989), and references therein.

91. R. L. Pruett and J. A. Smith, *J. Org. Chem.*, **34**, 327 (1969).
92. C. K. Brown and G. Wilkinson, *J. Chem. Soc. A*, 2753 (1970).
93. J. C. Bayon, J. Real, C. Claver, A. Polo, and A. Ruiz, *J. Chem. Soc., Chem. Commun.*, 1056 (1989).
94. M. M. Taqui Khan, S. B. Halligudi, and S. H. R. Abdi, *J. Mol. Catal.*, **48**, 313 (1988).
95. (a) T. Hayashi, Y. Kawabata, T. Isoyama, and I. Ogata, *Bull. Chem. Soc., Jpn.*, **54**, 3438 (1981). (b) E. Paumard, A. Mortreux, and F. Petit, *J. Chem. Soc., Chem. Commun.*, 1380 (1989). (c) A. Scrivanti, C. Botteghi, L. Toniolo, and A. Berton, *J. Organomet. Chem.*, **344**, 261 (1988). (d) A. Scrivanti, S. Paganelli, U. Matteoli, and C. Botteghi, *J. Organomet. Chem.*, **385**, 439 (1990). (e) L. Kollár, J. Bakos, B. Heil, P. Sándor, and G. Szalontai, *J. Organomet. Chem.*, **385**, 147 (1990).
96. S. C. Tang and L. Kim, *J. Mol. Catal.*, **14**, 231 (1982).
97. B. Heil and L. Marko, *Chem. Ber.*, **102**, 2238 (1969).
98. P. W. N. M. van Leeuwen and C. F. Roobeek, *J. Organomet. Chem.*, **258**, 343 (1983).
99. A. I. M. Keulemans, A. Kwantes, and T. van Bavel, *Recl. Trav. Chim. Pays-Bas*, **67**, 298 (1948).
100. J. M. Brown, S. J. Cook, and R. Khan, *Tetrahedron*, **42**, 5105 (1986).
101. R. Grigg, G. Reimer, and A. R. Wade, *J. Chem. Soc., Perkin Trans. (1)*, 1929 (1983).
102. P. Salvadori, G. Vitulli, A. Raffaelli, and R. Lazzaroni, *J. Organomet. Chem.*, **258**, 351 (1983).
103. Y. Ono, S. Sato, M. Takesada, and H. Wakamatsu, *J. Chem. Soc., Chem. Commun.*, 1255 (1970).
104. J. Falbe and F. Korte, *Chem. Ber.*, **97**, 1104 (1964).
105. J. Falbe, H. J. Schultze-Steinen, and F. Korte, *Chem. Ber.*, **98**, 886 (1965).
106. B. Fell and M. Barl, *Chem. Z.*, **101**, 343 (1977).
107. I. Ojima, *Chem. Rev.*, **88**, 1011 (1988).
108. C. Botteghi and F. Soccolini, *Synthesis*, 596 (1985).
109. H. Siegel and W. Himmele, *Angew. Chem., Int. Ed. Engl.*, **19**, 178 (1980).
110. G. Consiglio, P. Pino, L. I. Flowers, and C. U. Pittman, *J. Chem. Soc., Chem. Commun.*, 612 (1983).
111. L. Kollar, J. Bakos, I. Toth, and B. Heil, *J. Organomet. Chem.*, **370**, 257 (1989).
112. L. Kollar, G. Consiglio, and P. Pino, *J. Organomet. Chem.*, **330**, 305 (1987).
113. (a) G. Consiglio, F. Morandini, M. Scalone, and P. Pino, *J. Organomet. Chem.*, **279**, 193 (1985). (b) C. F. Hobbs and W. S. Knowles, *J. Org. Chem.*, **46**, 4422 (1981).
114. G. Parrinello, R. Deschenaux, and J. K. Stille, *J. Org. Chem.*, **51**, 4189 (1986).
115. G. Parrinello and J. K. Stille, *J. Am. Chem. Soc.*, **109**, 7122 (1987).
116. I. Matsuda, A. Ogiso, S. Sato, and Y. Izumi, *J. Am. Chem. Soc.*, **111**, 2332 (1989).
117. S. Murai and N. Sonoda, *Angew. Chem., Int. Ed. Engl.*, **18**, 837 (1979).
118. J. P. Collman, S. R. Winter, and D. R. Clark, *J. Am. Chem. Soc.*, **94**, 1788 (1972).
119. M. P. Cooke Jr. and R. M. Parlman, *J. Am. Chem. Soc.*, **99**, 5222 (1977).
120. W. H. Tamblyn and R. E. Waltermire, *Tetrahedron Lett.*, **24**, 2803 (1983).
121. G. Tanguy, B. Weinberger, and H. Des Abbayes, *Tetrahedron Lett.*, **25**, 5529 (1984).
122. Y. Sawa, M. Ryang, and S. Tsutsumi, *Tetrahedron Lett.*, **10**, 5189 (1969).
123. M. Yamashita and R. Suemitsu, *Tetrahedron Lett.*, **19**, 761 (1978).
124. R. C. Cookson and G. Farquharson, *Tetrahedron Lett.*, **20**, 1255 (1979).
125. T. Koga, S. Makinouchi, and N. Okukado, *Chem. Lett.*, 1141 (1988).
126. H. Alper and J. L. Fabre, *Organometallics*, **1**, 1037 (1982).
127. M. F. Semmelhack, J. W. Herndon, and J. P. Springer, *J. Am. Chem. Soc.*, **105**, 2497 (1983).
128. Y. Tamaru, H. Ochiai, Y. Yamada, and Z. Yoshida, *Tetrahedron Lett.*, **24**, 3869 (1983).

129. Y. Tamaru, H. Ochiai, and Z. Yoshida, *Tetrahedron Lett.*, **25**, 3861 (1984).
130. J. K. Stille, *Angew. Chem., Int. Ed. Engl.*, **25**, 508 (1986).
131. T. Kobayashi and M. Tanaka, *J. Organomet. Chem.*, **205**, C27 (1981).
132. J. H. Merrifield, J. Godschalx, and J. K. Stille, *Organometallics*, **3**, 1108 (1984).
133. F. K. Sheffy, J. P. Godschalx, and J. K. Stille, *J. Am. Chem. Soc.*, **106**, 4833 (1984).
134. M. Tanaka, *Synthesis*, 47 (1981).
135. A. M. Echavarren and J. K. Stille, *J. Am. Chem. Soc.*, **110**, 1557 (1988).
136. N. A. Bumagin, A. B. Ponomaryov, and I. P. Beletskaya, *Tetrahedron Lett.*, **26**, 4819 (1985).
137. T. Kondo, Y. Tsuji, and Y. Watanabe, *J. Organomet. Chem.*, **345**, 397 (1988).
138. N. A. Bumagin, P. G. More, and I. P. Beletskaya, *J. Organomet. Chem.*, **365**, 379 (1989).
139. R. F. Heck, *J. Am. Chem. Soc.*, **90**, 5546 (1968).
140. I. Rhee, M. Ryang, T. Watanabe, H. Omura, S. Murai, and N. Sonoda, *Synthesis*, 776 (1977).
141. R. C. Larock and S. S. Hershberger, *J. Org. Chem.*, **45**, 3840 (1980).
142. K. Kikukawa, K. Kono, F. Wada, and T. Matsuda, *Chem. Lett.*, 35 (1982).
143. M. Tanaka, T. Kobayashi, and T. Sakakura, *J. Chem. Soc., Chem. Commun.*, 837 (1985).
144. S. Aoki, E. Nakamura, and I. Kuwajima, *Tetrahedron Lett.*, **29**, 1541 (1988).
145. (a) E. A. Naragon, A. J. Millendorf, and J. H. Vergilio, U.S. Patent 2,699,453, (1955).
    (b) V. N. Zudin, G. N. Il'inich, V. A. Likholobov, and Y. I. Yermakov, *J. Chem. Soc., Chem. Commun.*, 545 (1984).
146. R. F. Heck, *J. Am. Chem. Soc.*, **85**, 3381 and 3383 (1963).
147. H. Alper and J. K. Currie, *Tetrahedron Lett.*, **20**, 2665 (1979).
148. L. S. Hegedus and Y. Inoue, *J. Am. Chem. Soc.*, **104**, 4917 (1982).
149. (a) T. Mise, P. Hong, and H. Yamazaki, *Chem. Lett.*, 401 (1982). (b) P. Hong, T. Mise, and H. Yamazaki, *J. Organomet. Chem.*, **334**, 129 (1987).
150. D. Seyferth and R. C. Hui, *J. Am. Chem. Soc.*, **107**, 4551 (1985).
151. T. W. Lai and A. Sen, *Organometallics*, **3**, 866 (1984).
152. (a) C. W. Bird, R. C. Cookson, and J. Hudec, *Chem. Ind. (London)*, 20 (1960).
    (b) J. Mantzaris and E. Weissberger, *J. Am. Chem. Soc.*, **96**, 1873 (1974).
153. J. Mantzaris and E. Weissberger, *J. Am. Chem. Soc.*, **96**, 1880 (1974).
154. J. E. McMurry and A. Andrus, *Tetrahedron Lett.*, **21**, 4687 (1980).
155. P. Eilbracht, M. Acker, and W. Totzauer, *Chem. Ber.*, **116**, 238 (1983).
156. I. U. Khand and P. Pauson, *J. Chem. Res. (S)* 9; *(M)* 0168 (1977).
157. D. C. Billington, W. J. Kerr, and P. L. Pauson, *J. Organomet. Chem.*, **341**, 181 (1988).
158. D. C. Billington, I. M. Helps, P. L. Pauson, W. Thompson, and D. Willison, *J. Organomet. Chem.*, **354**, 233 (1988).
159. (a) I. U. Khand, G. R. Knox, P. L. Pauson, W. E. Watts, and M. I. Foreman, *J. Chem. Soc., Perkin Trans. (I)*, 977 (1973). (b) S. E. MacWhorter, V. Sampath, M. M. Olmstead, and N. E. Schore, *J. Org. Chem.*, **53**, 203 (1988). (c) D. C. Billington, W. J. Kerr, P. L. Pauson, and C. F. Farnocchi, *J. Organomet. Chem.*, **356**, 213 (1988).
160. D. C. Billington and P. L. Pauson, *Organometallics*, **1**, 1560 (1982).
161. D. C. Billington and D. Willison, *Tetrahedron Lett.*, **25**, 4041 (1984).
162. D. C. Billington, *Tetrahedron Lett.*, **24**, 2905 (1983).
163. H. Alper and S. Gambarotta, *J. Organomet. Chem.*, **194**, C19 (1980).
164. (a) P. Magnus, C. Exon, and P. Albaugh-Robertson, *Tetrahedron*, **41**, 5861 (1985).
    (b) P. Magnus and L. M. Principe, *Tetrahedron Lett.*, **26**, 4851 (1985).
165. K. Doyama, T. Joh, T. Shiohara, and S. Takahashi, *Bull. Chem. Soc. Jpn.*, **61**, 4353 (1988).
166. E. Negishi and J. A. Miller, *J. Am. Chem. Soc.*, **105**, 6761 (1983).

167. J. M. Tour and E. Negishi, *J. Am. Chem. Soc.*, **107**, 8289 (1985).
168. R. C. Larock, K. Takagi, J. P. Burkhart, and S. S. Hershberger, *Tetrahedron*, **42**, 3759 (1986).
169. E. Amari, M. Catellani, and G. P. Chiusoli, *J. Organomet. Chem.*, **285**, 383 (1985).
170. L. S. Liebeskind and M. S. South, *J. Org. Chem.*, **45**, 5426 (1980).
171. A. S. Kende, R. Greenhouse, and J. A. Hill, *Tetrahedron Lett.*, **20**, 2867 (1979).
172. W. Best, B. Fell, and G. Schmitt, *Chem. Ber.*, **109**, 2914 (1976).
173. M. A. Sierra and L. S. Hegedus, *J. Am. Chem. Soc.*, **111**, 2335 (1989).
174. E. I. Negishi, Y. Zhang, I. Shimoyama, and G. Wu, *J. Am. Chem. Soc.*, **111**, 8018 (1989).
175. (a) G. P. Chiusoli and L. Cassar, *Angew. Chem., Int. Ed. Engl.*, **6**, 124 (1967). (b) L. Cassar, G. P. Chiusoli, and F. Guerrieri, *Synthesis*, 509 (1973). (c) I. Wender and P. Pino, *Organic Synthesis via Metal Carbonyls*, Vol. 2, Wiley, New York (1977). (d) J. Falbe (ed.), *New Syntheses with Carbon Monoxide*, Springer-Verlag, Berlin (1980). (e) P. W. Jolly, in *Comprehensive Organometallic Chemistry* (G. Wilkinson, F. G. A. Stone, and E. W. Able, eds.), Vol. 8, p. 773, Pergamon, Oxford (1982).
176. (a) L. Cassar and M. Foà, *J. Organomet. Chem.*, **134**, C15 (1977). (b) S. C. Shim, C. H. Doh, W. H. Park, Y. G. Kwon, and H. S. Lee, *J. Organomet. Chem.*, **382**, 419 (1990).
177. H. Alper and H. des Abbayes, *J. Organomet. Chem.*, **134**, C11 (1977).
178. H. des Abbayes and A. Buloup, *Tetrahedron Lett.*, **21**, 4343 (1980).
179. G. Tanguy, B. Weinberger, and H. des Abbayes, *Tetrahedron Lett.*, **24**, 4005 (1983).
180. G. C. Tustin and R. T. Hembre, *J. Org. Chem.*, **49**, 1761 (1984).
181. M. M. Taqui Khan, S. B. Halligudi, and S. H. R. Abdi, *J. Mol. Catal.*, **44**, 179 (1988).
182. L. Cassar, M. Foà, and A. Gardano, *J. Organomet. Chem.*, **121**, C55 (1976).
183. H. Alper, K. Hashem, and J. Heveling, *Organometallics*, **1**, 775 (1982).
184. M. Foà, F. Francalanci, A. Gardano, G. Cainelli, and A. Umani-Ronchi, *J. Organomet. Chem.*, **248**, 225 (1983).
185. S. Gambarotta and H. Alper, *J. Organomet. Chem.*, **194**, C19 (1980).
186. J. P. Collman, *Acc. Chem. Res.*, **8**, 342 (1975).
187. H. Arzoumanian, G. Buono, M. Choukrad, and J. F. Petrignani, *Organometallics*, **7**, 59 (1988).
188. T. Okano, I. Uchida, T. Nakagaki, H. Konishi, and J. Kiji, *J. Mol. Cat.*, **54**, 65 (1989).
189. F. Francalanci and M. Foà, *J. Organomet. Chem.*, **232**, 59 (1982).
190. F. Francalanci, A. Gardano, and M. Foà, *J. Organomet. Chem.*, **282**, 277 (1985).
191. H. des Abbayes and A. Buloup, *J. Chem. Soc., Chem. Commun.*, 1090 (1978).
192. R. Perron, U.S. Patent No. 4,152,352 (1979).
193. M. Foà and F. Francolanci, *J. Mol. Catal.*, **41**, 89 (1987).
194. H. Alper, *Adv. Organomet. Chem.*, **19**, 183 (1981).
195. F. Francalanci, A. Gardano, L. Abis, T. Fiorani, and M. Foà, *J. Organomet. Chem.*, **243**, 87 (1983).
196. H. Arzoumanian, G. Buono, M. Choukrad, and J. F. Petrignani, *Organometallics*, **7**, 59 (1988).
197. (a) J. Kiji, T. Okano, W. Nishiumi, and H. Konishi, *Chem. Lett.*, 957 (1988). (b) F. Joó and H. Alper, *Organometallics*, **4**, 1775 (1985).
198. I. Amer and H. Alper, *J. Am. Chem. Soc.*, **111**, 927 (1989).
199. E. R. H. Jones, G. H. Whitham, and M. C. Whiting, *J. Chem. Soc.*, 4628 (1957).
200. L. Cassar and M. Foà, *J. Organomet. Chem.*, **51**, 381 (1973).
201. L. Cassar, M. Foà, and A. Gardano, *J. Organomet. Chem.*, **121**, C55 (1976).
202. N. A. Bumagin, K. V. Nikitin, and I. P. Beletskaya, *J. Organomet. Chem.*, **358**, 563 (1988).
203. M. Tanaka, T. Kobayashi, and T. Sakakura, *J. Chem. Soc., Chem. Commun.*, 837 (1985).

204. J.-J. Brunet, C. Sidot, and P. Caubere, *Tetrahedron Lett.*, **22**, 1013 (1981).
205. T. Kashimura, K. Kudo, S. Mori, and N. Sugita, *Chem. Lett.*, 299 (1986).
206. K. Kudo, T. Shibata, T. Kashimura, S. Mori, and N. Sugita, *Chem. Lett.*, 577 (1987).
207. F. Francalanci, E. Bencini, A. Gardano, M. Vincenti, and M. Foà, *J. Organomet. Chem.*, **301**, C27 (1986).
208. M. Miura, K. Okuro, A. Hattori, and M. Nomura, *J. Chem. Soc., Perkin Trans. I*, 73 (1989).
209. I. Amer and H. Alper, *J. Org. Chem.*, **53**, 5147 (1988).
210. H. Alper, I. Amer, and G. Vasapollo, *Tetrahedron Lett.*, **30**, 2615 (1989).
211. H. Hohenschutz, N. von Kutepow, and W. Himmele, *Hydrocarbon Process.*, **45**, 141 (1966).
212. D. Forster, *Adv. Organomet. Chem.*, **17**, 255 (1979).
213. C. E. Hickey and P. M. Maitlis, *J. Chem. Soc., Chem. Commun.*, 1609 (1984).
214. T. W. Dekleva and D. Forster, *J. Am. Chem. Soc.*, **107**, 3565 (1985).
215. D. V. N. Hardy, *J. Chem. Soc.*, 1335 (1934).
216. H. Bahrmann, in Ref. 175d, p. 372, and references therein.
217. W. Haaf, *Chem. Ber.*, **99**, 1149 (1966).
218. Y. Takahashi, N. Yoneda, and H. Nagai, *Chem. Lett.*, 1733 (1985).
219. Y. Souma and H. Sano, *Bull. Chem. Soc. Jpn.*, **46**, 3237 (1973).
220. H. Koch, *Brennstoff. Chem.*, **36**, 321 (1955). See also Ref. 215.
221. G. G. Yakobson and G. G. Furin, *Synthesis*, 345 (1980) and references therein.
222. P. Pino, F. Piacenti, and M. Bianchi, in Ref. 175c, p. 233, and references therein.
223. C. A. Bertelo and J. Schwartz, *J. Am. Chem. Soc.*, **97**, 228 (1975).
224. H. C. Brown, *Acc. Chem. Res.*, **2**, 65 (1969).
225. F. Piacenti, M. Bianchi, and R. Lazzaroni, *Chim. Ind. (Milan)*, **50**, 318 (1968).
226. L. J. Kehoe and R. A. Schell, *J. Org. Chem.*, **35**, 2846 (1970).
227. H. C. Brown, J. L. Hubbard, and K. Smith, *Synthesis*, 701 (1979).
228. G. W. Kabalka, M. C. Delgado, U. Sastry, and K. A. R. Sastry, *J. Chem. Soc., Chem. Commun.*, 1273 (1982).
229. Y. Souma, H. Sano, and J. Iyoda, *J. Org. Chem.*, **38**, 2016 (1973).
230. Y. Souma and H. Sano, *J. Org. Chem.*, **38**, 3633 (1973).
231. N. Yoneda, Y. Takahashi, T. Fukuhara, and A. Suzuki, *Bull. Chem. Soc. Jpn.*, **59**, 2819 (1986).
232. S. Delavarenne, M. Simon, M. Fauconet, and J. Sommer, *J. Am. Chem. Soc.*, **111**, 383 (1989).
233. W. Reppe, *Justus Leibigs Ann. Chem.*, **582**, 1 (1953).
234. H. C. Clark and R. K. Mittal, *Can. J. Chem.*, **51**, 1511 (1973).
235. C. W. Bird and E. M. Briggs, *J. Chem. Soc. (C)*, 1265 (1967).
236. P. Pino and G. Braca, in Ref. 175c, p. 419.
237. L. Cassar, G. P. Chiusoli, and F. Guerrieri, *Synthesis*, 509 (1973).
238. J. C. Clark and R. C. Cookson, *J. Chem. Soc.*, 686 (1962).
239. G. N. Schrauzer, *Chem. Ber.*, **94**, 1891 (1961).
240. K. Nagira, K. Kikukawa, F. Wada, and T. Matsuda, *J. Org. Chem.*, **45**, 2365 (1980).
241. H. Wakamatsu, J. Uda, and N. Yamakami, *J. Chem. Soc., Chem. Commun.*, 1540 (1971).
242. J. J. Parnaud, G. Campari, and P. Pino, *J. Mol. Catal.*, **6**, 341 (1979).
243. (a) I. Ojima, K. Hirai, M. Fujita, and T. Fuchikami, *J. Organomet. Chem.*, **279**, 203 (1985). (b) I. Ojima, M. Okabe, K. Kato, H. B. Kwon, and I. T. Horváth, *J. Am. Chem. Soc.*, **110**, 150 (1988).
244. R. F. Heck and D. S. Breslow, *J. Am. Chem. Soc.*, **85**, 2779 (1963).
245. K. Shinoda and K. Yasuda, *Chem. Lett.*, 9 (1985).

246. K. Shinoda and K. Yasuda, *Bull. Chem. Soc. Jpn.*, **58**, 3082 (1985).
247. J. P. Collman, S. R. Winter, and R. G. Komoto, *J. Am. Chem. Soc.*, **95**, 249 (1973).
248. M. Rosenblum, *Acc. Chem. Res.*, **7**, 122 (1974).
249. S. G. Davies, *Pure Appl. Chem.*, **60**, 13 (1988), and references therein.
250. (a) A. Schoenberg, I. Bartoletti, and R. F. Heck, *J. Org. Chem.*, **39**, 3318 (1974). (b) J. K. Stille and P. K. Wong, *J. Org. Chem.*, **40**, 532 (1975). (c) T.-A. Kobayashi, F. Abe, and M. Tanaka, *J. Mol. Catal.*, **45**, 91 (1988). (d) S. R. Adapa and C. S. N. Prasad, *J. Chem. Soc., Perkin Trans. I*, 1706 (1989).
251. J. B. Woell and H. Alper, *Tetrahedron Lett.*, **25**, 3791 (1984).
252. H. Alper, N. Hamel, D. J. H. Smith, and J. B. Woell, *Tetrahedron Lett.*, **26**, 2273 (1985).
253. E. J. Corey and L. S. Hegedus, *J. Am. Chem. Soc.*, **91**, 1233 (1969).
254. L. Cassar and M. Foà, *J. Organomet. Chem.* **51**, 381 (1973).
255. H. Urata, O. Kosukegawa, Y. Ishii, H. Yugari, and T. Fuchikami, *Tetrahedron Lett.*, **30**, 4403 (1989).
256. R. Takeuchi, Y. Tsuji, and Y. Watanabe, *J. Chem. Soc., Chem. Commun.*, *351* (1986).
257. T. Kondo, Y. Tsuji, and Y. Watanabe, *Tetrahedron Lett.*, **29**, 3833 (1968).
258. W. R. Moser, A. W. Wang, and N. K. Kildahl, *J. Am. Chem. Soc.*, **110**, 2816 (1988).
259. J. K. Stille and P. K. Wong, *J. Org. Chem.*, **40**, 532 (1975).
260. K. E. Hashem, J. B. Woell, and H. Alper, *Tetrahedron Lett.*, **25**, 4879 (1984).
261. H. Alper, S. Antebi, and J. B. Woell, *Angew. Chem., Int. Ed. Engl.*, **23**, 732 (1984).
262. N. A. Bumagin, Y. V. Gulevich, and I. P. Beletskaya, *J. Organomet. Chem.*, **285**, 415 (1985).
263. J. B. Woell, S. B. Fergusson, and H. Alper, *J. Org. Chem.*, **50**, 2134 (1985).
264. R. A. Head and A. Ibbotson, *Tetrahedron Lett.*, **25**, 5939 (1984).
265. T. Takahashi, T. Nagashima, and J. Tsuji, *Chem. Lett.*, 369 (1980).
266. T. Takahashi, H. Ikeda, and J. Tsuji, *Tetrahedron Lett.*, **21**, 3885 (1980).
267. F. Ozawa, N. Kawasaki, H. Okamoto, T. Yamamoto, and A. Yamamoto, *Organometallics*, **6**, 1640 (1987).
268. M. Tanaka, T. Kobayashi, T. Sakakura, H. Itatani, S. Danno, and K. Zushi, *J. Mol. Catal.*, **32**, 115 (1985).
269. M. Tanaka, M. Koyonagi, and T. Kobayashi, *Tetrahedron Lett.*, **22**, 3875 (1981).
270. M. Foà, F. Francalanci, E. Bencini, and A. Gardano, *J. Organomet. Chem.*, **285**, 293 (1985).
271. (a) T. Kashimura, K. Kudo, S. Mori, and N. Sugita, *Chem. Lett.*, 851 (1986). (b) J.-J. Brunet, C. Sidot, and P. Caubere, *Tetrahedron Lett.*, **22**, 1013 (1981).
272. Y. Ben-David, M. Portnoy, and D. Milstein, *J. Am. Chem. Soc.*, **111**, 8742 (1989).
273. J. F. Knifton, *J. Org. Chem.*, **41**, 793 (1976).
274. J. F. Knifton, *J. Org. Chem.*, **41**, 2885 (1976).
275. E. E. Avery, M. F. Baevsky, S. J. Braswell, C. W. Duffus, G. O. Evans II, D. K. Rocha, and R. A. Wynne, *J. Mol. Catal.*, **53**, 179 (1989).
276. E. Drent, European Patent App., No. 218,282 A1 (1987), to Shell.
277. S. Hosaka and J. Tsuji, *Tetrahedron*, **27**, 3821 (1971).
278. A. Matsuda, *Bull. Chem. Soc. Jpn.*, **46**, 524 (1973).
279. J. Tsuji, Y. Mori, and M. Hara, *Tetrahedron*, **28**, 3721 (1972).
280. G. Cometti and G. P. Chiusoli, *J. Organomet. Chem.*, **236**, C31 (1982).
281. D. E. James and J. K. Stille, *J. Am. Chem. Soc.*, **98**, 1810 (1976).
282. J. K. Stille and R. Divakaruni, *J. Org. Chem.*, **44**, 3474 (1979).
283. L. de Picciotto, P. A. Carrupt, and P. Vogel, *J. Org. Chem.*, **47**, 3796 (1982).
284. G. Cometti and G. P. Chiusoli, *J. Organomet. Chem.*, **181**, C14 (1979).
285. (a) B. Despeyroux and H. Alper, *Ann. N.Y. Acad. Sci.*, **415**, 148 (1983). (b) H. Alper,

J. B. Woell, B. Despeyroux, and D. J. H. Smith, *J. Chem. Soc., Chem. Commun.*, 1270 (1983).

286. J. K. Stille and L. F. Hines, *J. Am. Chem. Soc.*, **92**, 1798 (1970).

287. D. Milstein, *Organometallics*, **1**, 888 (1982).

288. B. C. Söderberg, B. Åkermark, and S. S. Hall, *J. Org. Chem.*, **53**, 2925 (1988).

289. M. F. Semmelhack and A. Zask, *J. Am. Chem. Soc.*, **105**, 2034 (1983).

290. (a) M. F. Semmelhack and C. Bodurow, *J. Am. Chem. Soc.*, **106**, 1496 (1984). (b) M. McCormick, R. Monahan III, J. Soria, D. Goldsmith, and D. Liotta, *J. Org. Chem.*, **54**, 4485 (1989).

291. D. Lathbury, P. Vernon, and T. Gallagher, *Tetrahedron Lett.*, **27**, 6009 (1986).

292. J. S. Yadav, E. Sreenivasa Rao, C. S. Rao, and B. M. Choudary, *J. Mol. Catal.*, **49**, L61 (1989).

293. L. S. Hegedus and W. H. Darlington, *J. Am. Chem. Soc.*, **102**, 4980 (1980).

294. G. M. Wieber, L. S. Hegedus, B. Åkermark, and E. T. Michalson, *J. Org. Chem.*, **54**, 4649 (1989).

295. J. M. Tour and E. I. Negishi, *J. Am. Chem. Soc.*, **107**, 8289 (1985).

296. P. Pino and G. Braca, in *Organic Synthesis via Metal Carbonyls*, (I. Wender and P. Pino, eds.), Vol. 2, p. 419, Wiley, New York (1977).

297. J. F. Knifton, *J. Mol. Catal.*, **2**, 293 (1977).

298. Y. Tsuji, T. Kondo, and Y. Watanabe, *J. Mol. Catal.*, **40**, 295 (1987).

299. H. Alper, B. Despeyroux, and J. B. Woell, *Tetrahedron Lett.*, **24**, 5691 (1983).

300. L. Cassar, G. P. Chiusoli, and F. Guerreri, *Synthesis*, 509 (1973).

301. J. Tsuji and T. Nogi, *J. Am. Chem. Soc.*, **88**, 1289 (1966).

302. M. Foà and L. Cassar, *Gaz. Chim. Ital.*, **102**, 85 (1972).

303. G. P. Chiusoli, M. Dubini, M. Ferraris, F. Guerreri, S. Merzoni, and G. Mondelli, *J. Chem. Soc. C*, 2889 (1968).

304. H. Urata, H. Yugari, and T. Fuchikami, *Chem. Lett.*, 833 (1987).

305. C. Narayana and M. Periasamy, *Synthesis*, 253 (1985).

306. M. Yamashita and R. Suemitsu, *Tetrahedron Lett.*, 1477 (1978).

307. I. Rhee, M. Ryang, and S. Tsutsumi, *Tetrahedron Lett.*, 4593 (1969).

308. (a) P. M. Henry, *Tetrahedron Lett.*, 2285 (1968). (b) R. F. Heck, *J. Am. Chem. Soc.*, **90**, 5546 (1968).

309. W. C. Baird Jr., R. L. Hartgerink, and J. H. Surridge, *J. Org. Chem.*, **50**, 4601 (1985).

310. R. C. Larock, *J. Org. Chem.*, **40**, 3237 (1975).

311. R. C. Larock and K. Narayanan, *J. Org. Chem.*, **49**, 3411 (1984).

312. R. C. Larock and C. A. Fellows, *J. Am. Chem. Soc.*, **104**, 1900 (1982).

313. N. Miyaura and A. Suzuki, *Chem. Lett.*, 879 (1981).

314. K. Tamao, T. Kakui, and M. Kumada, *Tetrahedron Lett.*, **20**, 619 (1979).

315. (a) S. Antebi and H. Alper, *Tetrahedron Lett.*, **26**, 2609 (1985). (b) S. Uemura, H. Takahashi, K. Ohe, and N. Sugita, *J. Organomet. Chem.*, **361**, 63 (1989).

316. S. Cacchi, E. Morera, and G. Ortar, *Tetrahedron Lett.*, **25**, 2271 (1984).

317. S. Cacchi, P. G. Ciattini, E. Morera, and G. Ortar, *Tetrahedron Lett.*, **27**, 3931 (1986).

318. R. E. Dolle, S. J. Schmidt, and L. I. Kruse, *J. Chem. Soc., Chem. Commun.*, 904 (1987).

319. S. Cacchi, E. Morera, and G. Ortar, *Tetrahedron Lett.*, **26**, 1109 (1985).

320. E. Dalcanali and M. Foà, *Synthesis*, 492 (1986).

321. J. L. Eisenmann, R. L. Yamartino, and J. F. Howard Jr., *J. Org. Chem.*, **26**, 2102 (1961).

322. R. F. Heck, *J. Am. Chem. Soc.*, **85**, 1460 (1963).

323. A. Mullen, in *New Syntheses with Carbon Monoxide* (J. Falbe, ed.), p. 293, Springer-Verlag, Berlin (1980).

324. J. Tsuji, J. Kiji, S. Imamura, and M. Morikawa, *J. Am. Chem. Soc.*, **86**, 4350 (1964).

325. J. Kiji, T. Okano, I. Ono, and H. Konishi, *J. Mol. Catal.*, **39**, 355 (1987).
326. M. C. Bonnet, J. Coombes, B. Manzano, D. Neibecker, and I. Tkatchenko, *J. Mol. Catal.*, **52**, 263 (1989).
327. D. M. Fenton and P. J. Steinwand, *J. Org. Chem.*, **39**, 701 (1974).
328. F. Rivetti and U. Romano, *J. Organomet. Chem.*, **174**, 221 (1979).
329. Ube Industries Ltd., Belgian Patent No. 870,268 (1979).
330. J. E. Hallgren, G. M. Lucas, and R. O. Matthews, *J. Organomet. Chem.*, **204**, 135 (1981).
331. J. Tsuji, I. Shimizu, I. Minami, and Y. Ohashi, *Tetrahedron Lett.*, **23**, 4809 (1982).
332. J. Tsuji, K. Sato, and H. Okumoto, *Tetrahedron Lett.*, **23**, 5189 (1982); *J. Org. Chem.*, **49**, 1341 (1984).
333. J. Tsuji, T. Sugiura, and I. Minami, *Tetrahedron Lett.*, **27**, 731 (1986).
334. P. Pino, F. Piacenti, and M. Bianchi, in *Organic Synthesis via Metal Carbonyls* (I. Wender and P. Pino, eds.), Vol. 2, p. 233, Wiley, New York (1977), and references therein.
335. A. Schoenberg and R. F. Heck, *J. Org. Chem.*, **39**, 3327 (1974).
336. T. Kobayashi and M. Tanaka, *J. Organomet. Chem.*, **233**, C64 (1982).
337. F. Ozawa, H. Soyama, T. Yamamoto, and A. Yamamoto, *Tetrahedron Lett.*, **23**, 3383 (1982).
338. F. Ozawa, H. Soyama, H. Yanagihara, I. Aoyama, H. Takino, K. Izawa, T. Yamamoto, and A. Yamamoto, *J. Am. Chem. Soc.*, **107**, 3235 (1985).
339. F. Ozawa, T. Sugimoto, Y. Yuasa, M. Santra, T. Yamamoto, and A. Yamamoto, *Organometallics*, **3**, 683 (1984); F. Ozawa, T. Sugimoto, T. Yamamoto, and A. Yamamoto, *ibid.*, **3**, 692 (1984).
340. F. Ozawa, H. Yanagihara, and A. Yamamoto, *J. Org. Chem.*, **51**, 415 (1986).
341. Y. Ben-David, M. Portnoy, and D. Milstein, *J. Am. Chem. Soc.*, **111**, 8742 (1989).
342. F. Dany, R. Mutin, C. Lucas, V. Dufaud, J. Thivolle-Cazat, and J. M. Basset. *J. Mol. Catal.*, **51**, L15 (1989).
343. R. F. Heck and D. S. Breslow, *J. Am. Chem. Soc.*, **85**, 2779 (1963).
344. J. P. Collman, S. R. Winter, and R. G. Komoto, *J. Am. Chem. Soc.*, **95**, 249 (1973).
345. F. Merger, R. Kummer, and H. M. Hutmacher, German Patent No. 3,401,503 (1985), to BASF.
346. T. Hirao, Y. Harano, Y. Yamana, Y. Hamada, S. Nagata, and T. Agawa, *Bull. Chem. Soc. Jpn.*, **59**, 1341 (1986).
347. T. Hirao, S. Nagata, and T. Agawa, *Tetrahedron Lett.*, **26**, 5795 (1985).
348. S. Cacchi, E. Morera, and G. Ortar, *Tetrahedron Lett.*, **26**, 1109 (1985).
349. S. Cacchi, P. G. Ciattini, E. Morera, and G. Ortar, *Tetrahedron Lett.*, **27**, 3931 (1986).
350. R. E. Dolle, S. J. Schmidt, and L. I. Kruse, *J. Chem. Soc., Chem. Commun.*, 904 (1987).
351. J. Tsuji, K. Sato, and H. Okumoto, *Tetrahedron Lett.*, **23**, 5189 (1982); *J. Org. Chem.*, **49**, 1341 (1984).
352. S. I. Murahashi and Y. Imada, *Chem. Lett.*, 1477 (1985).
353. S. I. Murahashi, Y. Imada, and K. Nishimura, *J. Chem. Soc., Chem. Commun.*, 1578 (1988).
354. P. Pino and R. Magri, *Chim. Ind. (Milan)*, **34**, 511 (1952).
355. N. S. Imyanitov and D. M. Rudkovskii, *Zh. Prikl. Khim.*, **39**, 2335 (1966); *Chem. Abs.*, **66**, 10530 (1967).
356. H. J. Riegel and H. Hoberg, *J. Organomet. Chem.*, **260**, 121 (1984).
357. F. J. Fañanás and H. Hoberg, *J. Organomet. Chem.*, **275**, 249 (1984).
358. F. Ozawa, M. Nakano, I. Aoyama, T. Yamamoto, and A. Yamamoto, *J. Chem. Soc., Chem. Commun.*, 382 (1986).
359. W. Reppe, *Justus Liebigs Ann. Chem.*, **582**, 1 (1953).

360. H. Hoberg and F. J. Fañanás, *J. Organomet. Chem.*, **262**, C24 (1984).
361. D. G. Perez and N. S. Nudelman, *J. Org. Chem.*, **53**, 408 (1988), and references therein.
362. V. Rautenstrauch and M. Joyeux, *Angew. Chem., Int. Ed. Engl.*, **18**, 83 (1979).
363. M. R. Kilbourn, P. A. Jerabek, and M. J. Welch, *J. Chem. Soc., Chem. Commun.*, 861 (1983).
364. T. Tsuda, M. Miwa, and T. Saegusa, *J. Org. Chem.*, **44**, 3734 (1979).
365. C. M. Lindsay and D. A. Widdowson, *J. Chem. Soc., Perkin Trans. I*, 569 (1988).
366. F. Ozawa, I. Yamagami, N. Nakano, F. Fujisawa, and A. Yamamoto, *Chem. Lett.*, 125 (1989).
367. Y. Tsuji, M. Kobayashi, F. Okuda, and Y. Watanabe, *J. Chem. Soc., Chem. Commun.*, 1253 (1989).
368. V. H. Agreda, *Chem. Technol.*, 250 (1988).
369. N. Rizkalla, German Patent No. 2,610,036 (1975), to Halcon International.
370. M. Schrod and G. Luft, *Ind. Eng. Chem. Prod. Res. Dev.*, **20**, 649 (1981).
371. Y. Mori and J. Tsuji, *Bull. Chem. Soc. Jpn.*, **42**, 777 (1969).
372. W. Reppe, H. Kröper, H. J. Pistor, and O. Weissbarth, *Justus Liebigs Ann. Chem.*, **582**, 87 (1953).
373. J. Tsuji, J. Kiji, S. Imamura, and M. Morikawa, *J. Am. Chem. Soc.*, **86**, 4350 (1964).
374. K. Nagira, K. Kikukawa, F. Wada, and T. Matsuda, *J. Org. Chem.*, **45**, 2365 (1980).
375. K. Kikukawa, K. Kono, K. Nagira, F. Wada, and T. Matsuda, *J. Org. Chem.*, **46**, 4413 (1981).
376. I. Pri-Bar and H. Alper, *J. Org. Chem.*, **54**, 36 (1989).
377. T. Nogi and J. Tsuji, *Tetrahedron*, **25**, 4099 (1969).
378. E. R. H. Jones, G. H. Whitham, and M. C. Whiting, *J. Chem. Soc.*, 4628 (1957).
379. J. Tsuji, T. Sugiura, and I. Minami, *Tetrahedron Lett.*, **27**, 731 (1986).
380. D. Forster, *Adv. Organometallic Chem.*, **17**, 255 (1979).
381. W. T. Dent, R. Long, and G. H. Whitfield, *J. Chem. Soc.*, 1588 (1964).
382. R. F. Heck, *J. Am. Chem. Soc.*, **85**, 2013 (1963).
383. T. Sakakura, M. Chaisupakitsin, T. Hayashi, and M. Tanaka, *J. Organomet. Chem.*, **334**, 205 (1987).
384. W. W. Prichard, U.S. Patent No. 3,632,643 (1972), to Du Pont.
385. W. W. Prichard, U.S. Patent No. 3,560,553 (1971), to Du Pont.
386. J. A. Kampmeier and S. Mahalingam, *Organometallics*, **3**, 489 (1984); J. A. Kampmeier, S. Mahalingam, and T.-Z. Liu, *ibid.*, **5**, 823 (1986).
387. T. Susuki and J. Tsuji, *J. Org. Chem.*, **35**, 2982 (1970).
388. J. Tsuji, M. Morikawa, and J. Kiji, *J. Am. Chem. Soc.*, **86**, 4851 (1964).
389. J. Tsuji, M. Morikawa, and N. Iwamoto, *J. Am. Chem. Soc.*, **86**, 2095 (1964).
390. ICI, French Patent No. 1,379,231 (1964).
391. S. Cenini, M. Pizzotti, and C. Crotti, in *Aspects of Homogeneous Catalysis* (R. Ugo, ed.), Vol. 6, p. 97, Reidel, Dordrecht (1988).
392. V. I. Manov-Yuvenskii and B. K. Nefedov, *Russ. Chem. Rev.*, **50**, 470 (1981).
393. F. J. Weigert, *J. Org. Chem.*, **38**, 1316 (1973).
394. G. La Monica, C. Monti, and S. Cenini, *J. Mol. Catal.*, **18**, 93 (1983).
395. G. La Monica, C. Monti, S. Cenini, and B. Rindone, *J. Mol. Catal.*, **23**, 89 (1984).
396. T. Sakakibara and H. Alper, *J. Chem. Soc., Chem. Commun.*, 458 (1979).
397. H. Alper and C. P. Mahatantila, *Organometallics*, **1**, 70 (1982).
398. (a) ICI, Netherlands Patent Appl. No. 6,502,601 (1965). (b) ICI, Netherlands Patent Appl. No. 6,609,480 (1967). (c) G. A. Gamlen and A. Ibbotson, British Patent No. 1,092,157 (1967), to ICI.
399. Mitsui Toatsu, German Patent No. 2,555,557 (1976).

400. E. Alessio and G. Mestroni, *J. Organomet. Chem.*, **291**, 117 (1985).
401. Mitsui Toatsu, *Chemical Week*, March 9th, p. 43 (1977).
402. S. P. Gupte and R. V. Chaudhari, *J. Mol. Catal.*, **24**, 197 (1984).
403. Y. Watanabe, Y. Tsiji, R. Takeuchi, and N. Suzuki, *Bull. Chem. Soc. Jpn.*, **56**, 3343 (1983).
404. G. Besenyei, P. Viski, F. Nagy, and L. I. Simandi, *Inorg. Chim. Acta*, **90**, L43 (1984), and references therein.
405. S. Cenini, C. Crotti, M. Pizzotti, and F. Porta, *J. Org. Chem.*, **53**, 1243 (1988).
406. (a) E. Sappa and L. Milone, *J. Organomet. Chem.*, **61**, 383 (1973). (b) S. Bhaduri, K. S. Gopalkrishnan, G. M. Sheldrick, W. Clegg, and D. Stalke, *J. Chem. Soc., Dalton Trans.*, 2339 (1983). (c) A. Basu, S. Bhaduri, and H. Khwaja, *J. Organomet. Chem.*, **319**, C28 (1987).
407. G. D. Williams, R. R. Whittle, G. L. Geoffroy, and A. L. Rheingold, *J. Am. Chem. Soc.*, **109**, 3936 (1987).
408. H. Alper and G. Vasapollo, *Tetrahedron Lett.*, **28**, 6411 (1987).
409. S. Fukuoka, M. Chono, and M. Khono, *J. Org. Chem.*, **49**, 1458 (1984).
410. S. Fukuoka, M. Chono, and M. Khono, *Chem. Technol.*, 670 (1984).
411. H. Alper, G. Vasapollo, F. W. Hartstock, M. Mlekuz, D. J. H. Smith, and G. E. Morris, *Organometallics*, **6**, 2391 (1987).
412. J. K. Stille and R. Divakaruni, *J. Am. Chem. Soc.*, **100**, 1303 (1978).
413. J. K. Stille and D. E. James, *J. Am. Chem. Soc.*, **97**, 674 (1975); *J. Organomet. Chem.*, **108**, 401 (1976).
414. A. Cowell and J. K. Stille, *J. Am. Chem. Soc.*, **102**, 4193 (1980).
415. H. J. Nienburg and G. Elschnigg, German Patent No. 1,066,572 (1959).
416. M. Wang, S. Calet, and H. Alper, *J. Org. Chem.*, **54**, 20 (1989).
417. E. R. H. Jones, T. Y. Shen, and M. C. Whiting, *J. Chem. Soc.*, 230 (1950).
418. J. R. Norton, K. E. Shenton, and J. Schwartz, *Tetrahedron Lett.*, **16**, 51 (1975).
419. T. F. Murray, E. G. Samsel, V. Varma, and J. R. Norton, *J. Am. Chem. Soc.*, **103**, 7520 (1981).
420. T. F. Murray and J. R. Norton, *J. Am. Chem. Soc.*, **101**, 4107 (1979).
421. I. Matsuda, *Chem. Lett.*, 773 (1978).
422. M. F. Semmelhack and S. J. Brickner, *J. Org. Chem.*, **46**, 1723 (1981).
423. M. F. Semmelhack and S. J. Brickner, *J. Am. Chem. Soc.*, **103**, 3945 (1981).
424. L. D. Martin and J. K. Stille, *J. Org. Chem.*, **47**, 3630 (1982).
425. M. Mori, Y. Washioka, T. Urayama, K. Yoshiura, K. Chiba, and Y. Ban, *J. Org. Chem.*, **48**, 4058 (1983).
426. F. Henin and J. P. Pete, *Tetrahedron Lett.*, **24**, 4687 (1983).
427. H. Alper, J. K. Currie, and H. des Abbayes, *J. Chem. Soc., Chem. Commun.*, 311 (1978).
428. R. F. Heck, *J. Am. Chem. Soc.*, **86**, 2819 (1964).
429. J. C. Sauer, R. D. Cramer, V. A. Engelhardt, T. A. Ford, H. E. Holmquist, and B. W. Howk, *J. Am. Chem. Soc.*, **81**, 3677 (1959).
430. C. W. Bird, *J. Organomet. Chem.*, **47**, 295 (1973).
431. R. C. Larock, B. Riefling, and C. A. Fellows, *J. Org. Chem.*, **43**, 131 (1978).
432. J. X. Wang and H. Alper, *J. Org. Chem.*, **51**, 273 (1986).
433. H. Alper and D. Leonard, *Tetrahedron Lett.*, **26**, 5639 (1985).
434. J. Falbe, N. Huppes, and F. Korte, *Chem. Ber.*, **97**, 863 (1964).
435. Y. Tamaru, H. Higashimura, K. Naka, M. Hojo, and Z. Yoshida, *Angew. Chem., Int. Ed. Engl.*, **24**, 1045 (1985).
436. Y. Tamaru, T. Kobayashi, S. Kawamura, H. Ochiai, M. Hojo, and Z. Yoshida, *Tetrahedron Lett.*, **26**, 3207 (1985).
437. Y. Tamaru, M. Hojo, and Z. Yoshida, *J. Org. Chem.*, **53**, 5731 (1988).

438. H. Alper, H. Arzoumanian, J. F. Petrignani, and M. S. Maldonado, *J. Chem. Soc., Chem. Commun.*, **340** (1985).
439. M. F. Semmelhack, C. Bodurow, and M. Baum, *Tetrahedron Lett.*, **25**, 3171 (1984).
440. M. Foà, F. Francalanci, E. Bencini, and A. Gardano, *J. Organomet. Chem.*, **285**, 293 (1985).
441. R. C. Larock and C. A. Fellows, *J. Am. Chem. Soc.*, **104**, 1900 (1982).
442. H. Alper and G. Vasapollo, *Tetrahedron Lett.*, **30**, 2617 (1989).
443. P. G. M. Wuts, M. L. Obrzut, and P. A. Thompson, *Tetrahedron Lett.*, **25**, 4051 (1984).
444. R. Aumann and H. Ring, *Angew. Chem., Int. Ed. Engl.*, **16**, 50 (1977).
445. T. F. Murray, V. Varma, and J. R. Norton, *J. Org. Chem.*, **43**, 353 (1978).
446. Z. An, M. Catellani, and G. P. Chiusoli, *J. Organomet. Chem.*, **371**, C51 (1989).
447. T. Takahashi, H. Ikeda, and J. Tsuji, *Tetrahedron Lett.*, **21**, 3885 (1980).
448. T. Takahashi, T. Nagashima, and J. Tsuji, *Chem. Lett.*, 369 (1980).
449. For a review see A. G. M. Barrett and M. A. Sturgess, *Tetrahedron.*, **44**, 5615 (1988).
450. S. T. Hodgson, D. M. Hollinshead, and S. V. Ley, *Tetrahedron*, **41**, 5871 (1985).
451. G. D. Annis, E. M. Hebblethwaite, S. T. Hodgson, D. M. Hollinshead, and S. V. Ley, *J. Chem. Soc., Perkin Trans. (1)*, 2851 (1983).
452. S. R. Berryhill, T. Price, and M. Rosenblum, *J. Org. Chem.*, **48**, 158 (1983).
453. (a) M. Mori, K. Chiba, M. Okita, and Y. Ban, *J. Chem. Soc., Chem. Commun.*, 698 (1979). (b) M. Mori, K. Chiba, M. Okita, I. Kayo, and Y. Ban, *Tetrahedron*, **41**, 375 (1985).
454. (a) H. Alper, C. P. Perera, and F. R. Ahmed, *J. Am. Chem. Soc.*, **103**, 1289 (1981). (b) H. Alper and C. P. Mahatantila, *Organometallics*, **1**, 70 (1982).
455. (a) H. Alper, F. Urso, and D. J. H. Smith, *J. Am. Chem. Soc.*, **105**, 6737 (1983). (b) S. Calet, F. Urso, and H. Alper, *J. Am. Chem. Soc.*, **111**, 931 (1989).
456. D. Roberto and H. Alper, *Organometallics*, **3**, 1767 (1984).
457. M. A. McGuire and L. S. Hegedus, *J. Am. Chem. Soc.*, **104**, 5538 (1982).
458. D. Roberto and H. Alper, *J. Am. Chem. Soc.*, **111**, 7539 (1989).
459. J. F. Knifton, *J. Organomet. Chem.*, **188**, 223 (1980).
460. J. Falbe and F. Korte, *Chem. Ber.*, **98**, 1928 (1965).
461. G. P. Chiusoli, M. Costa, E. Masarati, and G. Salerno, *J. Organomet. Chem.*, **255**, C35 (1983).
462. J. M. Thompson and R. F. Heck, *J. Org. Chem.*, **40**, 2667 (1975).
463. M. Mori, K. Chiba, and Y. Ban, *J. Org. Chem.*, **43**, 1684 (1978).
464. C. W. Bird, *J. Organomet. Chem.*, **47**, 296 (1973).
465. S. Murahashi and S. Horiie, *J. Am. Chem. Soc.*, **78**, 4816 (1956).
466. L. S. Hegedus, G. F. Allen, and D. J. Olsen, *J. Am. Chem. Soc.*, **102**, 3583 (1980).
467. P. Braunstein, J. Kervennal, and J. L. Richert, *Angew. Chem., Int. Ed. Engl.*, **24**, 768 (1985).
468. Y. Uozumi, N. Kawasaki, E. Mori, and M. Shibasaki, *J. Am. Chem. Soc.*, **111**, 3725 (1989).
469. J. Falbe and F. Korte, *Chem. Ber.*, **95**, 2680 (1962).
470. L. S. Trifonov and A. S. Orahovats, *Tetrahedron Lett.*, **26**, 3159 (1985).
471. G. D. Pandey and K. P. Tiwari, *Tetrahedron*, **37**, 1213 (1981).
472. Y. Mori and J. Tsuji, *Tetrahedron*, **27**, 3811 (1971).
473. M. Mori, Y. Uozumi, M. Kimura, and Y. Ban, *Tetrahedron*, **42**, 3793 (1986).
474. M. Ishikura, M. Mori, M. Terashima, and Y. Ban, *J. Chem. Soc., Chem. Commun.*, 741 (1982).
475. J. N. Pitts and J. K. S. Wan, in *The Chemistry of the Carbonyl Group* (S. Patai, ed.), Chap. 16, Interscience, New York (1966).
476. R. Srinivasan, *Adv. Photochem.*, **1**, 83 (1963).

477. S. G. Cohen, J. D. Berman, and S. Orman, *Tetrahedron Lett.*, **3**, 43 (1962).
478. W. E. Bachmann, W. Cole, and A. L. Wilds, *J. Am. Chem. Soc.*, **62**, 824 (1940).
479. P. A. Leermakers, P. C. Warren, and G. F. Vesley, *J. Am. Chem. Soc.*, **86**, 1768 (1964).
480. J. D. Roberts and M. C. Caserio, *Basic Principles of Organic Chemistry*, p. 500, Benjamin, New York (1964).
481. B. P. Stark and A. J. Duke, *Extrusion Reactions*, Chap. 2, Pergamon Press, Oxford (1967).
482. S. C. Clarke and B. L. Johnson, *Tetrahedron*, **27**, 3555 (1971).
483. M. A. Ogliaruso, M. G. Romanelli, and E. I. Becker, *Chem. Rev.*, **65**, 261 (1965).
484. J. O. Hawthorne and M. H. Wilt, *J. Org. Chem.*, **25**, 2215 (1960).
485. J. Coca, E. S. Morrondo, and H. Sastre, *J. Chem. Technol. Biotechnol.*, **32**, 904 (1982).
486. R. D. Srivastava and A. K. Guha, *J. Catal.*, **91**, 254 (1985).
487. A. P. Dunlop and G. W. Huffman, U.S. Patent No. 3,257,417 (1966).
488. D. G. Manley and J. P. O'Halloran, U.S. Patent No. 3,223,714 (1965).
489. H. B. Coppelin and D. I. Garnett, U.S. Patent No. 3,007,941 (1961).
490. H. E. Eschinazi, *Bull. Soc. Chim. Fr.*, **19**, 967 (1952).
491. P. Lejemble, Y. Maire, A. Gaset, and P. Kalck, *Chem. Lett.*, 1403 (1983).
492. P. Lejemble, A. Gaset, and P. Kalck, *Biomass*, **4**, 263 (1984).
493. H. J. Anderson, C. E. Loader, and A. Foster, *Can. J. Chem.*, **58**, 2527 (1980).
494. C. E. Loader and H. J. Anderson, *Can. J. Chem.*, **59**, 2673 (1981).
495. B. J. Demopoulos, H. J. Anderson, C. E. Loader and K. Faber, *Can. J. Chem.*, **61**, 2415 (1983).
496. H. M. Colquhoun, J. Holton, D. J. Thompson, and M. V. Twigg, *New Pathways for Organic Synthesis*, pp. 188, 191, 192, 272, Plenum Press, New York (1984).
497. R. P. Linstead, K. O. A. Michaelis, and S. L. S. Thomas, *J. Chem. Soc.*, 1139 (1940).
498. M. C. Baird, in *The Chemistry of Acid Derivatives*, (S. Patai, ed.), Supplement B, Part 2, pp. 825–857, Wiley, New York (1979).
499. T. Nomoto, N. Nasui, and H. Takayama, *J. Chem. Soc., Chem. Commun.*, 1646 (1984).
500. R. McCague, C. J. Moody, and C. W. Rees, *J. Chem. Soc., Perkin Trans. I*, 165 (1984).
501. S. B. Laing and P. J. Sykes, *J. Chem. Soc. (C)*, 2915 (1968).
502. R. W. Fries and J. K. Stille, *Synth. Inorg. Metal-Organ. Chem.*, **1**, 295 (1971).
503. G. L. Geoffroy, D. A. Denton, M. E. Keeney, and R. R. Bucks, *Inorg. Chem.*, **15**, 2382 (1976).
504. J. Tsuji and K. Ohno, *Synthesis*, **1**, 157 (1969).
505. K. Ohno and J. Tsuji, *J. Am. Chem. Soc.*, **90**, 99 (1968).
506. J. K. Stille, M. T. Regan, R. W. Fries, F. Huang, and T. McCarley, *Adv. Chem. Ser.*, **132**, 181 (1974).
507. W. Strohmeier and P. Pfohler, *J. Organomet. Chem.*, **108**, 393 (1976).
508. D. H. Doughty and L. H. Pignolet, *J. Am. Chem. Soc.*, **100**, 7083 (1978).
509. D. H. Doughty, M. F. McGuiggan, H. Wang, and L. H. Pignolet, *Fundamental Research in Homogeneous Catalysis*, p. 109, Plenum Press, New York (1979).
510. A. R. Sanger, *J. Chem. Soc., Dalton Trans.*, 120 (1977).
511. M. C. Hall, B. T. Kilbourn, and K. A. Taylor, *J. Chem. Soc. (A)*, 2539 (1970).
512. M. D. Meyer and L. I. Kruse, *J. Org. Chem.*, **49**, 3195 (1984).
513. J. Tsuji and K. Ohno, *Tetrahedron Lett.*, **8**, 2173 (1967).
514. H. M. Walborsky and L. E. Allen, *Tetrahedron Lett.*, **11**, 823 (1970).
515. H. M. Walborsky and L. E. Allen, *J. Am. Chem. Soc.*, **93**, 5465 (1971).
516. D. E. Iley and B. Fraser-Reid, *J. Am. Chem. Soc.*, **97**, 2563 (1975).
517. B. M. Trost and M. Preckel, *J. Am. Chem. Soc.*, **95**, 7862 (1973).
518. P. D. Hobbs and P. D. Magnus, *J. Chem. Soc., Chem. Commun.*, 856 (1974).

519. M. Sergent, M. Mongrain, and P. Deslongchamps, *Can. J. Chem.*, **50**, 336 (1972).
520. J. Falbe, H. Tummes, and H.-D. Hahn, *Angew. Chem., Int. Ed. Eng.*, **9**, 169 (1970).
521. (a) M. A. Andrews and S. A. Klaeren, *J. Chem. Soc., Chem. Commun.*, 1266 (1988). (b) M. A. Andrews, G. L. Gould, and S. A. Klaeren, *J. Org. Chem.*, **54**, 5257 (1989).
522. C. F. Lochow and R. G. Miller, *J. Am. Chem. Soc.*, **98**, 1281 (1976).
523. K. S. Y. Lau, Y. Becker, F. Huang, N. Baenziger, and J. K. Stille, *J. Am. Chem. Soc.*, **99**, 5664 (1977).
524. For a list see F. H. Jardine, *Prog. Inorg. Chem.*, **28**, 109 (1981).
525. J. K. Stille and R. W. Fries, *J. Am. Chem. Soc.*, **96**, 1514 (1974).
526. J. K. Stille and K. S. Y. Lau, *Acc. Chem. Res.*, **10**, 434 (1977).
527. J. K. Stille, F. Huang, and M. T. Regan, *J. Am. Chem. Soc.*, **96**, 1518 (1974).
528. J. Blum, S. Kraus, and Y. Pickholtz, *J. Organomet. Chem.*, **33**, 227 (1971).
529. J. Tsuji and K. Ohno, *J. Am. Chem. Soc.*, **90**, 94 (1968).
530. J. G. Burr, *J. Am. Chem. Soc.*, **73**, 3502 (1951).
531. C. A. Rojahn and A. Seitz, *Justus Liebigs Ann. Chem.*, **437**, 297 (1924).
532. T. Nozoe and T. Kinugasa, *Nippon Kagaku Zasshi*, **59**, 772 (1938).
533. J. A. Kampmeier, S. H. Harris, and R. M. Rodehorst, *J. Am. Chem. Soc.*, **103**, 1478 (1981).
534. J. A. Kampmeier and T. Z. Liu, *Organometallics*, **8**, 2742 (1989).
535. J. A. Kampmeier, R. M. Rodehorst, and J. B. Philip, *J. Am. Chem. Soc.*, **103**, 1847 (1981).
536. J. A. Kampmeier and S. Mahalingam, *Organometallics*, **3**, 489 (1984).
537. H. U. Blaser and A. Spencer, *J. Organomet. Chem.*, **233**, 267 (1982).
538. C. M. Andersson and A. Hallberg, *J. Org. Chem.*, **53**, 235 (1988).
539. J. D. Rich, *J. Am. Chem. Soc.*, **111**, 5886 (1989).
540. J. D. Rich, *Organometallics*, **8**, 2609 (1989).
541. K. Kaneda, H. Azuma, M. Wakayu, and S. Teranishi, *Chem. Lett.*, 215 (1974).
542. M. A. Andrews, *Organometallics*, **8**, 2703 (1989).
543. J. Tsuji and K. Ohno, *Tetrahedron Lett.*, **7**, 4713 (1966); G. A. Olah and P. Kreienbühl, *J. Org. Chem.*, **32**, 1614 (1967); K. Ohno and J. Tsuji, *J. Am. Chem. Soc.*, **90**, 99 (1968).
544. J. Blum, H. Rosenman, and E. D. Bergmann, *J. Org. Chem.*, **33**, 1928 (1968).
545. G. A. Olah and P. Kreienbuhl, *J. Org. Chem.*, **32**, 1614 (1967).
546. R. E. Ehrenkaufer, R. R. MacGregor, and A. P. Wolf, *J. Org. Chem.*, **47**, 2489 (1982).
547. J. Blum, E. Oppenheimer, and E. D. Bergmann, *J. Am. Chem. Soc.*, **89**, 2338 (1967).
548. S. I. Murahashi, T. Naota, and N. Nakajima, *J. Org. Chem.*, **51**, 898 (1986).
549. A. Emery, A. C. Oehlschlager, and A. M. Unrau, *Tetrahedron Lett.*, **11**, 4401 (1970).
550. K. Osakada, T. Yamamoto, and A. Yamamoto, *Tetrahedron Lett.*, **28**, 6321 (1987).
551. H. Nakazawa, Y. Matsuoka, H. Yamaguchi, T. Kuroiwa, K. Miyoshi, and H. Yoneda, *Organometallics*, **8**, 2272 (1989).
552. P. Gilmont and A. A. Blanchard, *Inorg. Synth.*, **2**, 238 (1946).
553. P. Chini, M. C. Malatesta, and A. Cavalieri, *Chim. Ind. (Milan)*, **55**, 120 (1973).
554. P. Szabo, L. Marko, and G. Bor, *Chem. Tech. (Berlin)*, **13**, 549 (1961).
555. R. B. King, *Organometallic Syntheses*, Vol. 1, pp. 98–101, Academic Press, New York (1965).
556. J. K. Ruff and W. J. Schlientz, *Inorg. Synth.*, **15**, 84 (1974).
557. J. P. Collman and R. Winter, U.S. Patent No. 3,872,218 (1975).
558. R. J. Angelici, *Synthesis and Technique in Inorganic Chemistry*, pp. 131–136, W. B. Saunders, Philadelphia (1969).
559. R. B. King and F. G. A. Stone, *Inorg. Synth.*, **7**, 110 (1963).
560. (a) G. Booth and J. Chatt, *J. Chem. Soc.*, 3238 (1965). (b) M. J. Hudson, R. S. Nyholm,

and M. H. B. Stiddard, *J. Chem. Soc. A*, **40** (1968). (c) Y. Kiso, K. Tamao, N. Miyake, K. Yamamoto, and M. Kumada, *Tetrahedron Lett.*, **15**, 3 (1974).

561. (a) L. M. Venanzi, *J. Chem. Soc.*, 719 (1958). (b) F. A. Cotton, O. D. Faut, and D. M. L. Goodgame, *J. Am. Chem. Soc.*, **83**, 344 (1961).

562. T. A. Stephenson, S. M. Morehouse, A. R. Powell, J. P. Heffer, and G. Wilkinson, *J. Chem. Soc.*, 3632 (1965).

563. P. Fitton and E. A. Rick, *J. Organomet. Chem.*, **28**, 287 (1971).

564. F. Puche, *Ann. Chim.*, **9**, 233 (1938).

565. A. Yatsimirski and R. Ugo, *Inorg. Chem.*, **22**, 1395 (1983).

566. H. Schafer, U. Wiese, K. Rinke, and K. Brendel, *Angew. Chem. Int. Ed. Engl.*, **6**, 253 (1967).

567. *Gmelin Handbook of Inorganic Chemistry: Palladium*, 8th Edn., Suppl. Vol. B2 (W. P. Griffith and K. Swars, eds.), p. 66, Springer-Verlag, Berlin (1989).

568. S. E. Livingstone, in *Comprehensive Inorganic Chemistry*, Vol. 3, p. 1278, Pergamon Press, Oxford (1973).

569. R. F. Heck, *Palladium Reagents in Organic Synthesis*, p. 17, Academic Press, New York (1985).

570. W. T. Dent, R. Long, and A. J. Wilkinson, *J. Chem. Soc.*, 1585 (1964).

571. H. A. Tayim, A. Bouldoukian, and F. Awad, *J. Inorg. Nucl. Chem.*, **32**, 3799 (1970).

572. H. Itatani and J. C. Bailar, *J. Am. Oil Chem. Soc.*, **44**, 147 (1967).

573. J. R. Doyle, P. E. Slade, and H. B. Jonassen, *Inorg. Synth.*, **6**, 216 (1960).

574. D. R. Coulson, *Inorg. Synth.*, **13**, 121 (1972).

575. R. Mozingo, in *Organic Syntheses* (E. C. Horning, ed.), Coll. Vol. III, p. 685, Wiley, New York (1955).

576. A. I. Vogel, *A Text Book of Practical Organic Chemistry*, 3rd Ed., pp. 948–951, Longmans, New York (1964).

577. K. Brodersen, G. Thiele, and H. G. Schnering, *Z. Anorg. Allgem. Chem.*, **337**, 120 (1965).

578. A. J. Cohen, *Inorg. Synth.*, **6**, 209 (1960).

579. W. E. Cooley and D. H. Busch, *Inorg. Synth.*, **5**, 208 (1957).

580. H. L. Grube, in *Handbook of Preparative Inorganic Chemistry*, (G. Brauer, ed.), 2nd Ed., pp. 1572–1573, Academic Press, New York (1965).

581. J. C. Bailar and H. Itatani, *Inorg. Chem.*, **4**, 1618 (1965).

582. R. Ugo, F. Cariati, and G. la Monica, *Inorg. Synth.*, **11**, 105 (1968).

583. N. Ahmad, E. W. Ainscough, T. A. James, and S. D. Robinson, *J. Chem. Soc.*, *Dalton Trans.*, 1148 (1973).

584. K. A. Hofmann and G. Bugge, *Chem. Ber.*, **40**, 1772 (1907).

585. T. Uchiyama, Y. Toshiyasu, Y. Nakamura, T. Miwa, and S. Kawaguchi, *Bull. Chem. Soc. Jpn.*, **54**, 181 (1981).

586. R. Cramer, *Inorg. Synth.*, **15**, 14 (1974).

587. J. A. McCleverty and G. Wilkinson, *Inorg. Synth.*, **8**, 211 (1966).

588. D. Evans, J. A. Osborn, and G. Wilkinson, *Inorg. Synth.*, **11**, 99 (1968).

589. F. H. Jardine, *Prog. Inorg. Chem.*, **28**, 63 (1981).

590. J. A. Osborn, F. H. Jardine, J. F. Young, and G. Wilkinson, *J. Chem. Soc. A*, 1711 (1966).

591. G. A. Rempel, P. Legzdins, H. Smith, and G. Wilkinson, *Inorg. Synth.*, **13**, 90 (1972).

592. A. van der Ent and A. L. Onderlinden, *Inorg. Synth.*, **14**, 92 (1973).

593. B. R. James and D. Mahajan, *Canad. J. Chem.*, **57**, 180 (1979).

594. J. Chatt and L. M. Venanzi, *Nature*, **177**, 852 (1956); *J. Chem. Soc.*, 4735 (1957).

595. G. La Monica, C. Monti, and S. Cenini, *J. Mol. Catal.*, **18**, 93 (1983).

596. N. Ahmad, J. J. Levison, S. D. Robinson, and M. F. Uttley, *Inorg. Synth.*, **15**, 59 (1974).

597. P. E. Cattermole and A. G. Osborne, *Inorg. Synth.*, **17**, 115 (1977).
598. R. D. Lanam and E. D. Zysk, in *Kirk Othmer Encyclopedia of Chemical Technology*, 3rd Ed., Vol. 18, p. 228, Wiley, New York (1982).
599. M. J. Cleare, P. Charlesworth, and D. J. Bryson, *J. Chem. Technol. Biotechnol.*, **29**, 210 (1979).
600. S. E. Livingstone, The Platinum Metals, in *Comprehensive Inorganic Chemistry*, Vol. 3, pp. 1163–1370, Pergamon Press, Oxford (1973).
601. W. P. Griffith, *The Chemistry of the Rarer Platinum Metals*, Interscience, New York (1967).
602. N. V. Sidgwick, *The Chemical Elements and Their Compounds*, Vol. 2, pp. 1454–1628, Oxford University Press, Oxford (1950).
603. (a) N. I. Sax and R. J. Lewis, *Dangerous Properties of Industrial Materials*, 7th Ed., Vol. 3, p. 2810, Van Nostrand Reinhold, New York (1989). (b) M. J. Cleare, E. G. Hughes, B. Jacoby, and J. Pepys, *Clin. Allergy.*, **6**, 183 (1976).
604. G. B. Kauffman and L. A. Teter, *Inorg. Synth.*, **7**, 232 (1963).
605. W. E. Cooley and D. H. Busch, *Inorg. Synth.*, **5**, 208 (1957).
606. S. N. Anderson and F. Basolo, *Inorg. Synth.*, **7**, 214 (1963).
607. G. B. Kauffman and R. D. Myers, *Inorg. Synth.*, **18**, 131 (1978).
608. G. L. Silver, *J. Less-Common Met.*, **40**, 265 (1975); **45**, 335 (1976).
609. F. E. Beamish and J. C. Van Loon, *Analysis of Noble Metals*, Chap. 7, Academic Press, New York (1977).
610. G. A. Stein, H. C. Vogel, and R. G. Valerio, U.S. Patent No. 2,610,907 (1952).
611. J. Harkema, U.S. Patent No. 3,582,270 (1971).

Index